AF335226

polarized sources and targets

Proceedings of the Ninth International Workshop

polarized sources and targets

Nashville, Indiana, USA 30 September – 4 October 2001

editors

Vladimir P. Derenchuk
Barbara von Przewoski

Indiana University Cyclotron Facility, USA

World Scientific
New Jersey • London • Singapore • Hong Kong

Published by

World Scientific Publishing Co. Pte. Ltd.

P O Box 128, Farrer Road, Singapore 912805

USA office: Suite 1B, 1060 Main Street, River Edge, NJ 07661

UK office: 57 Shelton Street, Covent Garden, London WC2H 9HE

British Library Cataloguing-in-Publication Data
A catalogue record for this book is available from the British Library.

Cover: Photographic images are complements of the Brown County
Convention and Visitors Bureau, Nashville, Indiana.

**Proceedings of the Ninth Internaltional Workshop on
POLARIZED SOURCES AND TARGETS**

ISBN 981-02-4917-9

Printed in Singapore by Fulsland Offset Printing

SPONSORS

v

THE INDIANA UNIVERSITY CYCLOTRON FACILITY

VACUUMSCHMELZE CORPORATION

THE INTERNATIONAL COMMITTEE FOR SPIN PHYSICS SYMPOSIA

THE INDIANA UNIVERSITY CHANCELLOR'S OFFICE

PREFACE

The Workshop on Polarized Sources and Targets (PST 2001) was held at the Brown County Inn in Nashville, Indiana from September 30th to October 4th, 2001. It was organized by the Indiana University Cyclotron Facility in Bloomington, Indiana. The Workshop is the most recent of a series held at about two year intervals, the last of which was held in Erlangen, Germany in 1999. About 80 scientists attended the Workshop. There were 12 invited talks, 30 contributed talks and 15 posters. E. Steffens (Erlangen) gave the summary talk at the conclusion of the Workshop. The subjects addressed in the Workshop included atomic beam polarized H and D targets, solid polarized targets, spin polarized HD, polarized electron sources, polarized ion sources, hadron polarimetry at intermediate to high energies, electron polarimetry, polarized neutrons and the use of polarized noble gases in medical imaging.

The Workshop provided a forum to explore the many fascinating aspects of the physics of polarized beams and targets. In addition to the technical sessions, the Workshop participants enjoyed a barbecue, a reception followed by a tour of the Indiana University Cyclotron Facility and a banquet.

Our special thanks go to the Workshop secretary, Janet Meadows, who was indispensable in the preparation and organization of the Workshop. She also helped the participants cope with problems during the meeting. We would also like to thank T. Jones, R. Candelaria, H. Workman, R. Yoder and B. Craig for their assistance with various aspects of the Workshop. Credit for suggestions of topics and speakers go to the members of the program committee: D.G. Crabb (U. of Virginia), H.O. Meyer (IUCF), B. von Przewoski (IUCF), P. Schwandt (IUCF), C. Sinclair (TJNAF), W.M. Snow (IUCF), E.J. Stephenson (IUCF) and T. Wise (U. of Wisconsin). Local members of the program committee together with external advisors W. Haeberli (U. of Wisconsin) and E. Steffens (Erlangen) formed the organizing committee who developed the overall program. Finally, we thank the speakers, session chairs and the participants who made this Workshop a tremendous success. The Workshop was sponsored by IUCF (NSF grant PHY-9602872), Vacuumschmelze, the International Committee for Spin Physics Symposia, and the Indiana University Chancellor's office.

CONTENTS

I. Polarized Gas Targets

II. Polarized Solid Targets

III. Polarized Electron Sources

IV. Polarized Ion Sources

V. Sources of Polarized Neutrons

VI. Polarimetry

VII. Applications and New Techniques

VIII. Summary Talk

I. Polarized Gas Targets

OVERVIEW OF GAS TARGET DEVELOPMENT

FRANK RATHMANN

Institut für Kernphysik, Forschungszentrum Jülich, 52425 Jülich, Germany

Experimental setups utilizing polarized internal gas targets are presently used at various electron and proton machines with typical luminosities of 10^{31} cm$^{-1}s^{-1}$ and 10^{29} cm$^{-1}s^{-1}$, respectively. The paper focuses mainly on experimental aspects, the design and construction of storage cells, and different methods employed for polarimetry.

1 Introduction

Storage cells to enhance the target thickness provided by a source of polarized atoms have already been conceived some thirty years ago[1]. An application of this technique to nuclear and particle physics became possible only through the advent of storage rings. The stored beam passes through the *internal* target about a million times per second and thus compensates for the low target thickness. In order to obtain higher target thicknesses compared to a free beam of polarized atoms, the dwell time of atoms in the vicinity of the stored beam must be increased. For that purpose an open ended T-shaped storage cell is inserted into the ring through which the stored beam passes and into which the atomic beam is injected. Thereby the luminosity can be raised by about two orders of magnitude. Polarized internal targets nowadays provide densities around 10^{14} atoms/cm^2 and nuclear vector or tensor polarization close to the theoretical maximum at various storage rings.

2 Overview

The suggestion by Haeberli[1] to use a storage cell to enhance the available target thickness produced by an atomic beam source (ABS) in **1966** was motivated by the successful application of a storage container in the hydrogen maser[2]. The proof of principle, carried out by the Wisconsin Group in **1980**, used a 12 MeV beam of α-particles incident on a proton target[3]. The atomic beam entered the storage vessel through a 10 cm long tube of 10 mm diameter. The storage volume (0.26 l) was coated with Teflon to avoid depolarization. The measured polarization after background subtraction was compatible with the polarization of a free beam, thus essentially no depolarization occurred during an average of 900 wall collisions inside the vessel. An improvement in target thickness of a factor of six over that of a free beam was reported.

4

In **1985** the FILTEX group proposed an internal polarized hydrogen target for LEAR at CERN[4]. The objective was to polarize a stored beam of antiprotons by spin-dependent attenuation in the target (Spin-Filter-Method[5]) and to carry out $\bar{p}p$-interaction studies. Target densities of 10^{14} atoms/cm^2 in one substate were required.

Since **1988** an internal polarized deuterium target is operated at VEPP-3 in Novosibirsk for $e\vec{d}$-scattering[6,7] experiments. Substantial progress in atomic beam intensity could be achieved through superconducting magnets[8].

In **1992**, the FILTEX target was tested at TSR[a] [9,10]. A target thickness close to 10^{14} atoms/cm^2 in two substates, with a nuclear polarization of 0.8 in one substate was obtained (see also Sec. 4.2).

In **1992**, an optically pumped polarized $^3\vec{\mathrm{He}}$ target was installed at IUCF to study spin-dependent $\vec{p}^3\vec{\mathrm{He}}$-reactions[11]. This experiment constitutes the first measurement with a polarized beam incident on a polarized internal target in a storage ring. A target thickness of $1.5 \cdot 10^{14}$ atoms/cm^2 was reported[12]. In **1995**, the target was installed at the HERMES experiment at HERA/DESY to investigate the spin-structure of the nucleon[13].

In **1996**, the polarized source[9], initially developed for the FILTEX experiment, was installed at HERMES. The HERMES storage cell setup is discussed in Sec. 4.2.

Since **1994** the PINTEX[b] collaboration at IUCF[14] is operating a polarized hydrogen and deuterium gas target[15] (see also Sec. 4.2).

Until **1999** $e\vec{d}$-reaction studies have been carried out at NIKHEF with an electron beam incident on a tensor polarized internal deuterium gas target[16,17]. An ion-extraction system was developed to determine the target polarization (Sec. 5.1).

Among the operating polarized internal target experiments, EDDA at COSY[c] is the only one that utilizes a free atomic jet[18]. First tests with a storage cell[19] have been recently carried out (see also Sec. 4.2).

A new type of polarized hydrogen and deuterium source based on spin-exchange optical pumping is currently installed at IUCF. First results in $\vec{p}\vec{d}$ elastic scattering could be obtained[20] (see also Sec. 4.2).

Two **new facilities** employing polarized internal gas targets, are presently prepared for nuclear physics experiments. At the BLAST[d] experiment[21] the polarized source previously used at NIKHEF will be utilized

---

[a]Test Storage Ring in Heidelberg, Germany.
[b]Polarized INternal Target EXperiments, at IUCF, Bloomington, Indiana, USA.
[c]COoler SYnchrotron in Jülich, Germany.
[d]Bates Large Acceptance Spectrometer Toroid, Cambridge, Massachusetts, USA.

for studies in polarized ed reactions. At COSY a polarized internal storage cell gas target[22] is being developed to study the $\vec{p}d$-breakup reaction[23] at small scattering angles with the magnetic spectrometer ANKE[e][24].

3 Polarized Internal Targets for Storage Rings

The interaction region produced by a target source alone is typically of the order of a cm^3, given by the volume of overlap between stored and atomic beam. Target sources for this type of setup are optimized for *volume density* in the interaction zone, which increases with decreasing atomic velocity. When a storage cell is used, the *intensity* fed into the acceptance of the feed tube of the cell is optimized instead. In this case the interaction region is extended. Typical cell dimensions range from 250 to 600 mm in length, and from 10 to 30 mm in diameter.

The main advantages of polarized gas targets over solid targets are: *i)* **high isotopical purity**, *ii)* possibility of **rapid spin reversal** up to 100 Hz in $\vec{H}$ or $\vec{D}$ targets, *iii)* **low background**, since no container walls intercept the particle beam, *iv)* **no radiation damage**, because the target gas is replenished every few ms. Polarized gas targets combined with beams internal to storage rings are therefore ideally suited for high precision experiments.

4 Storage Cells for polarized internal Targets

The main objective in the development of storage cells for polarized internal targets was the identification of suitable wall coating materials that inhibit depolarization of atoms in the cell that are at the same time compatible with the ultra-high vacuum requirements of a storage ring.

4.1 Cell Coating

Studies of wall depolarization in storage cells were performed by Price and Haeberli with an apparatus described in Ref.[25]. A 50 keV beam of D^+ ions from a duoplasmatron source passes through a storage cell. Polarized $\vec{H}^0$ atoms are injected through a feed tube into the cell, over which a weak magnetic guide field of 5 mT is applied. In a weak field electrons and protons of hydrogen atoms in a state $|m_j = \frac{1}{2}, m_I = \frac{1}{2}\rangle$ carry the same polarization. The electron pickup reaction $D^+(50\,\text{keV}) + \vec{H}^0 \rightarrow \vec{D}^0(50\,\text{keV}) + H^+$ in the

---

[e]**A**pparatus for **S**tudies of **N**ucleon and **K**aon **E**jectiles

cell transfers the initial polarization of the electrons through hyperfine interaction to the deuterium atom, which acquires a nuclear tensor polarization. The neutral $\vec{D^0}$ atoms impinge on a tritiated foil located in a strong field of 30 mT. From the neutron asymmetry of the $^3\mathrm{H}(\vec{d},n)^4\mathrm{He}$ reaction the tensor polarization of the deuterons is deduced. The depolarization behaviour of various wall coating materials was investigated. Among the various materials studied, Teflon stands out for its low depolarization and good vacuum compatibility.

4.2 Design and Construction of Storage Cells

The design of a storage cell must be well adapted to the specific requirements of an experiment, taking into account properties of the stored beam, detector acceptance, possible background from the cell walls, etc. These design constraints for polarized internal targets have been reviewed by Meyer[26]. A few experiments employing storage cells are discussed in the next section.

FILTEX: A boundary condition initially imposed by the machine group during the measurements was to allow for injection of the beam without the storage cell acting as acceptance limiter. Therefore a storage cell that could be opened and closed like a clam-shell (Fig. 1) was developed[10]. After the

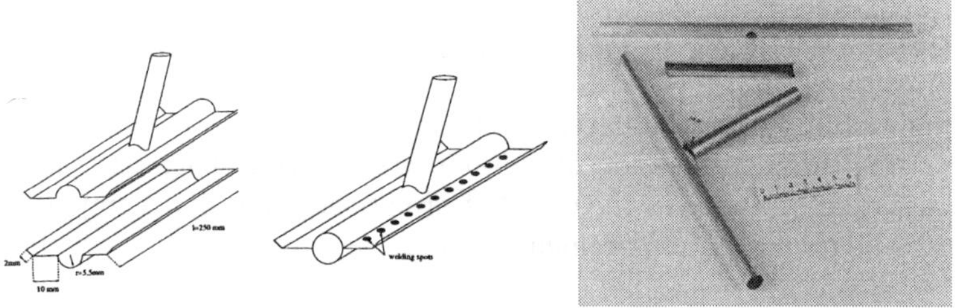

Figure 1. Storage cells used during the FILTEX test experiment at TSR[10]: Clam-shell cell (left), spot-welded clam-shell (center), and cell made from cylindrical tubing (right).

beam had been injected and cooled the cell was closed. The upstream end of the cell was attached to a cold head, thereby the cell center could be cooled to temperatures between 50-300 K. The rather low polarization observed for the clam-shell cell ($\approx$ 50% of the theoretical maximum) was probably caused by exposure of atoms to the badly-coated fins of the cell. Spot-welding the two half cells together (Fig. 1) substantially increased the polarization. Eventually, closed cells manufactured from cylindrical aluminum tubes were used. All cells were of the same dimensions, 250 mm length with a diameter of 11 mm and a wall thickness of 0.2 mm. The first two cell types (clam-shell, spot-welded clam-shell) were coated with Teflon FEP 120, while the final cells had Teflon TPFE 3170 coatings. All cells were equipped with small capillaries near the cell center to inject unpolarized gas for background studies. Results of the polarization measurements are presented in Sec. 5.1.

^{3}He-**Target at IUCF:** Depolarization in wall collisions in polarized ^{3}He gas targets is inhibited due to the absence of hyperfine interaction. Storage cells for experiments with ^{3}He (Fig. 2) therefore remained uncoated[11]. Aluminized mylar sheets of 1.7 μm thickness were attached to a support frame consisting of 0.2 mm thick aluminum pieces to form a 400 mm long cell of rectangular cross section ($h = 16.6$ mm, $w = 13.1$ mm). The experimental setup did not include any provision to cool the cell. A target thickness of $1.5 \cdot 10^{14}$ atoms/cm^2 was reported. Without restrictions due to deadtime, an order of magnitude higher target thickness would have been possible.

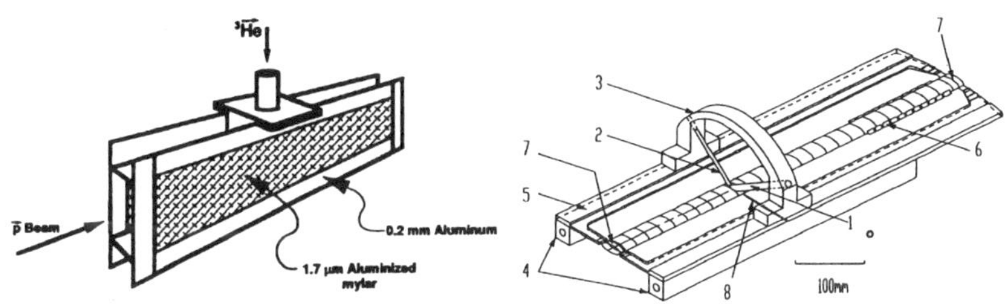

Figure 2. Storage cells used for the $\bar{p}^3\vec{\mathrm{He}}$ studies at IUCF[11] (left) and the storage cell for HERMES[27] (right): 1) H/D feed tube, 2) sample tube, 3) support arch, 4) cooling rails, 5) support plates, 6) cell extension, 7) end support 8) unpolarized feed tube.

HERMES The storage cell for HERMES (Fig. 2) is adapted to the beam distribution by an elliptical cross section ($w = 30$ mm, $h = 10$ mm). The cell walls consist of 75 μm thick aluminum, coated by drifilm to prevent depolarization[28]. The cell can be cooled to temperatures between 35-260 K by cold helium gas that passes through cooling rails alongside of the target cell. A thin layer of ice formed inside the cell was found to effectively suppress both depolarization and recombination. This effect was first observed at Heidelberg during the development of the target polarimeter[29]. The strong magnetic holding field must be rather uniform to prevent depolarization of target atoms by hyperfine transitions, induced by the periodic time structure of the HERA positron beam[30]. The HERMES target[31] is further discussed in Sec. 5.1.

8

PINTEX: For detection of low-energetic recoil particles, e.g. spectator protons from a deuteron target, thin cell walls are required. The PINTEX cell complies with this requirement (Fig. 3). Cells of this type were used for measurements of polarization observables in pp-elastic scattering at IUCF[32]. Teflon foil of $\approx$ 450 μg/cm^2 thickness, stretched over fins, form a quadratic channel of about 1 cm^2 cross section[33,34].

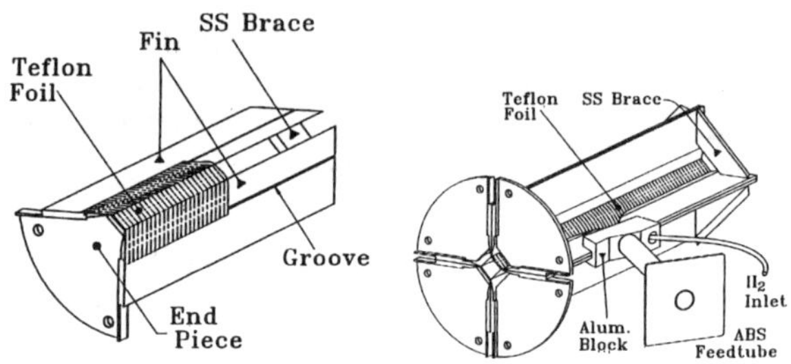

Figure 3. Storage cell for measurements of polarization observables in pp-elastic scattering[32]. One quadrant is depicted on the left, the fully assembled cell is shown on the right. (Figures from ref.[33].)

The cells for the pion production experiments[35] were made from 25 μm aluminum foil, coated by Teflon TPFE 3170. For the measurements of the nuclear polarization of molecular hydrogen formed by recombination, the cells had a teflon coated valve and a Cu-recombiner box attached at the center, whereby the atomic-to-molecular fraction inside the cell could be varied[36].

EDDA: Tests with a storage cell at COSY have been performed by the EDDA collaboration[19] Although the EDDA detector [18] was not designed to accomodate an extended target, employing a storage cell would lead to an enhancement in target thickness of about a factor ten[19], not taking into account the significant reduction in beam intensity because of acceptance limitations due to the cell during injection. For the time-reversal-invariance experiment[37] the full target thickness contributes and an enhancement factor of about 60 applies. The EDDA cell (Fig. 4) consists of a 300 mm long cell corpus ($h = 12$ mm, $w = 29$ mm) of 0.2 mm thick aluminum, coated by Teflon PTFE 3170, that can be cooled to 80-100 K. Polarized atoms are injected into the cell through a conical 10 cm long feed tube (entrance/exit diameter 20/12 mm, Fig. 4), not rigidly attached to the cell, in order to remove the cell from the beam position without breaking of vacuum.

Laser-driven Spin-Exchange Target: The storage cell of the first internal target experiment utilizing a laser driven spin-exchange source,

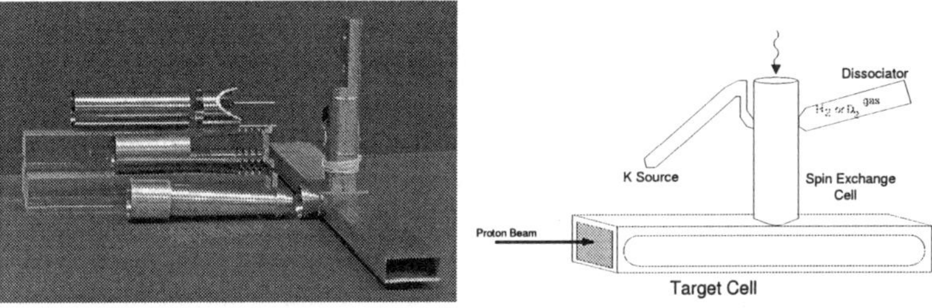

Figure 4. Storage cell used by EDDA at COSY (left) with cell body and mechanical fixture to hold the conical entrance tube in place[19]. Storage cell for the optically pumped spin-exchange target at IUCF[38] (right).

presently installed at IUCF[20], is shown in Fig. 4. The spin exchange cell is directly attached to the storage cell. Depolarization is inhibited by drifilm coating. Condensation of potassium vapour on the cell walls is prevented by operating the cell at 180 °C. The 400 mm long cell (wall thickness 3.2 mm ($h = 19.1$ mm, $w = 38.1$ mm) has thin windows of 0.3 mm thickness to the left and right to reduce multiple scattering of ejectiles. A flux of about 10^{18} polarized atoms/s generates a target thickness of $4 \cdot 10^{14}$ cm^{-2}, with vector polarization $P_z = 0.25$ for both hydrogen and deuterium targets[38].

5 Polarimetry of Internal Gas Targets

Various methods have been applied to measure the nuclear polarization of an internal gas target: *i)* Polarimetry through a **known reaction**, *ii)* **extraction of ions** formed inside the storage cell and subsequent polarization analysis, and *iii)* polarization analysis of an **extracted sample of neutral atoms** from the cell. These methods are discussed in more detail below.

Two **optical methods** were applied to determine the polarization of an internal gas target. At this stage both are suitable as a monitor only, and not for an absolute determination of the target polarization. The Balmer polarimeter measures the circular polarization of light from hydrogen atoms excited to $n = 3$ states by electron impact[39]. The initial electron polarization of the ground state atoms is partially transfered into angular momentum of the excited state. The emitted Balmer light is therefore circularly polarized and a measurement of the initial electron polarization is possible. The method is restricted to polarized targets operated in a weak guide field, where to a good approximation electron and nuclear polarization of hydrogen atoms are the same. At the ^{3}He target of HERMES an optical monitor measured the circular polarization of photons emitted in the $4^1D \to 2^1P$ transition[40].

It should be noted that a known (calibrated) reaction does not distinguish atoms from molecules, nor from any other contaminant in the cell. This method is therefore the first choice wherever feasible, because the target is sampled in the same way as in the experiment under consideration. Other methods determine the nuclear polarization of atoms, and, with additional instrumentation also the atomic-to-molecular fraction in the cell. Recently, the nuclear polarization of molecules was studied. A first measurement at NIKHEF[41] suggested a non-vanishing nuclear polarization of recombined deuterium molecules inside a storage cell $P_{\mathrm{molecule}}(\mathrm{D}_2^{\mathrm{rec}})/P_{\mathrm{atom}}(\mathrm{D}) = 0.81 \pm 0.32$. A more precise experiment with hydrogen molecules, reports $P_{\mathrm{molecule}}(\mathrm{H}_2^{\mathrm{rec}})/P_{\mathrm{atom}}(\mathrm{H}) = 0.42 \pm 0.02$ in a 0.66 T magnetic field[36].

5.1 Known Reactions

Suitable for the determination of the nuclear polarization is any *calibrated* reaction. Two examples are discussed below.

Example 1: 27 MeV αp scattering:
Analyzing power and cross section for 27 MeV α particles incident on a proton target are well known[42]. During the FILTEX test experiment at TSR (Fig. 5), polarization and density of the FILTEX polarized hydrogen target could be measured[10]. The target was operated in a weak vertical magnetic guide field. Two measurements, separated in time by about four weeks during which high proton currents were stored, revealed that the polarization of the target is very stable (Fig. 5). It was concluded that no damage to the wall coating occurred. The target was operated at a temperature of 125 K where the figure of merit is highest.

Example 2: pp elastic scattering: The PINTEX detector system at IUCF[14], is designed specifically for experiments involving both polarized beams and polarized targets. It enables a determination of target and beam polarization through the detection of pp elastic scattering. A precise analyzing power measurement in pp elastic scattering[43] provides the basis of a polarization calibration. It is of great importance that the polarization can be exported to serve as a standard at other energies[44]. Through the analysis of elastic pp events, it was possible to measure the target polarization along the cell axis, as shown in Fig. 6. It was also possible to deduce the effective reversal time (7 ± 1 ms), e.g. the time it takes to completely reverse or reorient the target polarization. High stability and high polarization characterize the longterm behaviour of the target, as illustrated in Fig. 6. Typical target

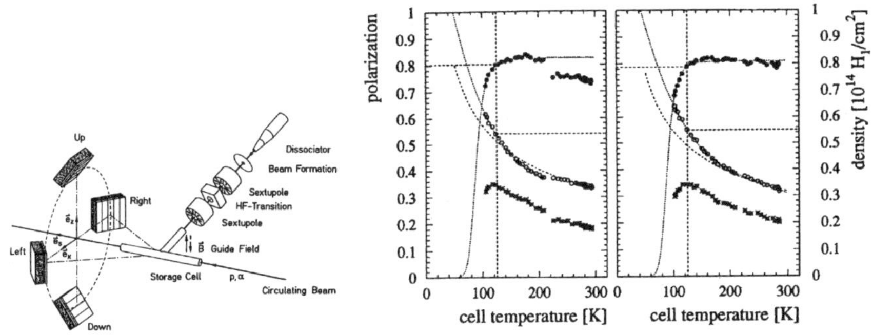

Figure 5. Polarized internal target setup of the FILTEX test experiment[10] (left). Four sets of scintillation counters are used to detect scattered α-particles. Polarization P ($\bullet$), density d_t ($\circ$), and figure of merit $P^2 \cdot d_t$ ($\star$) of the target as a function of the central cell temperature (right) for two measurements separated by four weeks.

thicknesses of $2 \cdot 10^{13}$ atoms/cm^2 with atomic beam intensities injected into the feed tube of the storage cell of about $3.6 \cdot 10^{16}$ atoms/s were obtained, with luminosities of about $5 \cdot 10^{28}$ cm^{-2}s^{-1} [32].

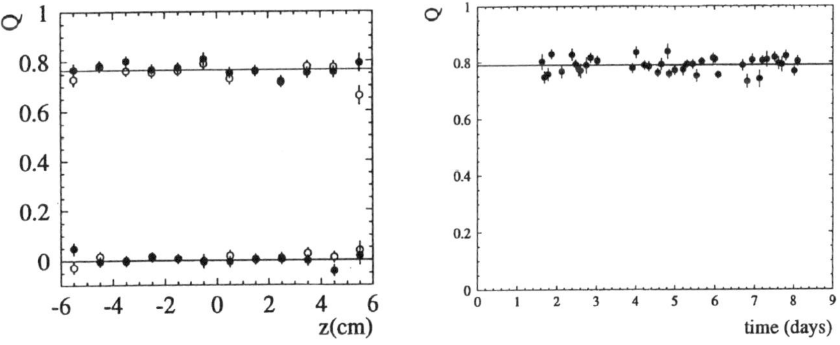

Figure 6. Target polarization as a function of the longitudinal position in the storage cell (left) and as a function of time during a run (right). (Figures from ref.[45].)

Ion-Extraction: A first experiment of this type used a low-energetic electron beam to ionize the target atoms, which were extracted, accelerated to 70 keV, and directed on a tritiated foil for polarization determination by the ^{3}H$(\vec{d}, n)^4$He reaction[46] (Sec. 4.1). The method can be directly applied to cell

12

targets internal to storage rings, because enough atoms are ionized by orbiting particles. The method is particularly useful in applications, where a strong longitudinal holding field is applied over the target which focuses the ions. The ion-extraction system of the internal target at NIKHEF [47,16] employed a spherical deflector (Fig. 7). Unfortunatley, this configuration is not directly

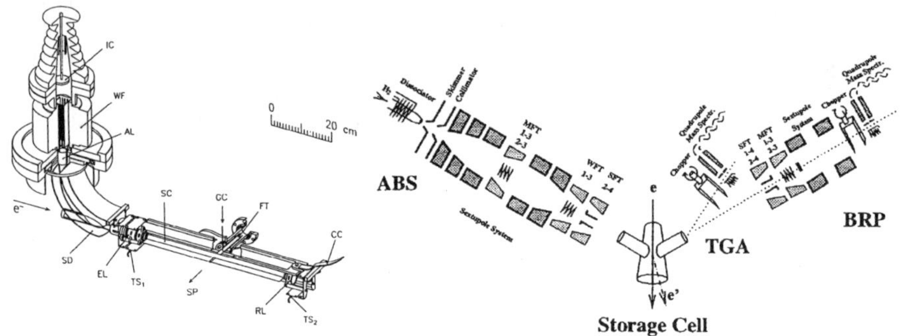

Figure 7. Left: Storage cell and ion-extraction system used by the NIKEF group[47]. SC: storage cell, FT: feed tube, SP: sample port, RL: repeller lens, EL: extraction lens, SD: spherical deflector, WF: Wien-filter. Right: The HERMES target[48]. The storage cell is fed by an atomic beam source (left). The composition of the target gas is measured via a target gas analyzer (TGA). The polarization of the target gas is measured by the Breit-Rabi Polarimeter (BRP).

applicable to polarized hydrogen targets, because of a lack of suitable low energetic reactions to analyze the polarization of protons. The ^{6}Li$(\vec{p},^3$He$)^4$He reaction, mentioned in ref.[46], requires energies of at least 260 keV, where the analyzing power becomes sufficiently large. A more promising method for both hydrogen and deuterium gas targets might be the combination of an ion-extraction system and a Lamb-shift polarimeter[49]. Such a polarimeter system is presently being developed at the University of Cologne for first use at the polarized internal gas target of ANKE at COSY[50].

Extraction of a Sample of Polarized Atoms: The HERMES target polarimeter determines the nuclear polarization of a small fraction of the target gas ($\approx 3\%$) extracted through a sample tube (Fig. 2). A polarization analysis is performed through a measurement of hyperfine state occupation numbers in a Breit-Rabi Polarimeter[29]. The polarimeter (Fig. 7) consists of a set of strong and weak-field transition units and a sextupole magnet system. A chopper is used to suppress background. A separate target gas analyzer measures the atomic-to-molecular fraction of the gas. Systematic errors of 0.01 for the nuclear polarization of the extracted sample of atoms are reported[48].

6 Conclusions and Outlook

Much experience has been gained in operating polarized internal targets in the environment of electron and ion storage rings. An impressive body of information on systematic effects arising from interactions of beam and target, such as unwanted reactions with the walls of the storage cell, radiation damage of the wall coating, longterm behaviour etc. has been obtained. The physics results cover a wide range from deep inelastic polarized lepton-nucleon scattering to measurements of spin correlation parameters and analyzing powers in elastic and inelastic pp collisions. Atomic physics experiments are carried out to investigate the nuclear polarization of recombined hydrogen and deuterium molecules. The experiments benefit from the high purity of the target material and the high degree of polarization. Fast polarization reversal and the alignment along different axes eliminates systematic asymmetries to a high degree. Furthermore, the low target thickness turns out to be actually a benefit, because lifetime and high momentum resolution of the stored beam are hardly altered by the thin target. Polarized targets internal to a storage ring can thus be regarded as an ideal tool for high precision experiments. They have just begun to open a new door to hadronic interaction studies and they will continue to play a major role in years to come.

The need for higher target thickness prevails, particularly for electronuclear experiments. Most experiments could digest a factor ten higher intensity from sources of polarized H and D atoms to feed storage cell targets. New developments in this field address a new dissociator based on a surface wave plasma discharge[51] and studies of a carrier jet system[52]. Higher fluxes of polarized atoms may be achieved through the application of new high field sextupole magnets. Considering the achievable fluxes from modern sources, a substantial increase in luminosity may also come from reduced transverse dimensions of the employed storage cells.

References

1. W. Haeberli, Proc. 2nd Int. Symp. on polarization Phenomena, eds. P. Huber and H. Schopper, Experientia Suppl. **12** (Birkhäuser, Basel, 1966), p. 64.
2. D. Kleppner *et al.*, Phys. Rev. A **138**, 972 (1965).
3. M.D. Barker *et al.*, AIP Conf. Proc. on Polarization Phenomena in Nuclear Physics, ed. G.G. Ohlsen, New York (1981), p. 931.
4. H. Döbbeling *et al.*, PROPOSAL CERN/PSSC/85-80 (1985) and Addendum (1986).

5. F. Rathmann *et al.*, Phys. Rev. Lett. **71**, 1379 (1993).

6. D. Toporkov, contribution to these proceedings.

7. R. Gilman *et al.*, Phys. Rev. Lett. **65**, 1733 (1990).

8. L.G. Isaeva *et al.*, Nucl. Instr. Meth. **A411**, 201 (1998).

9. F. Stock *et al.*, Nucl. Instr. Meth. **A343**, 334 (1994).

10. K. Zapfe, *et al.* Nucl. Instr. Meth. **A 368**, 293 (1996).

11. C. Bloch *et al.*, Nucl. Instr. Meth. **A354**, 437 (1995).

12. K. Lee *et al.*, Phys. Rev. Lett. **70**, 738 (1993); R.G. Milner, R.D. McKeown and C.E. Woodward, Nucl. Instr. Meth. **A274**, 56 (1989).

13. K. Ackerstaff *et al.*, Nucl. Instr. Meth. **A417**, 230 (1998).

14. T. Rinckel *et al.*, Nucl. Instr. Meth. **A439**, 117 (2000).
   http://www.iucf.indiana.edu/Experiments/PINTEX/pintex.html

15. B. von Przewoski, contribution to these proceedings; and T. Wise *et al.*, Nucl. Instr. Meth. **A336**, 410 (1993).

16. L.D. van Buuren, 4th Int. Conf., STORI 99, on Nuclear Physics at Storage Rings, eds. H.O. Meyer and P. Schwandt, AIP Conf. Proc. **512**, p. 362.

17. M. Bouwhuis *et al.*, Phys. Rev. Lett. **82**, 3755 (1999).

18. W. Scobel, p. 3 of Ref.[16]; and D. Albers *et al.*, Phys. Rev. Lett. **78**, 1652 (1997).

19. M. Glende, contribution to these proceedings.

20. R. V. Cadman *et al.*, Phys. Rev. Lett. **86**, 967 (2001).

21. H. Kolster, contribution to these proceedings.

22. M. Mikirtytchiants, contribution to these proceedings.

23. V.I. Komarov *et al.*, *Exclusive deuteron break-up study with polarized protons and deuterons at COSY*, COSY Proposal #20 (1992). well ad

24. S. Barsov *et al.*, Nucl. Instr. Meth. **A462**, 364 (2001).

25. J.S. Price and W. Haeberli, Nucl. Instr. Meth. **A349**, 321 (1994).

26. H.O. Meyer, Proc. Int. Workshop on Polarized Beams and Polarized Gas Targets, Cologne 1995, eds. H. Paetz gen Schieck and L. Sydow, World Scientific 1996, p. 355.

27. J. Stewart, Proc. 7th Int. Workshop on Polarized Gas Targets and Polarized Beams, eds. R.J. Holt and M.A. Miller, AIP Conf. Proc. **421**, p. 69.

28. G.E. Thomas *et al.*, Nucl. Instr. Meth. **A257**, 32 (1987).

29. B. Braun, *Spin Relaxation of Hydrogen and Deuterium in Storage Cells*, PhD Thesis, Fakultät für Physik, Ludwig-Maximilians-Universität, München (1995),
   http://www-library.desy.de/preprints.html#diss

30. K. Ackerstaff *et al.*, Phys. Rev. Lett. **82**, 1164 (1999).

31. P. Lenisa, contribution to these proceedings.
32. W. Haeberli *et al.*, Phys. Rev. C **55**, 597, 1997; F. Rathmann *et al.* Phys. Rev. C **58**, 658 (1998); B. von Przewoski *et al.*, Phys. Rev. C **58**, 1897 (1998); and B. Lorentz *et al.*, Phys. Rev. C 61, 054002 (2000).
33. M.A. Ross *et al.*, Nucl. Instr. Meth. **A344**, 307 (1994).
34. M.A. Ross *et al.*, Nucl. Instr. Meth. **A326**, 424 (1993).
35. H.O. Meyer *et al.*, Phys. Rev. C **63**, 064002, 2001.
36. T. Wise, contribution to these proceedings; and T. Wise *et al.*, Phys. Rev. Lett. **87**, 042741 (2001).
37. P.D. Eversheim, p. 224 of Ref.[16].
38. M.A. Miller, private communication.
39. J. Stenger, E. Steffens, and K. Zapfe, Nucl. Instr. Meth. **A330**, 21 (1993).
40. M.L. Pitt *et al.*, Proc. Int. Workshop on Polarized Beams and Polarized Gas Targets, Cologne 1995, eds. H. Paetz gen Schieck and L. Sydow, World Scientific 1996, p. 413.
41. J.F.J. van den Brand *et al.*, Phys. Rev. Lett. **78**, 1235 (1997).
42. P. Schwandt, T.B. Clegg and W. Haeberli, Nucl. Phys. **A163**, 432 (1971).
43. B. von Przewoski *et al.*, Phys. Rev. C **44**, 44 (1991).
44. R.E. Pollock *et al.*, Phys. Rev. E **55**, 7606 (1997).
45. W. Haeberli *et al.*, Phys. Rev. C **55**, 597, 1997.
46. J.S. Price and W. Haeberli, Nucl. Instr. Meth. **A326**, 416 (1993).
47. Z.L. Zhou *et al.*, Nucl. Instr. Meth. **A378**, 40 (1996).
48. B. Braun, Proc. 7th Int. Workshop on Polarized Gas Targets and Polarized Beams, eds. R.J. Holt and M.A. Miller, AIP Conf. Proc. **421**, p. 156.
49. J.E. Brolley, G.P. Lawrence and G.G. Ohlsen, Proc. 3rd Int. Symposium on Polarization Phenomena in Nucl. Reactions, Madison 1970, eds. H.H. Barschall and W. Haeberli, Univ. of Wisconsin Press Madison, p. 846 (1971).
50. R. Engels, contribution to these proceedings.
51. N. Koch and E. Steffens, Rev. Sci. Instrum. **70**, 1 (1999).
52. V.L. Varentsov *et al.*, Proc. 7th Int. Workshop on Polarized Gas Targets and Polarized Beams, eds. R.J. Holt and M.A. Miller, AIP Conf. Proc. **421**, p. 381.

NUCLEAR POLARIZATION OF RECOMBINED HYDROGEN ATOMS

T. WISE, W.HAEBERLI[1],B. LORENTZ,[1,2], P. A. QUIN[1], F. RATHMANN[1,2],
B. SCHWARTZ[1], T. G. WALKER[1], A. WELLINGHAUSEN[3], J. T. BALEWSKI[3], J. DOSKOW[3],
H.O. MEYER[3], R. E. POLLOCK,[3] B. V. PRZEWOSKI[3], T. RINCKEL[3], SWAPAN K. SAHA[4,5]
AND P.V. PANCELLA[6],

[1]*University of Wisconsin-Madison, WI 537061, USA*
[2]*Institure fürKernphysik, Forschungzentrum Jülich, 52425 Jülich, Germany*
[3]*Indiana University Cyclotron Facility, Bloomington, IN 47405*
[4]*University of Pittsburgh, Pittsburgh, PA 15260*
[5]*Bose Institute, Calcutta 700009, India*
[6]*Western Michigan University, Kalamazoo, MI 49008*

We have measured the nuclear polarization of hydrogen molecules formed by recombination of polarized atomic hydrogen gas. A polarized atomic hydrogen beam is incident upon a copper recombination zone and subsequently drifts into an internal target located in a straight section of the IUCF Cooler ring. The target contains an internal valve that isolates the recombination zone. With the valve closed the atomic beam enters the Teflon coated target cell directly. In this way we rapidly alternate between a mostly atomic and a mostly molecular target. A comparison of the target polarization for these two states can be used to determine the fraction of the initial atom polarization that survives recombination and subsequent wall collisions in the target. That fraction was studied for temperatures between 50K and 300K and for applied magnetic fields between 0.5mT and 0.6T. The target polarization was measured with a 200 MeV longitudinally polarized proton beam using the large p-p elastic spin correlation coefficient, A_{zz}. A simple model has been developed which relates the molecular depolarization to the randomization of the rotational states of the H_2 molecules after each wall collision.

1 Motivation

During the last few years increased use has been made of polarized hydrogen and deuterium gas targets, which are placed in the circulating beam of storage rings. In order to increase the target thickness over that obtained by a jet of polarized H atoms, the beam from atomic beam sources is directed into an open cell (``storage cell") in which the atoms make several hundred collisions before escaping from the target [1]. One example is the series of experiments by the HERMES collaboration, which has studied e.g. deep inelastic scattering of 27.6 GeV electrons from a polarized H target at DESY [2]. Other applications are measurements of spin correlation parameters in proton-proton scattering at IUCF [3], and studies of electronuclear form factors at Novosibirsk [4] and at NIKHEF [5].

The polarization of the H atoms in such a target can be monitored by an atomic polarimeter which determines the relative populations of the four hyperfine states of H. However, a fraction of the atoms recombine to form H_2 whose nuclear polarization is unknown, which causes a significant systematic error in some of the above measurements. The amount of recombination will depend strongly on the surface, its temperature, and on the number of wall collisions encountered by atoms before escape from the cell. We introduce the degree of dissociation, α, for gas in the target: $\alpha = n_{atom}/(n_{atom} + 2n_{molecule})$. The target polarization may then be expressed as: $P_{target} = \alpha P_{atom} + (1-\alpha)P_{molecule}$. For $\alpha = 0.95$, $\delta P/P = 5\%$ if no information about the molecular polarization is available.

This raises the interesting question to what extent the nuclear polarization of the atoms is preserved after recombination. A recent measurement with deuterium [6] reported that the molecules retain the tensor polarization of the atoms, but the statistical uncertainty of the measurements (fraction polarization retained = 0.81 $\pm$0.31 was very large.

Here we report measurements of the recombination of H atoms where we compare the nuclear polarization P_{atom} of the atoms from an atomic beam source [7] to the nuclear polarization $P_{molecule}$ of the molecules that are recombined on a copper surface. The temperature and applied magnetic field at the surface are systematically varied between 35-300K and 5mT-0.6T respectively. At the typical densities encountered in storage cells, three-body gas phase recombination may be ignored.

2   Measurement Method

The nuclear polarization is measured by passing a 203 MeV longitudinally polarized proton beam at the IUCF cooler ring through a target cell containing the gas to be analyzed. The polarization direction of the target is also longitudinal, defined by the longitudinal field of an open bore super-conducting magnet whose coils are shown in figure 1. We make use of the large spin correlation coefficient A_{zz} in pp elastic scattering near 45° in the lab [8]. With both target and beam polarized longitudinally the reaction cross-section may be written as: $\sigma(\theta) = \sigma_o(\theta)\{1+P_{beam}P_{target}A_{zz}(\theta)\}$, which for our conditions of $P_{beam} \approx 0.65$, $P_{target} \approx 0.75$ and $A_{zz} \approx 0.9$ gives a large spin dependent effect, $\sigma = \sigma_o(1\pm0.4)$. The choice of longitudinal target polarization was necessary to avoid large perturbations to the orbit of the circulating beam. The detector yields are, of course, proportional to the product of cross- section, target thickness, and integrated charge through the target. Throughout we factor out the charge dependence by dividing all yields by the integrated charge as measured by a current transformer.

The target cell is equipped with a remotely actuated internal valve with a Teflon gate as shown in figure 1. In the closed position atoms enter the cell and encounter only Teflon and aluminum coated by Teflon. Teflon is known to strongly inhibit both depolarization and recombination [9,10]. In the open position, atoms pass directly into the copper recombination zone where about 90% of them recombine and drift back into the cell as H_2. Inside the recombiner and cell they travel at a velocity $\sqrt{2}$ lower than the H atoms and therefore make the target thicker. The detector yields may then be expressed as $Y = Y_o t(\alpha)(1 + P_{beam}P_{target}A_{zz})$, where $t(\alpha)$ is the recombination dependent change in target thickness. For a purely atomic target $t(\alpha) = 1$ and for a purely molecular target $t(\alpha) = \sqrt{2}$. Neither condition is achieved in practice and therefore the ratio $t \equiv t(\alpha)_{open}/t(\alpha)_{closed} < \sqrt{2}$. By passing an unpolarized beam through the target it is possible to measure t. We obtain $t = 1.364 \pm 0.004$. The relationship between t, α_{open} and α_{closed} is

$$\alpha_{open} = \frac{1/t\left[1+\alpha_{closed}\left(\sqrt{2}-1\right)\right]}{\left(\sqrt{2}-1\right)}.$$
1)

Figure 2 shows the locus of pairs (α_{open}, α_{closed}) which satisfy equation 1. The locus appears as a diagonal line at nearly 45 degrees. The band surrounding the line is the statistical uncertainty of the data. Vertical and horizontal lines represent additional

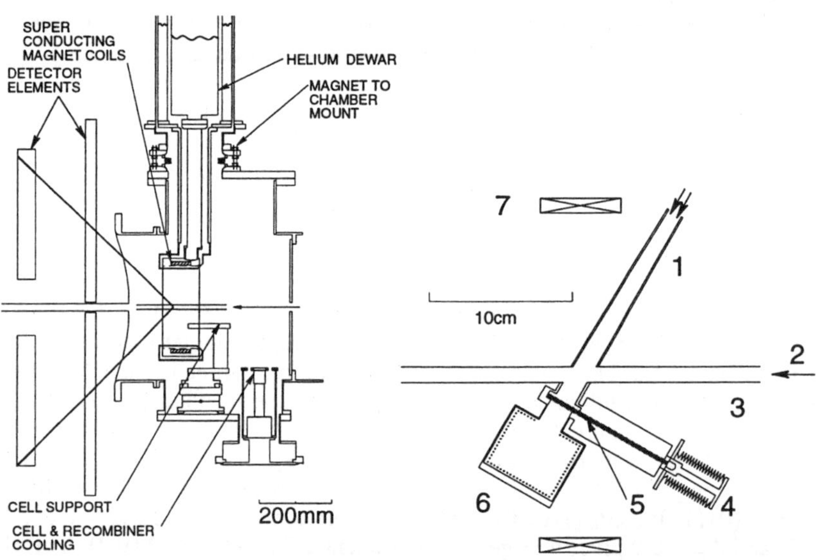

Figure 1. Left: side view showing the overall arrangement of target cell and detectors. The recombiner and storage cell which are mounted to the cell support are not shown in this view. Right: top view of cell and recombiner showing 1) atomic entrance tube, 2) 200 MeV proton beam, 3) beam tube, 4) pneumatic valve actuator, 5) Teflon valve gate, 6) copper recombiner with copper mesh, 7) Magnet coil.

physical limits. For example α_{closed} must be less than 1 because of ballistic and diffusive flow of molecules from the atomic beam source. This limit appears as the horizontal line near $\alpha_{closed} = 0.99$. A much stronger constraint to α_{open} and α_{closed} (represented by an additional nearly vertical band in figure 2) is applied in a later section.

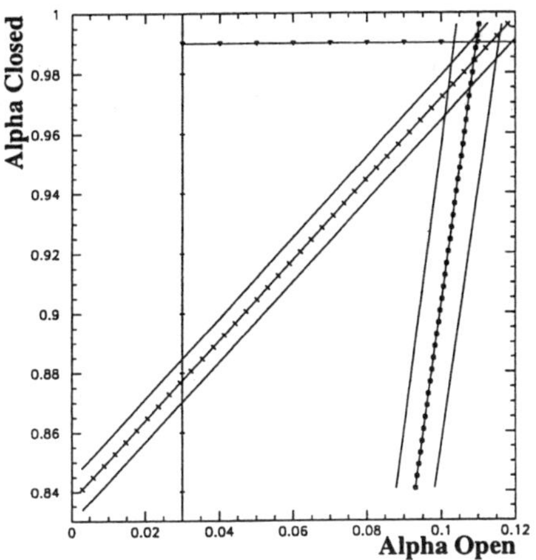

Figure 2. Allowed pairs of α_{open} and α_{closed} (~45° band). The horizontal and vertical lines are the weak physical limits. Adding the model dependent constraint described in section 6 restricts α_{open} and α_{closed} to the parallelogram near $\alpha_{closed} = 0.98$.

3   Measurement Protocol

Measurements are taken for valve closed and valve open with a polarized beam in the ring. The beam is dumped and the measurements are repeated with *unpolarized* beam. Figure 3 divides the two injections into four time periods A-D where the following relationships hold:

$$\alpha(A) = \alpha(D) \quad \text{(low alpha)}$$
$$\alpha(B) = \alpha(C) \quad \text{(high alpha)}$$
$$P_{beam}(A) = P_{beam}(B) = P_z$$
$$P_{beam}(C) = P_{beam}(D) = 0 \ .$$

We obtain the ratios of yields:

$$R_{open} = \frac{Y(A)}{Y(D)} = \left(1 + P_z P_{open}^{target} A_{zz}\right) \qquad 2)$$

$$R_{closed} = \frac{Y(B)}{Y(C)} = \left(1 + P_z P_{closed}^{target} A_{zz}\right)$$

and solve for

$$\frac{P_{open}^{target}}{P_{closed}^{target}} = \frac{R_{open} - 1}{R_{closed} - 1} . \qquad 3)$$

A significant simplification of the experiment results from the cancellation of P_z and A_{zz} in equation 3. We only require that P_z and A_{zz} be large in order to maintain the statistical significance of the data. The measurements we took consist of a series

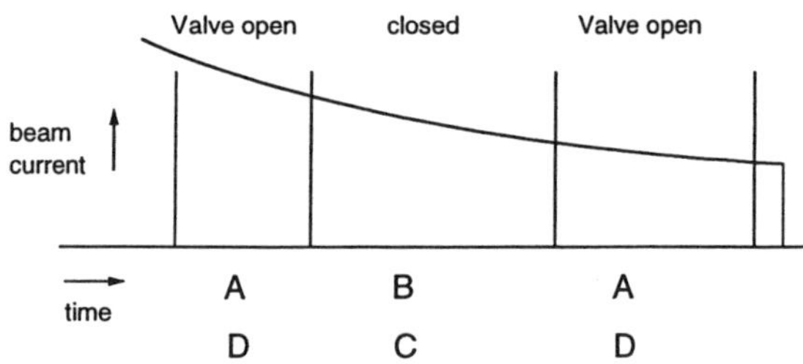

Figure 3. Division of the *unpolarized* injections into two time periods "A" (valve open) and "B" (valve closed) and the *polarized* injections into similar time periods "C" and "D". The relative duration of the periods is chosen to account for the decay of the beam current during the fill.

of determinations of the ratio in equation 3 as the B field and temperature at the recombiner were varied.

The aim of this experiment is to measure the ratio $R = P_{molecule}/P_{atom}$. In the ideal limit of $t = \sqrt{2}$, $R = P_{target\text{-}open}/P_{target\text{-}closed}$. Otherwise there is a more complicated relationship, with R depending additionally on t, α_{open}, and α_{closed}:

$$R = \frac{P_{molecule}}{P_{atom}} = \frac{\alpha_{closed}\left(P_{open}^{target}/P_{closed}^{target}\right) - \alpha_{open}}{1 - (1/t)(1 - \alpha_{closed}) - \alpha_{open}} \qquad 4)$$

4   Data Analysis and Results

Events were selected by requiring timing coincidence between the two out-going protons. The elastic nature of the beam-target reaction requires that the desired events have a ϕ separation of 180°. The delta ϕ spectrum has a strong peak at 180° with a somewhat wider background which we interpret as quasi-elastic scattering from higher z atoms contained in water, aluminum, and Teflon vapor. The subtraction of this background amounted to ~1% of the integrated peak. A small dead-time correction was also necessary. We corrected for a leak in the recombiner valve that was present only in the valve open position. The size of this effect (~2%) was measured by injecting bottled H_2 gas into the cell for a number of beam injection cycles.

Figure 4 shows R (Eq. 4) vs. B for the data we collected. The error bars combine the statistical uncertainty with the weak constraint on α_{open} and α_{closed} as represented in figure 2. We interpret the decrease in R at low B as arising from spin-relaxation of the ortho H_2 molecules during the time periods between wall collisions. In free flight, internal magnetic fields B_c from the spin-rotation interaction (B'=2.7 mT) and the direct dipole-dipole interaction (B''=3.4mT) cause the nuclei to rapidly precess about a direction that is skew to the external field by $\theta = B_c/B$. The orientation of B_c is randomized by each wall collision. Between successive wall collisions the component of the polarization along the external field decreases by an amount $(B_c/B)^2$. Thus after n wall bounces

$$R = R_0 \exp\{-n(B_c/B)^2\} \qquad\qquad 5)$$

where B_c is 6.1 mT as reported in [11]. Assuming $R_o = 0.46$, and an exponential distribution of n with a mean of 1000, gives the curve in Fig. 4.

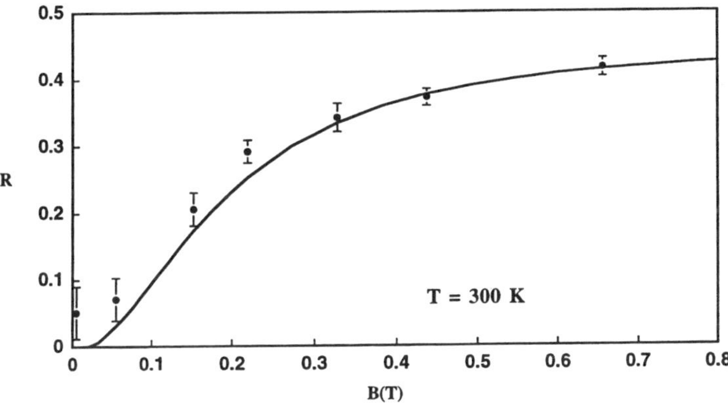

Figure 4. R vs B(Tesla) with the weak constraint on α_{open} and α_{closed} as shown in figure 2. The line represents the result of equation 5 with an exponential age distribution with a mean of 1000 wall bounces.

5   Further Analysis

A Monte Carlo simulation was developed where we modeled the evolution of atoms and molecules in the storage cell keeping track of their bounce age, recombination status, and their probability of detection after interacting with the circulating beam. The resulting molecular age distribution shown is clearly not a simple exponential and has $\langle n \rangle = 1600$. When this distribution is applied to Eq. 5 the model predicts *molecular* polarization essentially zero at the lowest two field values. In contrast, we measured $P_{target\text{-}open}/P_{target\text{-}closed} = 0.112$ at the lowest fields and we interpret this residual polarization as arising from polarized atoms which failed to recombine. This generates the relationship $\alpha_{open}/\alpha_{closed} = 0.112$ and the nearly vertical constraint band in figure 2. This additional constraint restricts the range of possible values for α_{open} and α_{closed} and the error bars are thereby greatly reduced. Figure 5 shows the B dependence of ($P_{target\_open}/P_{target\_closed}$) and R after a multi-parameter fit to the data that includes this new constraint. The solid curves are the prediction of the Monte Carlo. The fit is in reasonable agreement with the data but the χ^2 is rather large.

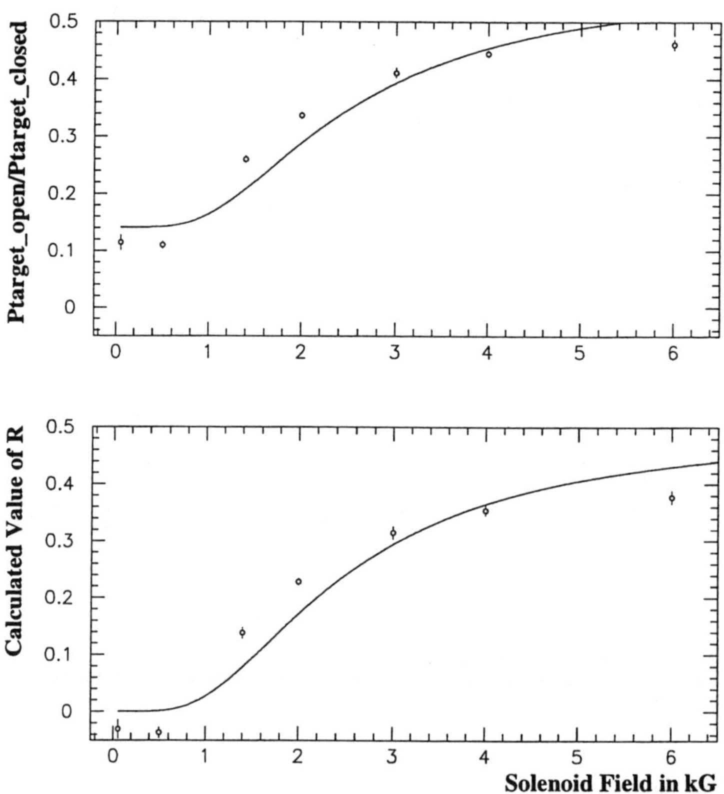

Figure 5. $P_{target\text{-}open}/P_{target\_closed}$ and R vs. B field using the tight constraint on α_{open} and α_{closed}.

In our model we assume that the rotational quanta are completely randomized by each wall collision and that rotational phase coherence is completely lost in the time between collisions. We can allow for the possibility that this is not the case by including a factor $0<\eta<1$ in the predicted depolarization per wall collision:

$$\frac{\Delta P}{P} = \eta\left(B_c/B\right)^2.$$

6)

Figure 6 shows the χ^2 of the fit of the Monte Carlo to the data vs. η which has a minimum at $\eta = 0.412$. The optimum fit to the data is shown figure 7. For the minimum $\chi2$ we obtain $\alpha_{open} = 0.102 \pm0.0055$, $\alpha_{closed} = 0.9791 \pm0.0119$, $R_o =$

0.42547 ±0.0067, and χ^2/df = 0.6. This analysis suggests that the molecular rotational quanta are not fully randomized by wall collisions.

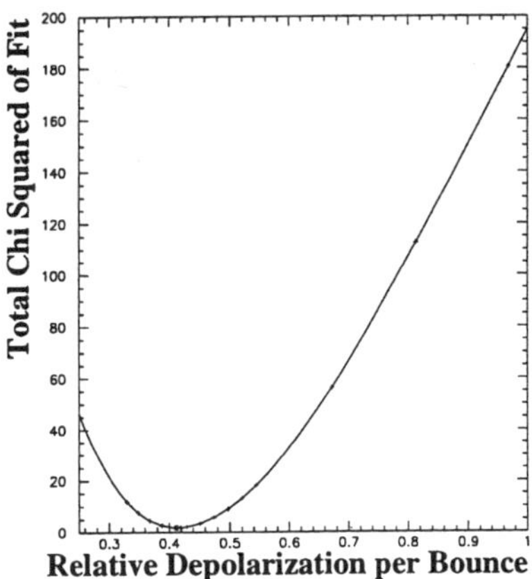

Figure 6. Total χ^2 of the fit of the Monte Carlo to the data vs. η.

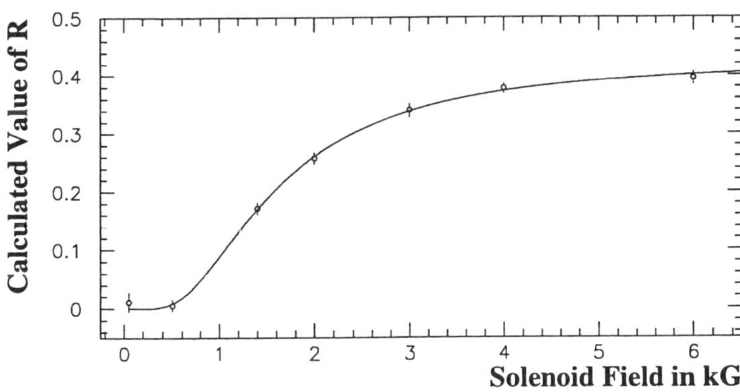

Figure 7. Fit to data for $\eta = 0.421$

The recombination proceeds either by an interaction between two atoms chemisorbed on the surface or between one chemisorbed atom and an atom from the gas phase. The number of available chemisorbed sites M is on the order of $10^{15}/cm^2$ on copper. At room temperature and below the average dwell time for chemisorbed hydrogen is long, and the dominant desorption mechanism is recombination. We have a recombining area $A \approx 200cm^2$ and a flux $N \approx 3 \times 10^{16}$ atoms/s resulting in an average dwell time on the surface $\langle \tau \rangle = 2M/N \approx 6s$. Thus we can safely assume the chemisorbed atoms have resided on our copper surface for a sufficient time to fully depolarize. The fraction of the atomic polarization retained by the molecules, R, for our situation is then expected to be limited to R<0.5 as the recombination proceeds between a polarized gas phase atom and a fully depolarized surface atom. In other situations such as high bombardment rates, low availability of chemisorb sites, or high temperatures where the chemisorbed dwell time is short, this assumption may not be valid.

Finally figure 8 shows our results for the temperature dependence of ($P_{target\_open}/P_{target\_closed}$). The points follow a curve similar to that measured by Price[12] for atom polarization on copper indicating that the loss at low temperatures is due to depolarization of atoms inside the recombiner.

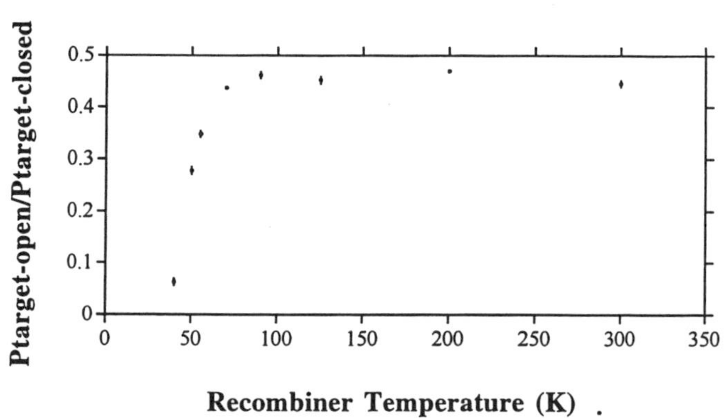

Figure 8. Temperature dependence of the target polarization.

6   Acknowledgements

This work was supported by the US National Science Foundation under Grants PHY-9602872, PHY-9722556, PHY-9901529 and by the US Department of Energy under Grant DOE-FG02-88ER40438.

References

1. W Haeberli, in *Free and Stored Atomic Beams as Internal Polarized Targets*, AIP Conf. Proc. No. 128 (AIP,New York, 1984) p. 251.
2. A. Airapetian *et al.*, *Phys. Lett. B* 442, 484 (1998).
3. F. Rathman *et al.*, *Phys. Rev. C* 58, 658 (1998); W. Haeberli *et al.*, *Phys. Rev C* 55, 597 (1997).
4. R. Gilman *et al.*, *Phys.Rev. Lett.* 65, 1733 (1990).

5. M. Ferro-Luzzi *et al.*, *Phys.Rev. Lett.* 77, 2630 (1996).

6. J. F. J. van den Brand *et al.*, *Phys.Rev. Lett.* 78, 1235 (1997).

7. T. Wise, A. D. Roberts, and W. Haeberli, *Nucl. Instrum. Methods Phys. Res,*. Sect. A 336, 410 (1993).

8. B. Lorentz et *al.*, *Phys. Rev. C* 61 54002 (2000).

9. J. S. Price and W. Haeberli, *Nucl. Instrum. Methods Phys. Res.* Sect. A 349, 321 (1994).

10. K. Zapfe et al., *Phys. Rev. C* 61, 28 (1995).

11. T. Wise et al., *Phys.Rev. Lett.* 87, 042701 (2001).

12. J. S. Price, Ph.D thesis, University of Wisconsin 1993. Available as Dissertation No. 9318638, from Dissertation Express (http://wwwlib.umi.com/dxweb/).

THE GASEOUS POLARIZED H-D TARGET OF THE HERMES EXPERIMENT

P. LENISA

ON BEHALF OF THE HERMES COLLABORATION

Università di Ferrara and INFN,
44100 Ferrara, ITALY
E-mail: lenisa@hermes.desy.de

The HERMES target makes use of a storage cell internal to the HERA lepton ring in which polarized D/H atoms from a Stern-Gerlach Atomic Beam Source (ABS) are injected. A sample beam is extracted from the center of the storage cell to determine the atomic fraction and the atomic polarization. The analysis of the data and the performance with Deuterium gas in 2000 are presented.

1 Target Setup

The HERMES experiment in the HERA electron storage ring at DESY is studying the spin structure of the proton and neutron via deep inelastic scattering of polarized electrons off the nucleons in a highly polarized internal gas target. The target was operated with Hydrogen gas in 1996 and 1997 and has been modified to use Deuterium gas since 1998. A new run with Hydrogen will start in 2002 when HERMES is planning to study transversity. A schematic diagram of this target, is shown in Figure 1

A beam of H/D atoms is generated RF or MW dissociator which forms part of the atomic beam source (ABS) [1]. These atoms are electron polarized by means of Stern-Gerlach separation in a sextupole magnet system. The electron polarization is transferred to the nucleons by means of adiabatic high frequency transitions that interchange the occupations of two different hyperfine states.

The beam of nuclear polarized atoms is injected into the center of the thin-walled storage cell via a side tube and the atoms then diffuse to the open ends of the cell where they are removed by a high speed pumping system. The storage cell acts to confine the polarized atoms near the beam and therefore increases the target areal density 150 times the free beam. The cell is coated with Drifilm in order to minimize wall interaction effects.

In the 1997-2000 running a longitudinal magnetic field of approximately 330 mT was maintained over the cell volume by a superconducting magnet coil. For the Hydrogen running starting 2002 the transverse field produced by a conventional dipole magnet will be used. The field provides a quantisation

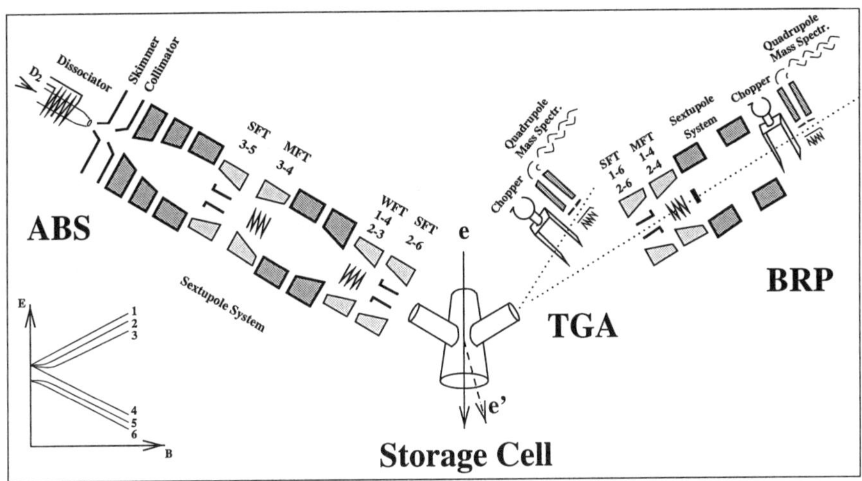

Figure 1. Schematic picture of the HERMES target.

axis for the spins and inhibits nuclear spin relaxation by decoupling nucleon and electron spins.

A second side tube is provided to sample the gas within the target cell. The beam emerging from this tube is analysed with a Target Gas Analyser (TGA) [2] to determine its atomic fraction and a Breit-Rabi polarimeter (BRP) [3] to measure its hyperfine occupation in order to calculate the atomic polarization. During the atom diffusion process relaxation by wall and spin exchange collisions and wall recombination changes the polarization and the atomic fraction of the target gas. The atom polarization and atomic fraction values measured by the BRP and TGA must be corrected for these effects to obtain the absolute target polarization.

2 Data analysis

2.1 Basic Formulas for the Target Polarization

The average target polarization is described by the following expression:

$$P = \alpha_0 \left[\alpha_r P_a + (1 - \alpha_r) P_m \right], \tag{1}$$

where P_a, P_m are the nuclear polarization of atoms and molecules; α_0

and α_r are the initial atomic fraction and the atomic fraction surviving recombination and they can be expressed by:

$$\alpha_0 = \frac{\bar{n}_a + 2\bar{n}_r}{\bar{n}_a + 2\bar{n}_m} \qquad \alpha_r = \frac{\bar{n}_a}{\bar{n}_a + 2\bar{n}_r} \tag{2}$$

$\bar{n}_a$, $\bar{n}_m$ are the atomic and the molecular density averaged over the cell; $\bar{n}_r$ is the density of the molecules produced by recombination in the cell.

The necessity of distinguishing between the two factors α_0 and α_r arises from the different origin of the molecules in the cell which can than present or not polarization. Molecules with polarized nuclei can originate only in the recombination of polarized atoms and the molecular polarization can be rapresented as: $P_m = \beta P_a$ [a].

2.2 From the measured fluxes to the averaged densities in the cell

The quantities in Equation (2) are average densities in the cell and they have to be related with the countrates measured in the detection system which are proportional to the flux of the sampled gas. The total atomic (ϕ_a) and molecular (ϕ_m) fluxes from the sample of the cell gas are continously monitored by the TGA: the detected count rates are proportional to the fluxes through the acceptance of the detector and the detection efficiencies.

Three different sources of molecules are present in the cell and the total molecular flux can then be written as:

$$\phi_m = \phi_{rg} + \phi_{bl} + \phi_r, \tag{3}$$

where: ϕ_{rg} is molecular flux due to the rest gas in the ABS and the target chamber; ϕ_{bl} is the contribution of molecules balistically injected into cell from the ABS and ϕ_r is the molecular flux originated from atom recombination in the cell.

The separate contributions to the molecular flux can be estimated by performing regular target calibrations [4,5].

Once the different contributions to the fluxes have been introduced, α_0 can be rewritten as:

$$\alpha_0 = \frac{\phi_a + \phi_r}{\phi_a + \phi_r + (\phi_{bl} + q\phi_{rg})[\sqrt{2} + (1 - \sqrt{2})(1 - \alpha_r)]}, \tag{4}$$

---

[a] As the mechanims of polarization transfer in the recombination process are not known, β in the Deuterim analysis has been considered to present an uniform distribution in the interval 0 - 1, i.e. $\beta = 0.5 \pm 0.5$.

In the previous formula, the parameter q has been introduced to take into account the different spatial distributions in the cell tube of $n_a + \sqrt{2}n_r$ and n_{bl} from n_{rg}, while the first factor $\sqrt{2}$ inside the parenthesis accounts for the relative weight of molecules and atoms when passing from fluxes to densities.

After the different fluxes have been defined, also the quantity α_r^{TGA} (expressing possible recombination in the cell) can be introduced:

$$\alpha_r^{TGA} = \frac{\phi_a}{\phi_a + \phi_r}. \tag{5}$$

α_r^{TGA} is then related to the average value in the cell α_r by a *sampling correction* $k_{\alpha r}$ which can be evaluted both analytically through the diffusion equation and numerically by means of Monte Carlo molecular flow simulations 6

$$1 - \alpha_r = k_{\alpha r}(1 - \alpha_r^{TGA}), \tag{6}$$

3   2000 running with Deuterium

The 2000 running with Deuterium has shown an outstanding smooth and stable performance of the overall target. This is confirmed if looking at the behaviour of the quantity α^{TGA} over the year (Fig. 2):

α^{TGA} is defined as

$$\alpha^{TGA} = \frac{\phi_a}{\phi_a + \phi_m} \tag{7}$$

α^{TGA} is a *self-normalized* quantity as it overcomes for possible systematics in the TGA performance due to variations in emission current and working temperature. α_{TGA} can thus be considered as on online monitoring of any systematic in the complete system: ABS + cell + TGA. Figures 2 evidences how smooth the behaviour of α^{TGA} was all over the year (the total dispersion less than 1.5%) reflecting the behaviour of the target.

The good performance of the target can be largely attributed to the almost ideal behaviour of the storage cell: no recombination and depolarization effects have ever been detected by means of the periodically perfomed calibrations during all the running. An independend confirmation of this assumption comes from the temperature scans of the storage cell. Both the atomic fraction and the atomic polarization showed no temperature dependence trough the year 2000. Since both the recombination and depolarization mechanisms

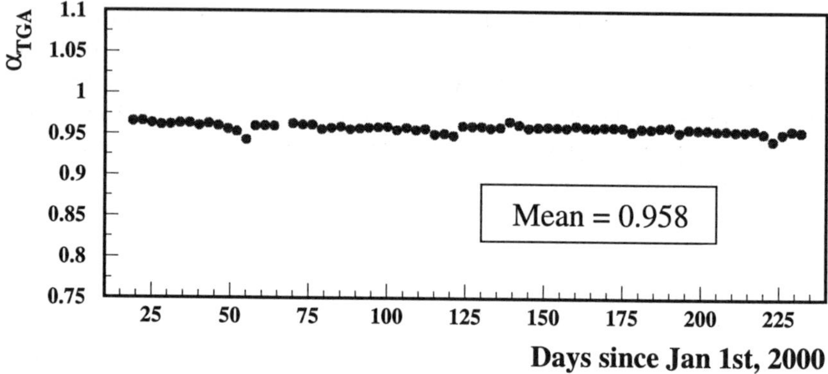

Figure 2. α^{TGA} for the 2000 Deuterium running.

active in the cell depend exponentially on the temperature [4,5], this can be considered as an indication that the storage cell used in 2000 caused no measurable recombination and depolarization on the cell surface.

The smooth and stable running during all the period allow to give an average value for the target polarization for all the period [7]:

$$P_{z+} = +0.851 \pm 0.031 \qquad P_{z-} = -0.840 \pm 0.028 \tag{8}$$

4 Conclusions

The HERMES target ran reliably with very stable operation during the 2000 running period. The collected data are consistent with having little or no recombination and depolarization inside the storage cell. The average value of the polarization has been given.

Starting 2002 the target will run transversally polarized Hydrogen.

References

1. N. Koch and E. Steffens Rev. Sci. Instrum. 70, 1999, 1631
2. M.C. Simani *et al The Gas Analyzer of the Target of the HERMES Experiment* submitted.

3. C. Baumgarten *et al The HERMES Breit-Rabi Polarimeter* to appear on NIM-A.
4. H. Kolster Ph. D Thesis, LMU München 1998
5. C. Baumgarten Ph. D Thesis, Univeristät München 2000
6. C. Baumgarten *et al Molecular Flow and Wall Collision Age Distribution* submitted.
7. HERMES Target Group; HERMES Int. Note

STATUS OF THE MICHIGAN ULTRA-COLD POLARIZED HYDROGEN JET TARGET*

V.G. LUPPOV, B.B. BLINOV, J.M. DENBOW, T. KAGEYA, M.C. KANDES,
A.D. KRISCH, D.A. KULKARNI, M.A. LEHMAN, M.A. LEONOVA,
V.S. MOROZOV, C.C. PETERS, R.S. RAYMOND, M.R. ROSS, K. YONEHARA

Spin Physics Center, University of Michigan, Ann Arbor, MI - 48109-1120, USA

N.S. BORISOV, V.V. FIMUSHKIN

Joint Institute for Nuclear Research , RU-141980, Dubna, Russia

A.F. PRUDKOGLYAD

Institute for High Energy Physics, RU-142284, Protvino, Russia

D. KLEPPNER

*Department of Physics, Massachusetts Inst. of Technology, Cambridge,
Massachusetts 02139, USA*

Progress on the Michigan ultra-cold proton-spin-polarized hydrogen Jet target is presented. We describe the present status of the Jet and some beam test results.

To study spin effects in high energy collisions, we are developing an ultra-cold high-density jet target of proton-spin-polarized hydrogen atoms (Michigan Jet). The Jet first uses a high magnetic field and an ultra-cold separation cell coated with a superfluid helium-4 film to produce a slow monochromatic electron-spin-polarized atomic hydrogen beam; an rf transition unit then converts this into a proton-spin-polarized beam[1].

A layout of the Michigan Jet is shown in Fig. 1. Atomic hydrogen is produced in a room-temperature rf dissociator and guided to the ultra-cold separation cell coated with superfluid helium-4 to suppress the surface recombination of atoms (see Fig. 2). The double walls of the cell form the mixing chamber of the dilution refrigerator. The cell's entrance and exit apertures are respectively located at about 95% and 50% of the superconducting solenoid's 12 Tesla central magnetic field. After the hydrogen atoms are thermalized by collisions with the cell surface, the magnetic field gradient physically separates the atoms according to their electron-spin states. The atoms in the two lowest hyperfine states ($|3>$ and $|4>$) are attracted toward the high field region and escape from the cell. They quickly recombine on bare surfaces and are

---

* Supported by a research grant from the U.S. Department of Energy.

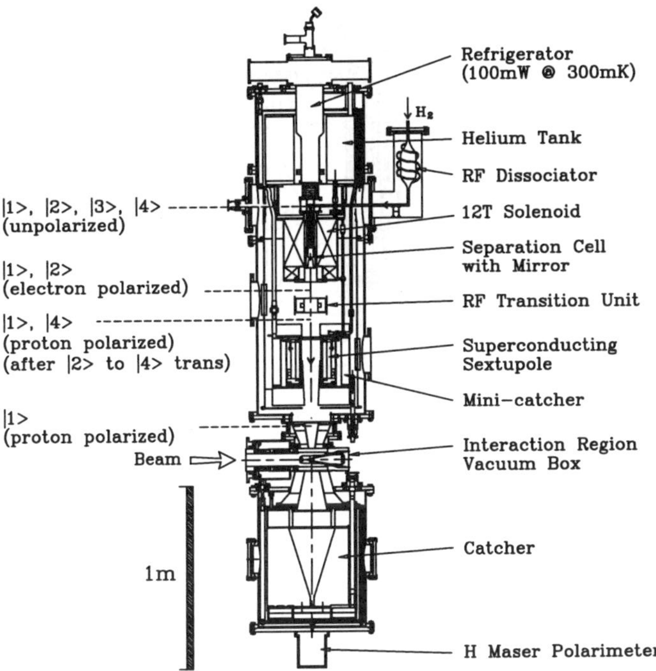

Figure 1. The layout of the Michigan ultra-cold Jet.

cryopumped. The atoms in the two higher hyperfine states ($|1>$ and $|2>$) are repelled toward the low field region and effuse from the exit aperture, forming a rather monochromatic electron-spin-polarized beam. To increase the jet density, we use a gold-coated copper focusing mirror with a polished surface covered with a helium-4 superfluid film similar to the prototype mirror[2].

After an rf transition unit, which changes state $|2>$ atoms into state $|4>$ atoms, the beam passes through a superconducting sextupole. The sextupole selects atoms in electron spin state $+1/2$, by focusing atoms in state $|1>$ into the interaction region and defocusing atoms in state $|4>$, which are then cryopumped. The proton-spin-polarized beam then passes through the interaction region and is caught below by a cryopumping catcher. A maser polarimeter below the catcher monitors the beam proton polarization.

Most of the Michigan Jet parts have been fabricated and successfully

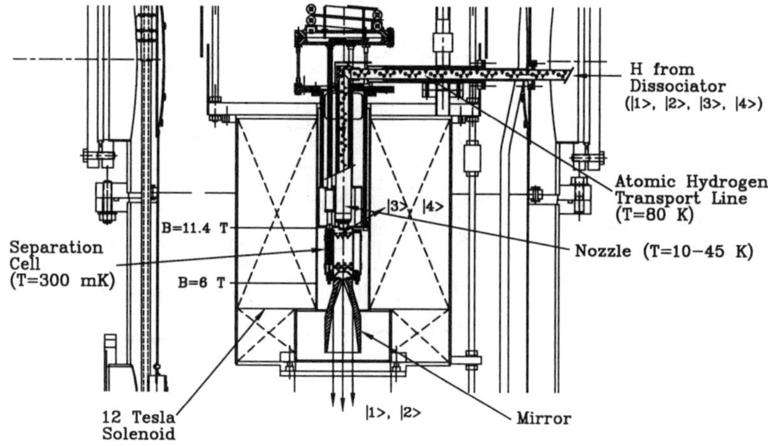

Figure 2. The layout of the Michigan Jet's electron-spin separation region.

tested. This hardware includes a 12 Tesla superconducting solenoid, a dilution refrigerator with a cooling power of about 80 mW at 300 mK, a 20 cm long superconducting sextupole magnet with iron poles and a 10.5 cm diameter bore, a cryocondensation pump with a measured pumping speed of about 1.2×10^7 liters s^{-1} (4.2×10^{26} atoms Torr^{-1} sec^{-1}), and a hydrogen maser polarimeter capable of monitoring the polarization to about $\pm$ 2% in a few minutes.

We studied a polarized beam of hydrogen atoms focused by the superconducting sextupole into a compression tube detector which measured the polarized atoms' intensity. The measured compression tube signal versus the focusing sextupole current corresponds to the calculated beam velocity distribution[1,3]. According to our Monte-Carlo simulations[3], the ultra-cold beam is very monochromatic due to the strong magnetic field gradient acceleration along the solenoid axis. Thus, we expected a small size of the focused beam; indeed, the measured radial beam distribution[1] has been as small as 4 mm FWHM. The superfluid-^{4}He-coated parabolic mirror, attached to the separation cell, appeared to increase the beam intensity by a factor of about 3, as expected.

The highest measured spin-polarized atomic hydrogen jet intensity, into the 11 mm by 1.4 mm compression tube slot, was about 2.2×10^{15} H s^{-1}. This intensity corresponds to jet thickness of about 1.1×10^{12} H cm^{-2}. So far, the

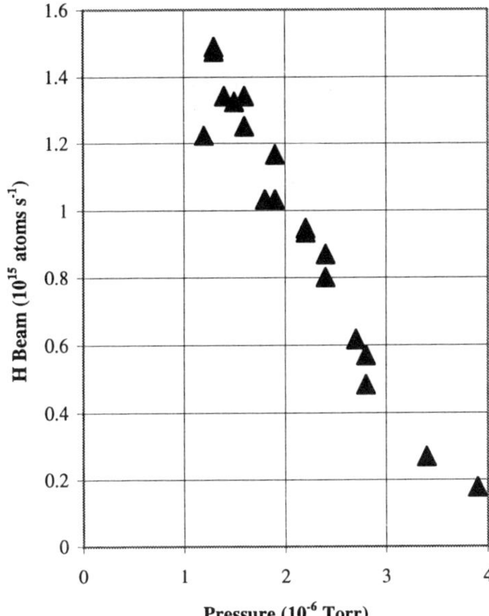

Figure 3. The measured beam intensity into the Compression Tube detector versus gas pressure in the cryostat.

intensity seems limited by the high vacuum pressure due to the evaporation of the separation cell's helium film. The measured beam intensity versus helium gas pressure in the cryostat is shown in Fig. 3. There is no sign of any saturation of the intensity at low pressure, which might indicate intrinsic beam scattering. Thus, we plan to increase the Jet's cryopumping capacity to farther decrease beam scattering from residual helium gas and thus to increase the beam intensity.

The electron-spin polarized beam has a proton polarization of about 50%. Most of the electron polarization will soon be converted into proton polarization by adiabatic passage through an rf transition unit with a novel ring dielectric resonator that accepts the 6 cm diameter beam. A room temperature prototype rf unit was built and tested; its maximum measured transition efficiency was 97%. A preliminary design and some tests of the cryogenic rf unit were recently made[4].

References

1. B.B. Blinov , *et al.*, Michigan Ultra-Cold Polarized Atomic Hydrogen Jet, in *Proceedings of the 14th International Spin Physics Symposium*, Osaka, Japan, October 2000, edited by K. Hatanaka, T. Nakano, K. Imai, H. Ejiri, AIP Conf. Proc. **570**, pp 856-860 (2001)
2. V.G. Luppov, W.A. Kaufman, K.M. Hill, R.S. Raymond, and A.D. Krisch, *Phys. Rev. Lett.* **71**, 2405 (1993)
3. V.G. Luppov , *et al.*, Status of the Mark-II Polarized Hydrogen Jet Target, in *Proceedings of the 12th International Symposium on High-Energy Spin Physics*, Amsterdam, Netherlands, September 1996, edited by C.W. de Jager *et al.*, World Scientific, pp 434-437 (1997)
4. R.S. Raymond, Development of a Large-bore Cryogenic 2-4 Transition Unit, in *Proceedings of the International Workshop on Polarized Sources and Targets*, Erlangen, Germany, September 1999, edited by A. Gute *et al.*, pp 494-495 (1999)

THE BLAST POLARIZED H/D TARGET

H. Kolster[1], C. Crawford[1], K. McIlhany[1,3], N. Meitanis[1], R. Milner[1,2], T. Wise[4],
V. Ziskin[1],
and the *BLAST–Collaboration*

*(1) Laboratory for Nuclear Science, Massachusetts Institute of Technology,
Cambridge, MA 02139, USA*
(2) MIT–Bates Linear Accelerator Center, Middleton, MA 01949, USA
*(3) Current address: Department of Physics, US Naval Academy, Annapolis,
MD 21402, USA*
(4) Physics Department, University of Wisconsin, Madison, WI 53706, USA

The Bates Large Acceptance Spectrometer Toroid (BLAST) [1] is a detector designed to study the spin dependent electromagnetic response of few–body nuclei at momentum transfers up to 1 $(GeV/c)^2$. The strength of the experiment lies in its unique combination of large acceptance spectrometer, highly polarized electron beam, and dilution free internal gas targets. The Atomic Beam Source (ABS) [2,3] which operated in the NIKHEF AmPS electron ring throughout the 1990's was delivered to Bates to be installed in the BLAST experiment. The ABS at Bates will be able to deliver a polarized hydrogen or deuterium target. The high magnetic field of the toroid magnet of up to 0.18Tin at the location of the ABS demands a redesign of crucial components of the ABS to ensure high performance. We report on recent developments on the ABS target and the future setup within the BLAST spectrometer.

1 The Bates Large Acceptance Spectrometer Toroid (BLAST)

Scattering of polarized circulating electrons from polarized internal targets is a significant new capability for intermediate energy nuclear physics. It makes possible the full exploitation of spin observables in electronuclear processes. This new technology has recently become available to experimentalists in the US after the completion of the MIT-Bates South Hall Ring.

The detector, called the Bates Large Acceptance Spectrometer Toroid (BLAST) [1], consists of an eight sector copper coil array producing a toroidal magnetic field, instrumented with two opposing wedge-shaped sectors of wire chambers, scintillation detectors, Cerenkov counters, lead glass calorimeter and neutron detectors. The open geometry maximizes acceptance while allowing good momentum and angular resolution and has a luminosity capability matched to the projected densities of the polarized internal targets.

The initial scientific program has been approved by the Bates Program Advisory Committee for a total of 3200 hours. This program addresses issues of central importance to intermediate energy nuclear physics: the structure of the nucleon, the spin structure of the weakly bound deuteron system and the three-body ^{3}He system, the nature of multinucleon absorption mechanisms in electromagnetic reactions.

1.1 The Internal Targets for BLAST

Several internal targets will be used for data taking with the BLAST detector which operate with a variety of target gas species and target densities:

• The Atomic Beam Source [2,3] (ABS) for hydrogen and deuterium polarized targets with an average target thickness of 5×10^{13} nucl cm^{-2} and average target polarization of 80%.

• The Laser Driven Target [4] (LDT) for hydrogen with an average target thickness of 2.4×10^{15} nucl cm^{-2} and average target vector polarization of 50%.

• The ^{3}He target with a target thickness of 2×10^{15} nucl cm^{-2} and average target polarizations of 50%. The ^{3}He target is currently being developed at MIT-Bates.

• Unpolarized target gases of different types: hydrogen, deuterium, helium, oxygen, and nitrogen.

Available for target diagnostics are a Breit-Rabi Analyzer [2] , an analysing sextupole combined with a QMA for measurements of the ABS beam intensity, and an ion polarimeter [5] to measure the tensor polarization of the deuterium target and the degree of dissociation in the hydrogen and deuterium targets.

2   The Atomic Beam Source Target at Bates

The Atomic Beam Source was operating in the NIKHEF AmPS electron ring throughout the 1990's was delivered to Bates to be installed in the BLAST experiment. The new facility and the environment at the BLAST experiment imposed changes to the ABS because it is operated inside the spectrometer toroidal field with field values up to 0.2 T. An improvement of the figure of merit, a new and different data acquisition system, and the change from European to American standards requires further effort. Another important motivation for improved diagnostics as well as transparent operation and maintenance procedures is the decision to have Bates staff personnel operating the ABS during the run of the experiment which ensures continuous reliable operation of the system throughout the run of the experiments.

Figure 1 shows a drawing of the ABS, the target chamber and the analyzing sextupole in a Breit-Rabi configuration. Molecular hydrogen or deuterium is dissociated in the dissociator and forms a super sonic atomic beam in a cooled nozzle. The beam passes through a sequence of sextupole magnets and RF-transition units until it enters the target cell. The atomic beam is electron spin polarized by Stern-Gerlach separation in the sextupole magnets. The RF-transition units allow for an additional selection of hyper-fine states resulting in vector or tensor polarized targets. A fraction of this beam passes straight through the exit tube in the cell and enters into the Breit-Rabi Analyzer which is used to measure the ABS intensity.

Large pump stations form a differential pumping system with pumping speeds of $4000\,\mathrm{mbar\,l\,s^{-1}}$ from turbo pumps in the dissociator and skimmer chamber (first stage), $3500\,\mathrm{mbar\,l\,s^{-1}}$ from a cryo pump in the sextupole chamber (second stage), and $500\,\mathrm{mbar\,l\,s^{-1}}$ from a getter pump in the appendix (third stage). The third pumping stage was added at Bates to reduce the high background of unpolarized molecules entering as diffusive flow from the ABS into the target cell that was present at NIKHEF. The new design of the medium field transition (MFT) magnet made it necessary to stretch the ABS below the first stage and regroup the first four sextupole magnets to add more space between the first and second sextupole systems. Tracking calculation showed that the intensity of the ABS will not be affected.

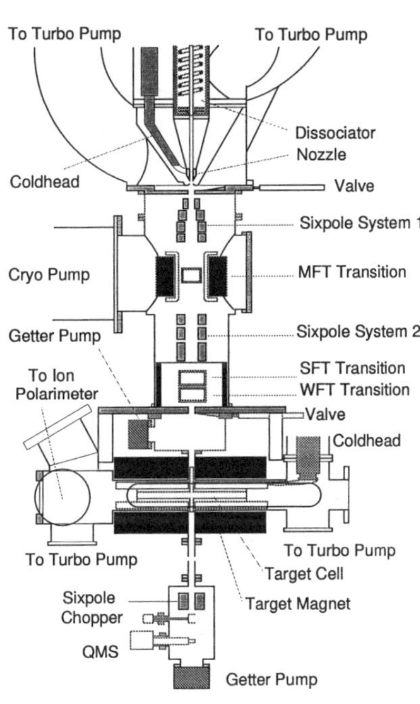

Figure 1: Drawing of the ABS with target chamber and analyzing sextupole for intensity measurements at the ABS.

New developments for the dissociator include an improved resonant circuit and additional matching network. Figure 2 shows a typical circuit diagram for a dissociator with matching network. The plasma couples to the RF by capacitive coupling mainly in the capacitor C with voltage V, and also along the coil with n turns which represents a series of capacitors with voltage $V\cdot n^{-1}$ across each turn.

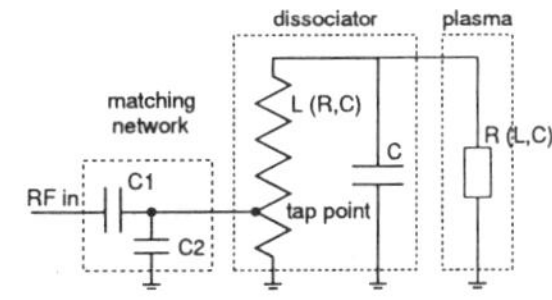

Figure 2: Circuit diagram for the dissociator.

The inductor L is needed to adjust to the resonant frequency of the circuit $\omega = \sqrt{LC}^{-1}$. A commercial monitor was added to measures the in and outgoing power and phase relation of the system including the matching network.

It determines the impedance of the system and displays the information in a Smith-chart, a polar graph of the complex reflection coefficient γ (see Fig 3).

The optimization of the dissociator depends on several factors which are strongly interdependent: The amount of RF power coupling into the resonant circuit is determined by the tap position along the coil of the inductor. The power stored in the resonant circuit determines the strength of the plasma. The resistance of the plasma in turn changes the resonance condition of the resonant circuit. A series of measurements with different tap positions and variety in the input power is shown in Fig. 3. No matching network was used for these measurements. The resonant frequency of the L-C circuit was set to 27.14 MHz the frequency of the RF power supply with a Q factor of 150. A change in tap position results in change of the position on the Smith chart but never gets close to the optimum value of 50 Ω. A change in the power results in a shift in the Smith chart along a straight line. Adding a matching network helps to adjust the starting point such that at a certain power level the optimum of 50 Ω is reached. This procedure is schematically shown in Fig. 4. Operation of the dissociator in this optimized mode results in full length of the plasma inside the disociator tube up to the nozzle at an input power of 60 W, with 0 W reflected power, at a gas flow of 80 sccm.

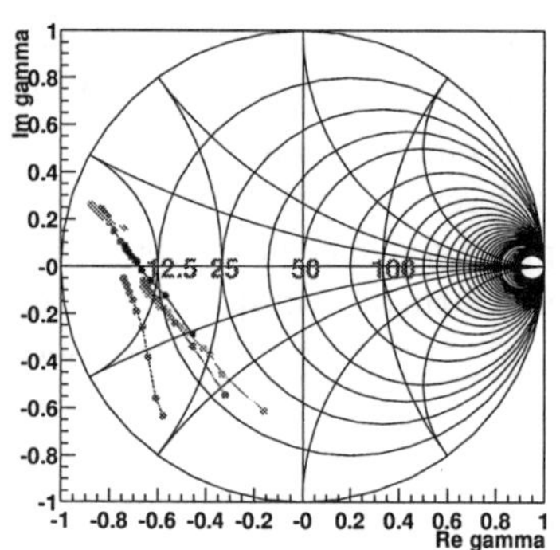

Figure 3: Measurements at different tap-points and different power levels plotted in a Smith chart. Each series of measurements is connected by a line, starting at low power from below right and moving upwards and left with increase in power.

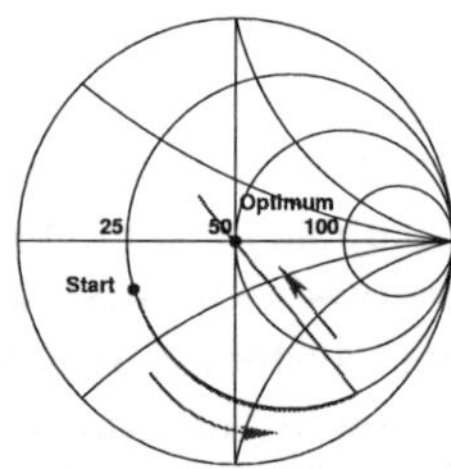

Figure 4: Matching procedure with matching network.

The high magnetic field of the spectrometer of up to 0.2 T at the location of the medium field transition (MFT) inside the ABS requires a new design of the MFT magnet. Typical field values for the operation of the MFT are 2.5 mT for the static field and 0.2 mT cm^{-1} for the gradient field. Figure 5, top, shows the new design of the MFT magnet as it was developed at Bates. The superposition of the spectrometer field and the compensating field of the magnet, which results in the static field used for the RF-transition, was simulated with TOSCA from the OPERA3D software package. The resulting magnetic field along the atomic beam axis is shown in Fig. 5, bottom. No gradient field was applied in the calculation.

The calculations show a sufficiently uniform field within a range of 60 mm inside the magnet with a slope of less than $0.1\,\mathrm{mT\,cm^{-1}}$ at a static field value close to the operation point. The transition units for the weak field and strong field transition also received additional shielding and show good expected field values and gradient in the calculations. The strong field transitions were equipped with new phase loop controllers to lock the cavity like resonators in resonance.

The target cell and cooling frame were newly designed as a 400 cm long cell tube with 15 mm diameter and $100\,\mu$m thin cell walls. It is supported by a 1 cm thick aluminum bar on the top and bottom for good thermal conduction to reduce temperature fluctuations during filling and decay of the electron beam.

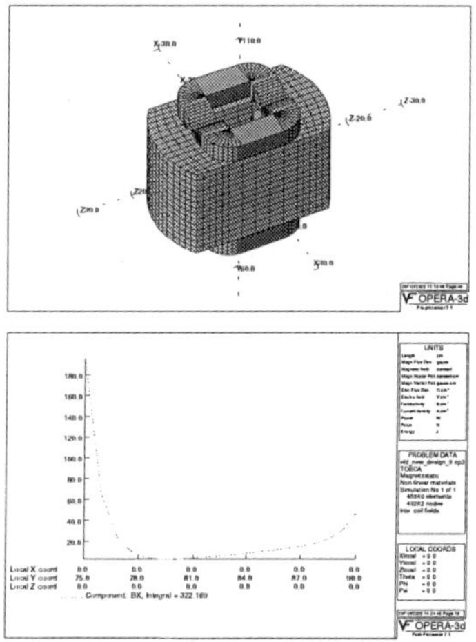

Figure 5: Drawing of the new MFT magnet (top) and plot of the magnetic field distribution along the atomic beam axis (y) (bottom).

3 Conclusion

The transfer of the NIKHEF ABS to the Bates Laboratory for installation in the BLAST experiment resulted in a variety of modifications to make the ABS operational at Bates. The changes applied to the setup in different areas such as dissociator, RF-transition units, vacuum system, and target cell should sufficiently improve the figure of merit of the target. The projected target operational parameters are a degree of dissociation in the target cell higher than 90% and a target polarization of up to 80%. These values are expected at flow rates of $3.5 \times 10^{16}\,\mathrm{atoms\,s^{-1}}$ for hydrogen single state injection which have been confirmed in measurements at Bates.

References

1. R. Arlacon *et al.*, *Nucl. Phys.* A **663-664**, 1111 (2000).
2. D. Szczerba *et al.*, *Nucl. Instrum. Methods* A **455**, 769 (2000).
3. L.-D. van Buuren *et al.*, *Nucl. Phys.* A **663-664**, 1049 (2000).
4. C. Crawford, *contribution to this workshop*.
5. Z.-L. Zhou, *et al.*, *Nucl. Instrum. Methods* A **379**, 212 (1996).

STUDIES ON BEAM FORMATION IN THE HERMES-ABS

A. NASS, N. KOCH, M. RAITHEL AND E. STEFFENS

Physikalisches Institut der Universität Erlangen - Nürnberg,
E. Rommel - Str. 1, 91058 Erlangen, Germany

The HERMES experiment is installed in the HERA electron ring at DESY (Hamburg) and taking data with longitudinally polarized deuterium since 1998. The target consists of an Atomic Beam Source (ABS), injecting into a cooled storage cell, and a Breit–Rabi Polarimeter. The flow rate of polarized deuterium[1] in three substates is about $4.5 \cdot 10^{16} \vec{D}/s$.

A new dissociator based on a microwave discharge at 2.45 GHz has been developed[2] and installed into the HERMES–ABS. The long term stability was found to be excellent compared with the Radio–Frequency dissociator (RFD) used so far. The intensity increased by 15 % although the sextupole system has been optimized for the RF–dissociator. Since the velocity distribution of the microwave based dissociator differs from that of the RF–dissociator the intensity could be increased further with a modified sextupole magnet system. For that purpose the way for a new start generator for sextupole Monte–Carlo simulations was opened.

Furthermore a new method of beam formation was tested, the Carrier Jet method[3], based on an expanded beam surrounded by an overexpanded carrier jet. Since there was a large deviation between the measurements and the results of the Navier–Stokes calculations Monte–Carlo calculations[4] were performed. The results agree well with the measurements, showing that the method used in ref.[3] does not provide a realistic description of beam formation in this flow range, in particular concerning the diffusion between the two components.

In addition a new type of beam monitor[5] was used to study the beam formation after the conventional sonic nozzle. Again there was a good agreement between the measured intensity profiles and the predictions of the Monte–Carlo calculations[4] of the beam formation.

1 Beam Formation and its Simulation

1.1 Formation of a Free Jet Molecular Beam

A free jet molecular beam is formed if a gas is flowing through a conical nozzle from a high pressure source (p_0) into a low pressure background (p_b). The gas is accelerated and reaches the speed of sound (Machnumber $M = 1$) at the exit of the nozzle. If the exit pressure $p_e = p_0/G$ with $G = ((\gamma + 1)/2)^{\gamma/\gamma+1}$ exceeds the background pressure p_b an expansion occurs and a supersonc beam is formed. At low p_b a transition from continuum to molecular flow takes place due to the lack of collisions. Downstream of this point all beam parameters are frozen. Only background scattering due to restgas occurs.

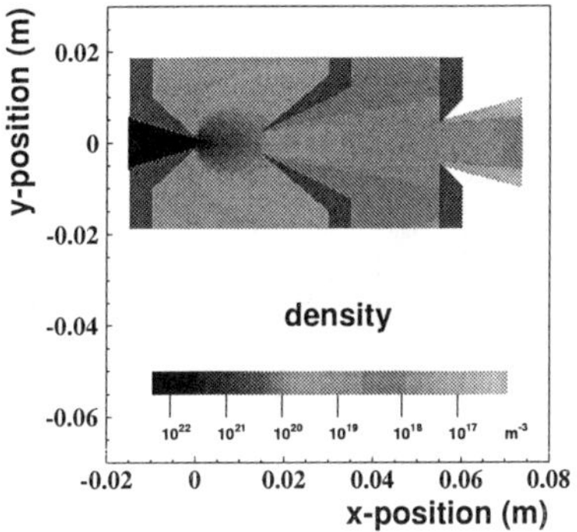

Figure 1. Simulated atomic density distribution of an expansion of hydrogen (flux: 1mbarl/s, degree of dissociation: 67%, nozzle temperature: 100 K).

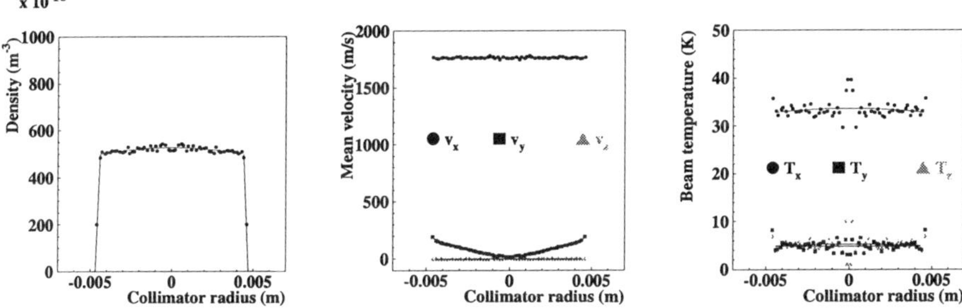

Figure 2. Simulated values of density, beam velocity and temperature of an expansion of hydrogen (x - beam direction, y - radial direction, z - azimuthal direction).

1.2 Monte - Carlo Simulation of an Expansion

To describe this type of expansions a code (DS2G) of G.A. Bird [4] was used. It simulates binary (hard sphere) collisions between particles characterized by their effective diameter $d_{part}^{eff}(T)$, mass m_{part} and for diatomic molecules by the rotational collision number c_{coll}^{rot}. The input parameters are the geometry (e.g. nozzle, skimmer, collimator), the stream input (flux) and the specified flow due to the chamber pressure. The space is divided into regions and small cells in which the sampling occurs. The results are the density and velocity distribution averaged over every cell and the flux inbetween the regions (e.g. nozzle throat, skimmer ...). Additionally a molecule output file could be created on any desired boundary. Figure 1 shows the calculated density distribution using ABS-geometry at usual running conditions. An atomic beam is created with a small divergence after the collimator. If the simulations describe the reality good they could be used to create new geometries

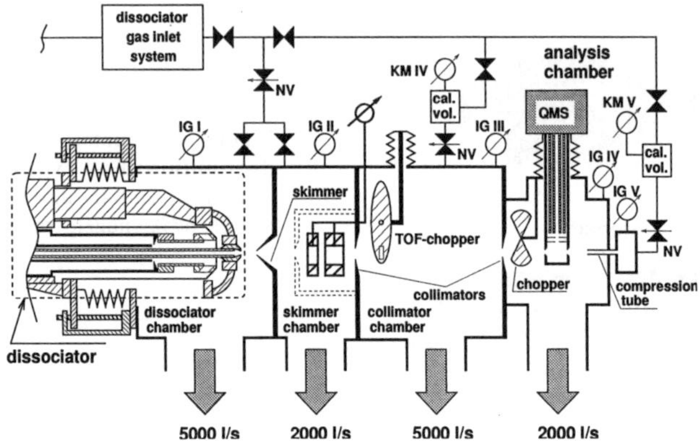

Figure 3. The teststand and its diagnostics, the pumping speeds are denoted with the arrows and the numbers below every chamber.

to increase the atomic intensity on the collimator surface. On the other hand a new starting generator for sextupole Monte-Carlo simulations could be created. It would be the most realistic starting generator ever used independent of any model. Figure 2 shows the set of beam parameters on the collimator surface that could be used to create such a generator.

2   The Teststand and Diagnostics

To verify if the simulations describe the reality good a lot of measurements were done on the atomic beam teststand at DESY/ Hamburg. In figure 3 all components of the teststand are shown including the compression tube (intensity measurements), the QMS (measurements of the degree of dissociation), the Time-Of-Flight-Chopper (TOF-measurements) and the dummy magnets (attenuation studies). Additionally a beam profile monitor is installed in the first chamber between the nozzle and the skimmer. It consits of 32 gold plated tungsten wires on which the hydrogen atoms recombine. The recombination heat load leads to a increasing resistance of the wires.

3   Measurements

First of all, TOF-measurements were performed and the results agreed well with the simulated values. Even the mean velocity was reproduced in an excellent manner (table 1).

Table 1. Comparision between simulated and measured mean velocity of an expansion of hydrogen (degree of dissociation: 64 %, flux: 1 mbarl/s, nozzle temperature: 100 K).

Species	Measurement	Simulation
H_1	(1750 ± 47) m/s	(1760 ± 20) m/s
H_2	(1579 ± 51) m/s	(1590 ± 33) m/s

3.1 The Beam Profile Monitor and the Simulations

The wires can be calibrated using U - I characteristics to get the resistance of the wire R_w as a function of the incident power $P_w = U \cdot I$:

$$R_w = f(P_w). \tag{1}$$

Assuming a homogeneous wire with $R_w/l = dR/dy$ and $P_w/l = dP/dy$ the resistance is:

$$R = \frac{1}{l} \int_{y_a}^{y_e} f\left(\frac{dP}{dy} \cdot l\right) dy, \tag{2}$$

with the length of the wire l and the wire ends y_a and y_e. This equation always holds even in the case of inhomogeneous power distribution. From the simulation one can the density $n(y)$ and velocity distribution $v(y)$. So the power distribution on the wire is:

$$\frac{dP}{dy} = d\frac{\epsilon_r}{2}\alpha_{Au}n(y)v(y), \tag{3}$$

where $d = 5$ μm is the diameter of the wire, $\epsilon_r = 4.75$ eV is the recombination energy of hydrogen and $\alpha_{Au} = 0.5$ [5] is the accomodation coefficient of gold. With eq. 2 one can get a prediction of the resistances of the wires. In figure 4 these values are compared with measured values at two different distances from nozzle to monitor. The agreement is again very good.

3.2 The Carrier Jet Method

The carrier jet method was proposed in 1997 [3] to increase the phase space density of the atomic beam and to get a higher intensity through the colli- mator of the ABS. An overexpanded carrier jet surrounding the inner atomic beam should cool and confine this inner beam. There were a lot of mea- surements done in the past [6] but there was a big discrepancy between these measurements and the predictions. Again Monte-Carlo simulations were car- ried out and they verified the measurements. So it can be concluded that the Navier-Stokes equations used for the predictions are an invalid assumption in this flow range.

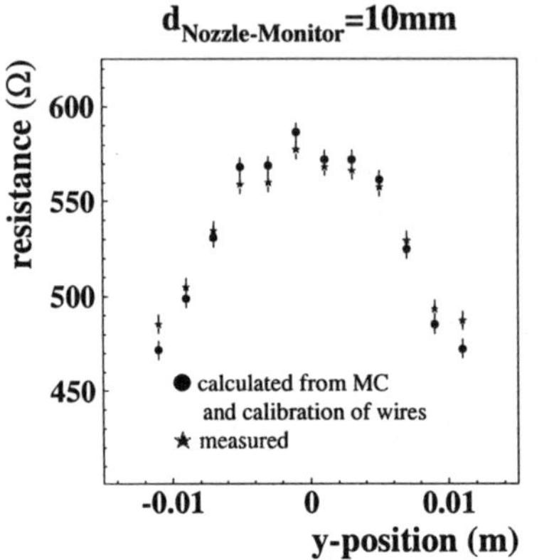

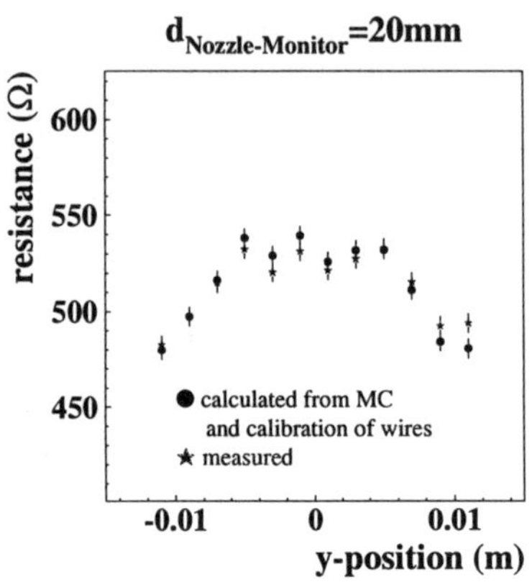

Figure 4. Calculated and measured wire resistances for 2 different distances between monitor and nozzle (flux: 1mbarl/s, degree of dissociation: 80%, nozzle temperature: 100 K.

4 Conclusions

It has been found that the Monte-Carlo simulations (DS2G) are an excellent tool to describe the formation of the atomic beam in this flow region. The results were verified by many measurements including velocity (TOF) and beam profile measurements. The reason of the discrepancy between the prediction (increase of the phase space density of the atomic beam) and the carrier jet experiment was found. Now the simulations can be used to optimize the geometry of the beam formation system and to create a starting generator for sextupole Monte-Carlo simulations and hence improve the lay-out of sextupole systems of atomic beam sources.

References

1. C. Baumgarten, PhD theses (2000).
2. N. Koch, E. Steffens, Review of Scientific Instr. **70**, 1631 (1999).
3. V.L. Varentsov *et al*, PST 1997 Urbana, AIP Conf. Proc. **421**, 381 (1997).
4. G.A. Bird, Molecular Gas Dynamics and the direct Simulation of Gas Flows, Oxford (1998).
5. A. Vassiliev *et al*, PST 1999 Erlangen, AIP Conf. Proc. , 200 (1999).
6. A. Nass *et al*, PST 1999 Erlangen, AIP Conf. Proc. , 482 (1999).

THE POLARIZED GAS TARGET FOR THE ANKE SPECTROMETER AT COSY/JÜLICH

M. MIKIRTYTCHIANTS[1,2], R. BRÜGGEMANN[3], R. EMMERICH[3],
R. ENGELS[3], H. KLEINES[4], V. KOPTEV[2], P. KRAVTSOV[2], S. LEMAÎTRE[3],
J. LEY[3], B. LORENTZ[1], S. LORENZ[5], M. NEKIPELOV[1,2], V. NELYUBIN[2],
H. PAETZ GEN. SCHIECK[3], F. RATHMANN[1], J. SARKADI[4], H. SEYFARTH[1],
E. STEFFENS[5], H. STRÖHER[1], A. VASSILIEV[2], K. ZWOLL[4]

(1) Institut für Kernphysik, Forschungszentrum Jülich, 52425 Jülich, Germany
(2) High Energy Phys. Department, Petersburg Nuclear Physics Institute, 188300
Gatchina, Russia
(3) Institut für Kernphysik, Universität zu Köln, 50937 Köln, Germany
(4) Zentrallabor für Elektronik, Forschungszentrum Jülich, 52425 Jülich, Germany
(5) Physikalisches Institut II, Friedrich-Alexander-Universität, 91058 Erlangen,
Germany

The polarized atomic beam source (ABS) will be utilized to feed the storage cell gas target in future experiments at the magnetic spectrometer ANKE. A state-of-the-art system of sextupole magnets has been developed for the source. An intensity of $(6.9 \pm 0.3) \cdot 10^{16}$ atoms/s into a compression tube has been achieved for two hyperfine states of hydrogen. The medium field rf-transition (MFT) unit has been installed. An efficiency of 0.90 ± 0.05 has been measured for the 2-3 transition. For future experiments at COSY with polarized internal targets a Lamb-shift polarimeter, built at the University of Cologne, has been installed at the ABS. First measurements of the nuclear polarization of the atomic hydrogen beam, produced by the MFT unit, yield a value of $|P| = 0.889 \pm 0.009$. First studies of the properties of the COSY beam have been carried out with an aperture at the ANKE target position.

A polarized atomic beam source (ABS) will be utilized to feed a gas-storage cell at the magnetic spectrometer ANKE[1] with polarized hydrogen or deuterium atoms. By the use of a storage cell the luminosity will be increased by about two orders of magnitude compared to an atomic gas jet crossing the stored, circulating COSY beam[2]. The ABS has been assembled and tested.

1 Layout of the ABS

The layout of the ABS is shown in Fig. 1. A powerful pumping system yields pressures of about $10^{-4}, 10^{-6}, 10^{-7}$, and $5 \cdot 10^{-8}$ mbar in chamber I, II, III, and IV, respectively, at a primary H_2 inlet flux of 1.5 mbar·l/s.

48

In the dissociator (labeled 1 in the figure) a plasma discharge is maintained by coupling of 13.56 MHz. The nozzle at the lower end can be cooled to about 65 K by a cryocooler via a flexible Cu connection and a Cu bridge (label 2). The details at the lower end of the dissociator are shown in Fig. 2. The inner discharge tube (Duran 8330 glass of 14 mm outer diameter and 1.5 mm wall thickness) is cooled by a water flow between coaxial glass tubes. The transition piece from the cold nozzle to the outer dissociator tube avoids forces on the discharge tube. Thermal contact to the lower part of the discharge tube is provided by a modified sliding rf connector. This special construction allows to extract and reinstall the glass tubes and the rf-generating elements without disassembling the heat bridge and nozzle.

The upper baffle, separating the chambers I and II, carries the beam skimmer. The nozzle-to-skimmer distance can be optimized from outside without breaking the vacuum.

Two groups of permanent sextupole magnets (labels 3 and 5), delivered by Vacuumschmelze[a], have been installed. The excellent properties of the magnets are described in ref.[3]. Magnetic field distributions, calculated with the computer code MAFIA, are in perfect agreement with the results of the measurements. Protection of the magnets from hydrogen was done by enclosing them in stainless steel cans with inner tubes of 0.2 mm thickness in order to maximize the free magnet apertures.

In order to achieve vector and tensor nuclear polarization of hydrogen and deuterium, medium- and weak-field rf transition units have been designed and built at Universität Erlangen[4]. The latter unit and the strong-field units for H and D, man-

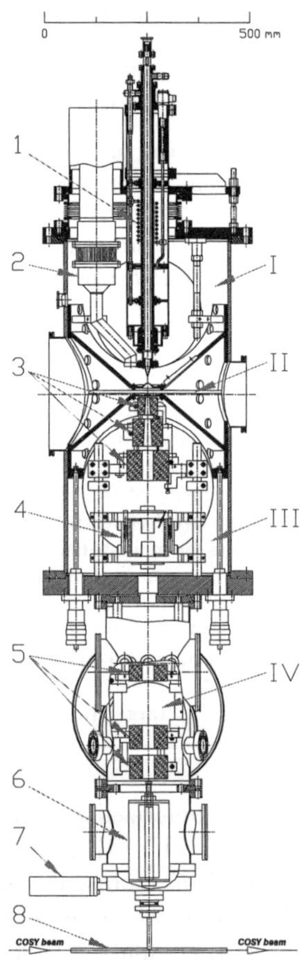

Figure 1. Layout of the ANKE polarized atomic beam source (for details see text).

---

^aVacuumschmelze GmbH, D-63450 Hanau, Germany.

ufactured following an earlier design[5], are combined into one unit (label 6).

A special vacuum gate valve[b] (label 7) allows one to disconnect the ABS chamber from the target chamber with the storage cell (label 8), i.e. from the COSY beam line system.

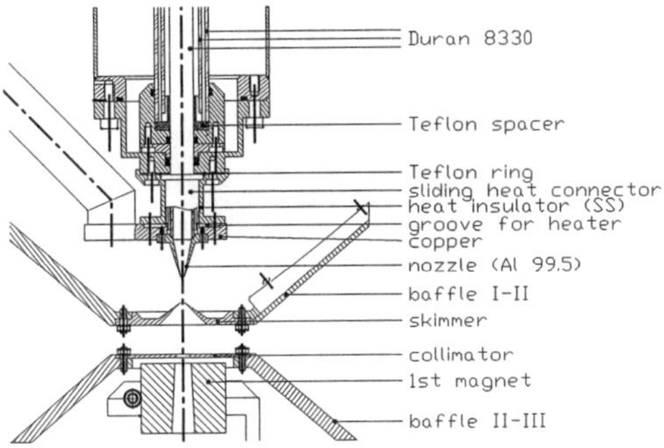

2 Results of the beam studies

The intensity of the polarized beam, as it will feed the storage cell, has been measured with a compression tube, having the dimensions of the feeding tube of the storage cell (10 mm diameter, 100 mm length) and installed at its position (entrance 300 mm behind the exit of the last magnet).

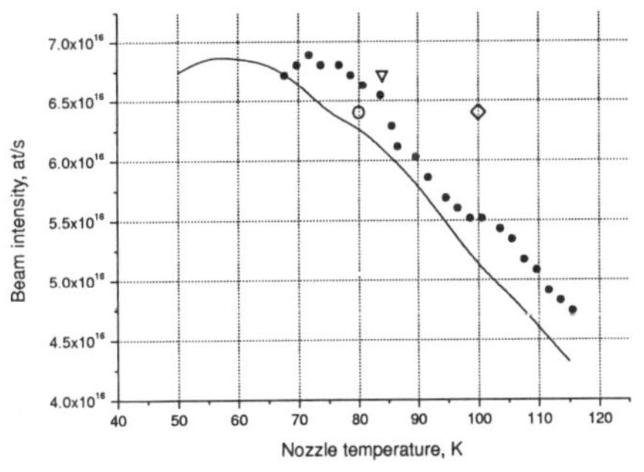

Figure 2. The region between the lower end of the dissociator and the first sextupole magnet.

The obtained absolute beam intensity as a function of the nozzle temperature is shown in Fig. 3. The maximum of $(6.9 \pm 0.3) \cdot 10^{16}$ H atoms per second is achieved around 7? K at primary H_2 flow of 1 mbar·l/s, O_2 admixture of 0.1% and dissociator power 300 W. The measured slope is reproduced well by the result of a trajectory calculations, which assumed a constant degree of dissociation, verified through earlier measurements[7]. The discrepancy at lower temperatures may be explained by increasing intra-beam and rest-gas scattering, which is not taken into account in the simulations.

Figure 3. Absolute beam intensity as a function of the nozzle temperature. For the comparison, data from the HERMES ABS($\diamond$) and PINTEX ($\triangledown$) ABS and Munich ion source ($\bigcirc$) are shown on the figure. The solid line represents results of simulations.[6]

The present results are compared with the maximum intensities given for the sources operated at HERMES (DESY)[8], PINTEX (IUCF)[9], and the polarized ion source at tandem accelerator in Munich, Germany[10].

---

[b]VAT Deutschland GmbH, D-85630, Grasbrunn, Germany.

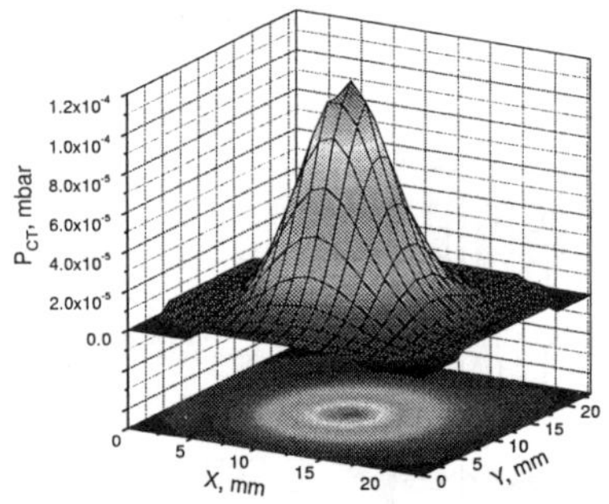

Figure 4. xy-distribution of the pressure in the compression volume measured with narrow (5 mm diameter) compression tube 300 mm behind the last magnet.

In addition to the absolute beam intensity the atomic beam profile has been measured. These studies were carried out with a compression tube of 5 mm diameter and 50 mm length at two positions: 300 mm (1) and 337 mm (2) behind the last magnet. Figure 4 shows the xy-distribution of the pressure in the compression volume measured 300 mm behind the last magnet. Both measured distributions (1) and (2) can be reproduced by calculations performed under the assumption of the gaussian beam profile with $\sigma_1 = 2.85 \pm 0.42$ mm and $\sigma_2 = 2.95 \pm 0.43$ mm, respectively.

At present only the medium-field rf transition unit has been installed. Its efficiency has been measured with the compression-tube set-up. Preliminary results show that the efficiency of the medium-field rf transition unit is 0.90 ± 0.05, which corresponds to a polarization larger than 85%.

This result is confirmed by measurements with the recently installed Lamb-shift polarimeter (LSP), which yields the polarization of the H beam behind the second group of magnets to be $|P| = 0.889 \pm 0.009$. The polarimeter has been designed and built at University of Cologne[11]. At present its commissioning is completed and it is used for further studies of the ABS beam properties and for fine-tuning of the transition units. Later it will serve for investigations of the polarization of the gas in the storage cell.

The construction of the LSP necessitates its horizontal position. Thus a $90°$ electrostatic spherical deflector[c] was installed between the vertical ABS and the ionizer and the subsequent components of the LSP: Wien filter, Cs cell for charge exchange and excitation of atoms into the metastable 2S state, spin filter as the heart of the set-up to select the atoms in the different nuclear spin states (magnetic quantum numbers $m = \pm 1/2$ for H and $m = +1, 0, -1$ for D) for D), and finally the quenching chamber with the photomultiplier to measure the light from the Lyman-α transitions into the ground state.

---

[c]on loan from University of Wisconsin, Madison (W. Heberli)

3 Future developments

Since the future experiments (e.g. pd breakup[12]) at ANKE will require the use of a storage cell it is very important to study the properties of the COSY beam and to define the dimensions of the storage cell. These studies have been started with simple rectangular and elliptical diaphragms, 10 mm high and 30 mm wide, placed at the ANKE-target position. With the available vertical steerers and the rectangular diaphragm $9.6 \cdot 10^9$ protons could be accelerated and stored, whereas in the same steering mode but without the diaphragm this value is about $\sim 1.2 \cdot 10^{10}$. Without this additional steering a higher number of $\sim 1 \cdot 10^{11}$ protons could be stored which indicates beam losses elsewhere in the COSY ring. Thus additional vertical steerers are needed and will be installed for the next beam studies. In parallel to these activities the design work on a new large target chamber (length 800 mm, width 600 mm and height 400 mm) is in progress. The chamber will allow us to carry out beam studies with storage cell prototypes next year. It will also house new beam-position monitors and the set-up of silicon vertex detectors.

References

1. S. Barsov *et al*, *Nucl. Instrum. Methods* A **462**, 364 (2001).
2. W. Haeberli, Proc. 2nd Int. Symp. On Polarization Phenomena, eds. P. Huber and H. Schopper, Experientia Suppl. **12** (Birkhauser, Basel, 1966), p. 64.
3. A. Vassiliev *et al*, *Rev. Sci. Instrum.* **71** (2000) 3331.
4. S. Lorenz, Diploma thesis, Universität Erlangen-Nürnberg (1999).
5. F.Stock, PhD thesis, MPI Heidelberg (1994).
6. Based on a code obtained from D. Toporkov, BINP Novosibirsk.
7. M. Mikirtytchiants, Diploma thesis, St. Petersburg State Technical University (1999).
8. J. Stewart *et al*, AIP Conf. Proc. **421**, 69 (1998).
9. F. Rathmann *et al*, AIP Conf. Proc. **421**, 89 (1998).
10. R. Hertenberger *et al*, Proc. 8th Int. Workshop on Polarized Sources and Targets (PST99), Erlangen, 1999. Edt. A. Gute, S. Lorenz, E. Steffens (Universität Erlangen-Nürnberg, 1999) , p. 52.
11. R. Engels, contribution to these proceedings.
12. S.V. Dshemuchadze *et al*, COSY Exp. Proposal No. 20 (1991); update V. Abazov *et al* (1999), spokesperson V. Komarov. available via `http://www.fz-juelich.de/ikp/anke`

A LAMB-SHIFT POLARIMETER FOR THE POLARIZED TARGET AT ANKE

R. ENGELS, R. EMMERICH, J. LEY AND H. PAETZ GEN. SCHIECK

*Institut für Kernphysik, Universität zu Köln, Zülpicher Str. 77,
50937 Köln, Germany
E-mail: engels@ikp.uni-koeln.de*

M. MIKIRTYTCHIANTS, F. RATHMANN AND H. SEYFARTH

*Institut für Kernphysik, Forschungszentrum Jülich, Leo-Brandt-Str.
52425 Jülich, Germany*

A. VASSILIEV

*Petersburg Nuclear Physics Institute
188350 Gatchina St. Petersburg, Russia*

With a Lamb-shift polarimeter (LSP) it is possible to measure the occupation numbers of the individual hyperfine states in a beam of hydrogen or deuterium. Therefore one can calculate the nuclear polarization of the atomic beam in magnetic fields of varying strength. The polarization of a slow (500 -2000 eV) ion beam can be measured too. Since effects in the ionizer, the Wien filter, the cesium cell, and the spin filter influence the measured polarization P_{Ly} value correction factors for calculating the *absolute* polarization p_z of the beam as a function of P_{Ly} have to be applied.

1 The Lamb-Shift Polarimeter

1.1 The Principle

The atomic beam of hydrogen or deuterium is ionized in an electron-impact ionizer [3],[4] with an efficiency of about 10^{-3}. The ionization volume is located in a strong magnetic field ($\sim 160\,\mathrm{mT}$) to decouple the nuclear and the electron spins. At the same time it increases the collision probability of the accelerated electrons and the beam particles. The vertical ion beam is deflected into the horizontal beamline by an electric deflector. The necessary precession of the polarization back onto the beamline direction is performed by a Wien filter. This Wien filter also acts as a mass filter and therefore only protons (deuterons) arrive in the cesium cell. By charge exchange with the cesium metastable atoms in the 2S state are produced. To define the polarization of the metastable hydrogen (deuterium) a strong magnetic field ($\sim 55\,\mathrm{mT}$) in the cesium cell is required. The spin filter [5] transmits only metastable atoms in different single hyperfine states depending on the magnetic field. All other atoms are quenched into the ground state. In the vacuum chamber at the end of the polarimeter the residual metastable atoms are then quenched by an electric field (Stark effect). The Lyman-α photons are registered by a photomultiplier [6] sensitive only to photons with a wavelength of 121 nm (Fig. 1). The number of photons counted in one hyperfine state is proportional to the number of atoms with the same nuclear spin in the primary beam. From the ratio of the photons which are produced at 53.5 mT or 60.5 mT in the spinfilter the polarization of the atomic beam or a slow ion beam of 500-2000 eV is calculated. With an overall efficiency of the polarimeter of $\sim 10^{-10}$ a beam of 10^{16} atoms/s will produce some 10^6 photons/s in the photomultiplier. So the statistical error $\Delta P_{Ly} \sim \pm 0.003$ obtained in 30 s is very small.

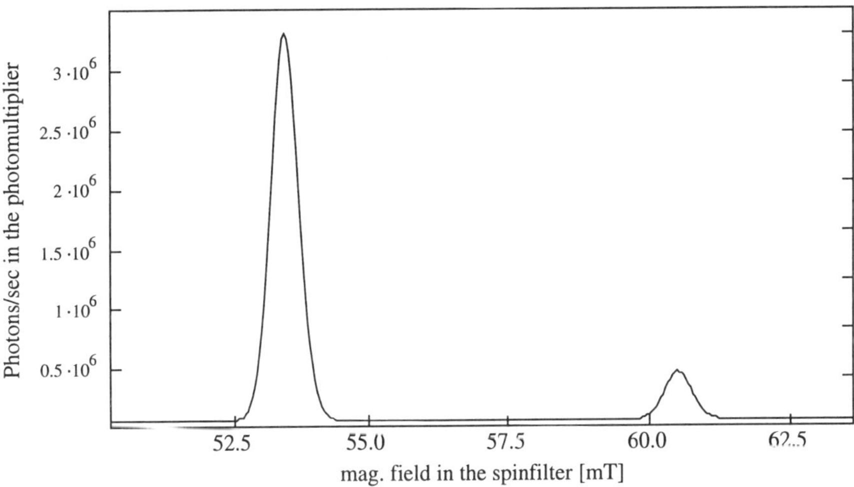

Figure 1. The Lyman-α spectrum of a polarized hydrogen beam in the hyperfine state 1. The polarization $P_{Ly} = 0.780 \pm 0.002$ was measured in less than 30 seconds.

2   The Corrections

The measured polarization P_{Ly} of the Lyman-α spectrum is different from the polarization p_z of the atomic beam. Some corrections are necessary. The knowledge of the correction factors determines the absolute accuracy which can be obtained with the LSP.

2.1 The Ionizer

The ionizer can produce protons from residual gas molecules like H_2O, H_2, C_nH_m or H_2 after the recombination of atoms in the ionizer. With atomic beam entering the ionizer an H_2^+ ion current of the same order of magnitude as the H^+ ions will be produced. So up to 10% of the proton beam is made from unpolarized particles and has to be subtracted in the Lyman-α spectrum. This will increase the polarization value by up to 10%.

The polarization of the hyperfine states 2 and 4 is a function of the magnetic field in the ionization volume. Only for $B \rightarrow \infty$ the produced protons will have the same polarization as the atoms of the beam. By varying the magnetic field in the ionizer it is possible to claculate the correction to the polarization. At 18 A in the solenoid coil only 93.5% of p_z will be measured. With this procedure it is even possible to indentify the single hyperfine substates or some mixtures of this states in the beam (Fig. 2).

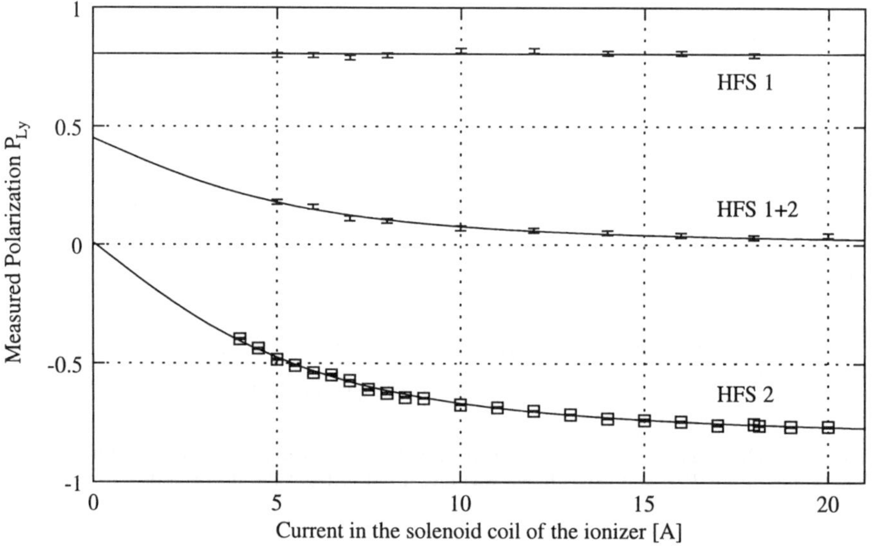

Figure 2. The polarization of an atomic hydrogen beam in the hyperfine states 1, 2 or 1+2 as a function of the magnetic field in the ionizer.

2.2 The Wien Filter

With the Wien filter the polarization vector is rotated into the horizontal direction. During this process 0.8% of the polarization will be lost for protons. For deuterons this loss will grow up to $\sim 1\%$.

2.3 The Cesium Cell

Like in the ionizer the magnetic field in the cesium cell influences the measured polarization because the production of the α_2 metastable atoms is a function of the magnetic field (Fig. 3). Due to the small critical field (6.34 mT) of the metastable atoms 99.4% of the polarization value is conserved at 55 mT on the beam axis.

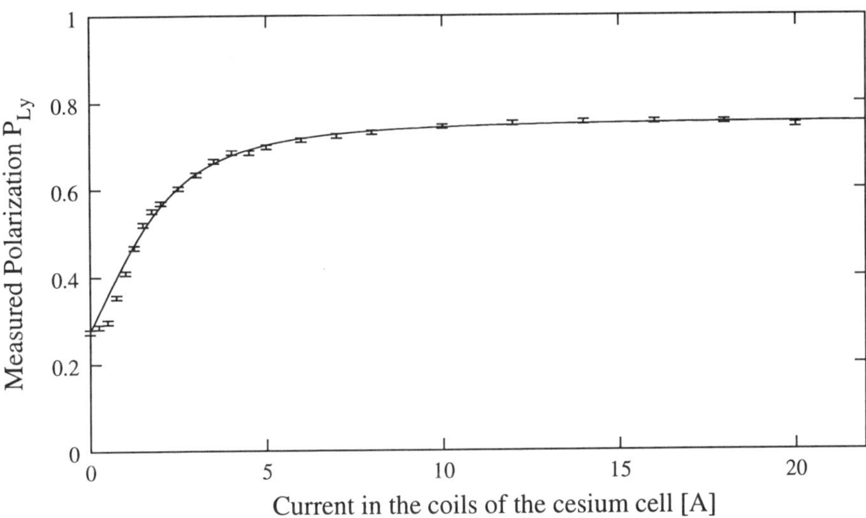

Figure 3. The measured polarization P_{Ly} as a function of the current in the magnetic-field coils of the cesium cell.

2.4 The Spin Filter

The transmission of the metastables through the spin filter depends on its magnetic field: at 53.5 mT it is nearly 1% better than at 60.5 mT. Therefore the measured polarization of the hyperfine states 1 and 4 will be higher by 0.2% and decreased for the states 2 and 3.

Thompson[7] explained that different magnetic gradients between the cesium cell and the spin filter will produce some depolarization. Therefore both gradients were designed to be the same in our LSP. However, after some small modifications, this effect will be investigated in the near future.

3   Discussion

The measurements on the LSP have been done with use of an ABS [8]. The LSP ist an excellent *relative* polarimeter with a statistical error of 0.3% within 30 seconds. To derive the *absolute* polarization of the atomic beam to high precision some known correction factors have to be applied. The known correction factors increase the measured polarization P_{Ly} of the hyperfine states 1 and 3 by about 11% and for the states 2 and 4 by about 19% with an error of 1% within 30 seconds. These corrections are dominated by effects in the ionizer. A new ionizer with a stronger magnetic field and a getter pump inside was designed to reduce these overall corrections to 5% at an error of 0.5%. This ionizer will be finished at the end of 2001.

Measurements with deuterium which will be easier due to the smaller critical field and lower background of the residual gas will be carried out in the fall of 2001.

With the present LSP it will be possible to measure the polarization of the primary atomic beam with intensities down to 1% of the present beam. It is planned to measure the polarization of atoms in a storage cell by extracting a small amount of the atoms. First tests will start in the spring of 2002.

Acknowledgments

This work has been supported by BMBF and FFE Jülich.

References

1. R. Engels, Proc. 8th Int. Workshop on Polarized Sources and Targets (PST99), Erlangen (1999)150.
2. S.K. Lemieux et al., Nucl. Instr. Meth. **A333**(1993)434.
3. H.F. Glavish et al., Nucl. Instr. Meth. **65**(1968)1.
4. R. Emmerich, Diploma-Thesis, Universität zu Köln (2000).
5. J.L. McKibben et al., Phys. Lett. **28B**(1969)594.
6. J. Ley, Diploma-Thesis, Universität zu Köln (2000).
7. W.J. Thompson, Nucl. Instr. Meth. **A333**(1993)443.
8. M. Mikirtytchiants et al., Proc. 9th Int. Workshop on Polarized Sources and Targets (PST01), Nashville, Indiana (2001), beeing published.

A VECTOR AND TENSOR POLARIZED INTERNAL DEUTERIUM TARGET FOR THE IUCF COOLER

B. v.Przewoski[1], J. Balewski[1], W.W. Daehnick[2], J. Doskow[1], W. Haeberli[3],
R. Ibald[1], B. Lorentz[4], H.O. Meyer[1], P.V. Pancella[5], R.E. Pollock[1],
F. Rathmann[4], T. Rinckel[1], Swapan K. Saha[2], B. Schwartz[3],
P. Thörngren-Engblom[6], A. Wellinghausen[1], T.J. Whitaker[1], and T. Wise[3]

(1) IUCF, Milo B. Sampson Lane, Bloomington, IN 47405, USA
(2) University of Pittsburgh, Pittsburgh, PA 15260, USA
(3) University of Wisconsin, Madison 53706, USA
(4) Forschungszentrum Jülich, Jülich, Germany
(5) Western Michigan University, Kalamazoo, MI 59008, USA
(6) The Svedberg Laboratory, Uppsala, Sweden

The PINTEX atomic beam source, which was previously used to produce a polarized, internal hydrogen target for measurements of spin correlation coefficients in pp elastic scattering and pion production, was modified to also produce a polarized deuterium target. The target thickness obtained with a 12mm diameter and 25cm long storage cell is on the order of 10^{13} atoms/cm^2. Two new transition units with variable gradient were installed to facilitate easy changeover from hydrogen to deuterium target. The ABS is capable of producing pure vector polarization with a theoretical maximum value of +2/3 and pure tensor polarization of theoretical maximal values of +/-1. The direction of the vector polarization can be reversed by reversing the direction of the holding field at the target. To date, this internal deuterium target has been used for a measurement of spin correlation coefficients in pd elastic scattering.

1 Introduction

The PINTEX[1] atomic beam source[2] has been used at the IUCF Cooler since 1993 to produce a beam of polarized, atomic hydrogen for injection into a storage cell. In 2000 it was upgraded to also produce either vector or tensor polarized deuteron beams.

Atoms from an 18 MHz dissociator emerge through an aluminum nozzle which is kept at liquid nitrogen temperature. The atoms then pass along the axis of a set of sextupole magnets where they are separated according to their electron polarization. In the following first medium field transition unit (MF1), transitions between hyperfine states are induced. The atoms pass along the axis of a second set of sextupole magnets whereby an atomic beam in a pure spin state is prepared. At last, depending which polarization state is desired, another transition between hyperfine states is induced in a second medium field transition unit (MF2). The atoms are then injected into the storage cell which is located in a weak holding field generated by a set of Helmholtz coils.

2 Medium Field Transitions

An MFT operates in magnetic fields of $B \sim 0.1B_c$ to $B \sim 0.2B_c$, where B_c is the hyperfine interaction field of 50.7 mT for hydrogen and 11.7 mT for deuterium. A field gradient along the beam direction is required to satisfy the condition of adiabatic passage. Multiple transitions can be made by adjusting the static field so that the beam passes in sequence through field regions where the populations of different pairs of hyperfine states are interchanged at a given, fixed RF frequency.

Previously, the atomic beam source was equipped with a single, fixed-gradient MF transition unit located after the first set of sextupole magnets. In order to facilitate easy switching between proton and deuteron target two new transition units with variable gradient and variable static field were installed. The linearity of the gradient field over the transition region as well as the homogeneity of the static field were determined prior to installation of the units in the ABS. For deuterium the gradient field is set to +0.2 mT/cm. The RF coil of each MF unit consists of 12 turn solenoids of 1.6 mm diameter wire with a length of 70 mm and an I.D. of 34 mm. The units are water cooled and operated at 60.5 MHz for hydrogen and 30 MHz for deuterium. The currents used to drive the offset and gradient coils are remotely controlled. This makes it possible to quickly change, during the experiment, between vector, + tensor and − tensor polarization.

3 Principle of Operation

Since operation with hydrogen has been discussed extensively elsewhere, we will limit the following discussion to deuterium. After the first set of sextupoles the atomic beam consists of states 1+2+3, where the states are labeled in order of decreasing energy in a non-zero magnetic field[3]. One or more transitions can be made sequentially in MF1. Figs.1a-d show the energy differences in MHz between pairs of hyperfine states with $|\Delta m_f| = 1$ as a function of magnetic field. It is graphically shown which transitions are selected as the static field is increased while the gradient field is kept constant. The transition unit is operated at 30 MHz, indicated by the vertical position of the horizontal bar. The length of the bar represents the range of the magnetic field over the RF region. For small static fields (Fig.1a) no transitions are made. When the static field is increased, the atoms undergo a $3 \leftrightarrow 4$ transition (Fig.1b). When the field is further increased the atoms pass through the $3 \leftrightarrow 4$ transition followed by the $2 \leftrightarrow 3$ transition (Fig.1c). If the static field is increased even further, the atoms undergo the $3 \leftrightarrow 4$, $2 \leftrightarrow 3$ and $1 \leftrightarrow 2$ transitions sequentially (Fig.1d). The second set of sextupoles eliminates state

4, so that one is left with states 1+2+3, 1+2, 1+3 and 2+3 for the cases shown in Fig.1a-c. The corresponding maximum polarizations of the atomic beam, before entering MF2, are $(P_z, P_{zz}) = (+1/3, -1/3)$, $(P_z, P_{zz}) = (+2/3, 0)$, $(P_z, P_{zz}) = (+1/3, 0)$ and $(P_z, P_{zz}) = (0, -1)$, where P_z, P_{zz} are vector and tensor polarization. MF2 is only needed to produce positive tensor polarization. If the transition parameters are set as shown in Fig.1b, atoms in states 1+3 with polarizations $(P_z, P_{zz}) = (+1/3, 0)$ undergo the $3 \leftrightarrow 4$ transition. Consequently, the atomic beam contains states 1+4 with polarizations $(P_z, P_{zz}) = (0, +1)$.

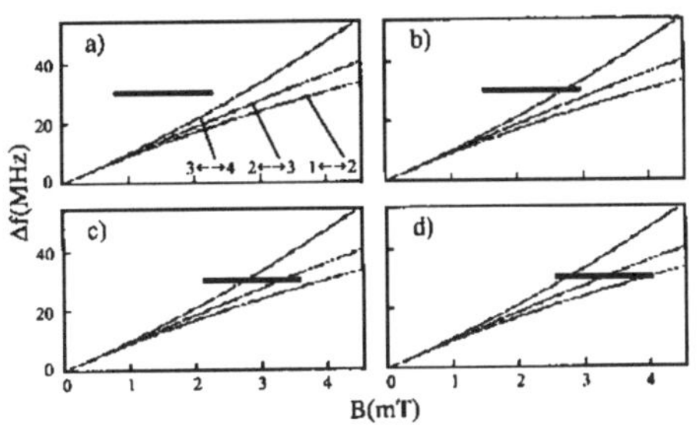

Figure 1: Sequence of deuterium transitions made as the MFT magnetic field is increased. See text for details and also ref.4.

4   Optimization Procedure

Although each MF unit is equipped with two Hall probes to monitor the field along the interaction region, a more direct way to assess the performance of MF1 and MF2 is desirable. The transition units are optimized by measuring the target vector asymmetry as a function of static field using the PINTEX forward detector located in the A-region of the Indiana Cooler. The detector consists of a thin start scintillator, a pair of wire chambers and a stopping scintillator. Fig.2a shows the vector asymmetry as a function of static field current for MF1. For currents $I_{MF1} \leq 2.8\text{A}$ no transitions are made, corresponding to $(P_z, P_{zz}) = (+1/3, -1/3)$. The theoretical polarizations for $2.8\text{A} \leq I_{MF1} \leq 3.4\text{A}$, $3.4\text{A} \leq I_{MF1} \leq 4.2\text{A}$ and $4.2\text{A} \leq I_{MF1} \leq 4.8\text{A}$ are $(P_z, P_{zz}) = (+2/3, 0)$, $(P_z, P_{zz}) = (+1/3, 0)$ and $(P_z, P_{zz}) = (0, -1)$ respectively. At higher MF1 currents the static field is too large and no transitions are made again, resulting in $(P_z, P_{zz}) = (+1/3, -1/3)$. It can be seen from Fig.2a that the vector asymmetry approximately doubles where the polarization is expected to increase from $P_z = +1/3$ to $P_z = +2/3$, indicating that the transition unit operates efficiently. Likewise, the vector asymmetry

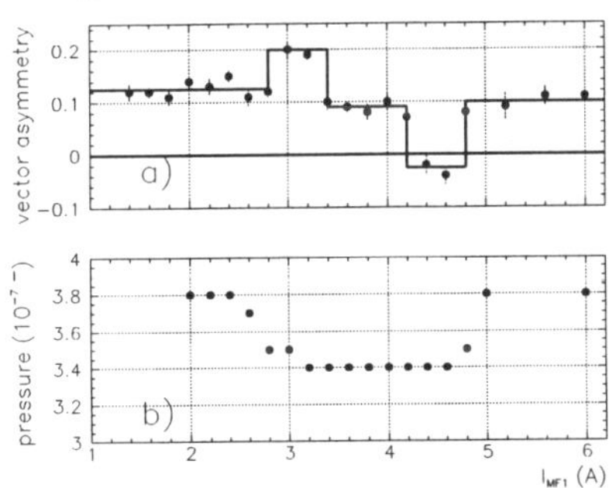

Figure 2: Vector asymmetry (a) and pressure in target chamber (b) as function of the MF1 static current.

vanishes where pure tensor polarization is expected. After setting the current in MF1 to 3.8A, MF2 was optimized for positive tensor polarization in a similar fashion. In addition to the measured asymmetry, the pressure in the target chamber provides some information about the performance of the transition units. Fig.2b shows the pressure as a function of current in MF1. When transitions are made, the pressure in the target chamber decreases by a third over the base pressure, because in this case the second set of sextupoles defocusses one of the three states.

5 Operation

In order to minimize systematic errors, it is desirable to be able to alternate between target states frequently, preferably while beam is stored in the Cooler so that data in different target states are taken with the same stored beam. This was accomplished by remotely controlling the currents to the transition units. A sinusoidal function with exponentially decreasing amplitude was programmed to reproducably degauss each unit with every current change. Fig.3 shows the static field current through MF1 (trace a), the beam current (trace b) and the trigger rate (trace c) as a function of time. Trace (a) shows how MF1 is degaussed prior to each new current setting. Trace (b) shows the Cooler being filled, a data taking flattop at 135 MeV, a ramp to 200 MeV (which corresponds to an increase in beam current), another data taking flattop and the reset of all Cooler magnets during which the stored beam is discarded before the ring is filled with protons of the opposite spin state. The trigger rate (trace (c) in Fig.3 increases during degaussing, because the target density increases by 1/3 (see Sec.4). Events during degaussing are ignored in the final analysis.

6 Spin Exchange

When we began using the deuterium target, we noted that the tensor polarization was significantly lower than expected. We investigated obvious causes for the deficiency, such as insufficient background subtraction, incomplete rejection of unwanted states by the sextupoles, wall depolarization and inefficiency of the transition units. None of these mechanisms quantitatively explained the low tensor polarization.

Spin exchange between deuterium atoms would have a significant effect on the tensor polarization only. It would also depend on the target density, in such a way that the polarization grows larger as the target density is decreased. We performed a series of measurements where we determined the tensor polarization as a function of target thickness by reducing the gas flow in the dissociator compared to its normal level. For these measurements we operated the ABS

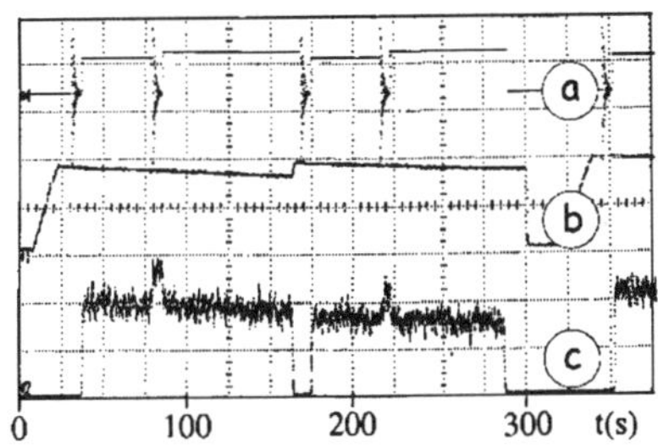

Figure 3: MF1 static current (a), stored beam current (b) and trigger rate (c) as function of time.

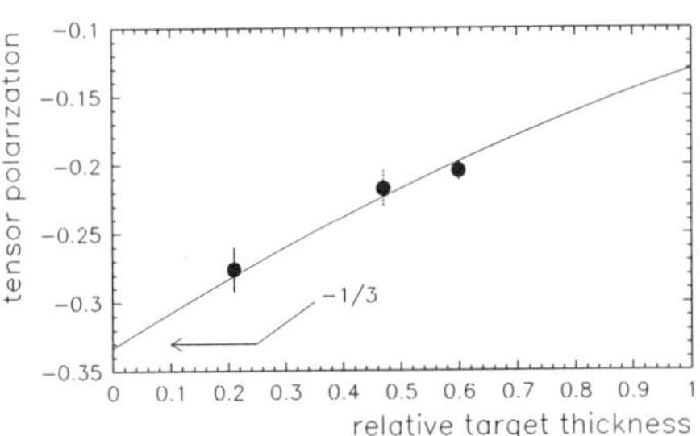

Figure 4: Target tensor polarization as a function of target thickness. The curve is the calculated effect from spin exchange.

without MF transitions in order to be independent from inefficiencies of the transition units. In this case the target polarization is a mixture of vector and tensor polarization $(P_z, P_{zz}) = (+1/3, -1/3)$. The tensor component is shown in Fig.4 as a function of relative target thickness. The data are normalized with a common, arbitrary factor. The curve shown is from Walker at al.[5] Note that the tensor polarization is negative and that polarizations of larger magnitude are indeed found at lower target densities. It can be seen from Fig.4 that the calculation explains the observed relative change in target tensor polarization with target density.

Conclusions

We have commissioned a vector and tensor polarized deuterium target for the IUCF Cooler. To date, we have used this target to measure spin correlation coefficients in pd elastic scattering at 135 and 200 MeV.

Acknowledgments

This work was supported by NSF grants PHY-9602872, PHY-9722556, PHY-9901529 and DOE grant DOE-FG02-88ER40438.

References

1. Polarized INternal Target EXperiments, www.iucf.indiana.edu/~pintex
2. T. Wise *et al*, *Nucl. Instrum. Methods* A **336**, 410 (1993).
3. W. Haeberli, *Ann. Rev. Nucl. Sci.* **17**, 373 (1967).
4. A.D. Roberts *et al.*, *Nucl. Instrum. Methods* A **322**, 6 (1992).
5. T. Walker *et al.*, *Nucl. Instrum. Methods* A **334**, 313 (1993).

STATUS OF THE NOVOSIBIRSK POLARIZED DEUTERON TARGET

M.V.DYUG[1], B.A.LAZARENKO[1], D.M.NIKOLENKO[1], A.N.OSIPOV[2],
I.A.RACHEK[1], R.SH.SADYKOV[1], YU.V.SHESTAKOV[1], A.A.SIDOROV[2],
V.N.STIBUNOV[2], D.K.TOPORKOV[1] AND S.A.ZEVAKOV[1]

(1)Budker Institute for Nuclear Physics, Novosibirsk 630090, Russia
(2) Institute for Nuclear Research, Tomsk 634050, Russia

Status of the polarized deuterium gas target used at the VEPP-3 electron storage ring is given. The main parameters of the target measured during the experiment are presented. Further improvements of the target and possible limitation of the beam intensity from atomic beam source (ABS) are discussed.

A storage cell polarized deuteron target was successfully used at the VEPP-3 electron storage ring in experiment on measurement of the components of tensor analyzing power in elastic *ed* scattering during the run 1999-2000[1]. The general view of the experimental straight section with polarized target is shown in Fig. 1.

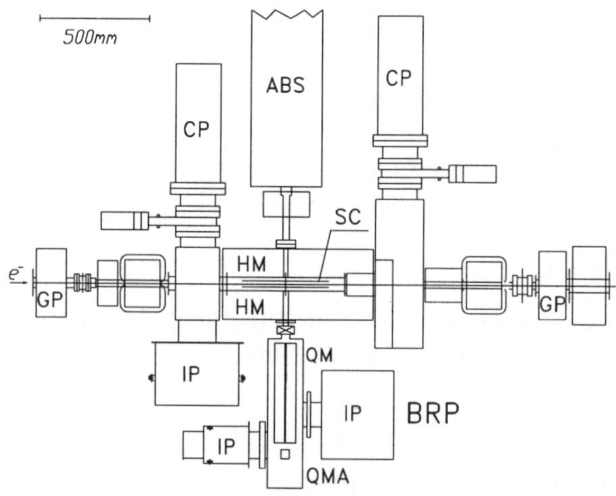

Figure 1. Top view of the internal target at VEPP-3. ABS - atomic beam source; SC - storage cell; BRP - Breit Rabi polarimeter; HM - holding field magnet; QM - quadrupole magnet; GP - getter pump; CP - cryo pump; IP - ion pump; QMA - quadrupole mass analyzer.

To determine the polarization of the target a special detector was introduced. It covered of the small momentum transfer range where theoretical predictions for the asymmetry in elastic *ed* scattering are reliable and also the absolute measurement of the component T_{20} has been performed earlier[2]. The average tensor polarization of the target was found to be $P_{zz} = 0.420 \pm 0.015 \pm 0.030$. Somehow the measured polarization was less than the expected one.

One considers two effects may reduce the polarization in the target. The high intensity bunches of the circulated electron beam produce a magnetic field transverse to the beam direction. The interaction of this field with the magnetic moments of atoms can cause depolarization of the atoms inside the cell[3]. This effect depends from the current of the electron beam and more important at higher value of the current. However the analysis of the measured polarization of the target versus the current of the electron beam has shown that this effect is very small in our case. The second one is depolarization of atoms during the collisions with the cell wall. The dependence of the polarization versus the position along the cell was not found that shows this effect small too. A drifilm coated aluminum storage cell used during the

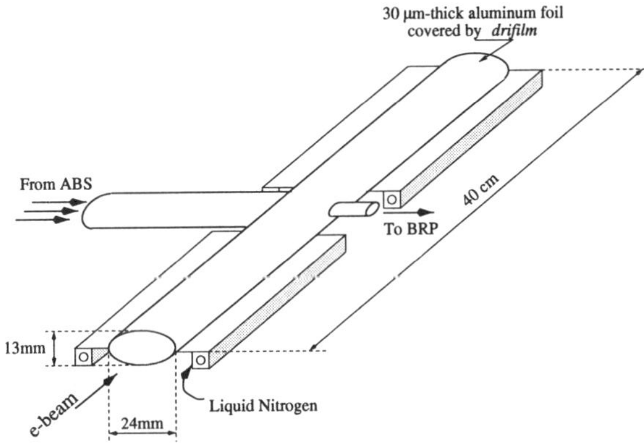

Figure 2. Storage cell for polarized atoms

experiment was consisted from the body of the cell (cooled by liquid nitrogen) and injection tube composed from aluminum tube (160 mm in length)and teflon tube (182 mm) passing through the SFT unit, see Fig. 2. After the run when the composite injection tube was replaced by a pure aluminum one coated with drifilm and cooled additionally the ratio of atoms to molecules leaving the cell and measured by QMA was increased about one order of

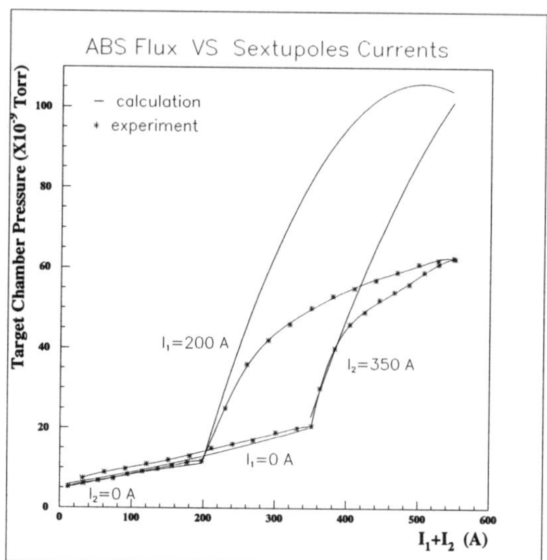

Figure 3. Intensity of the focused deuterium beam versus the currents through the coils of the magnets. I_1 - is a current through the coils of the first and second magnets; I_2 - is a current through the coils of the third last magnets. The right point of the curve with $I_1 = 0$ was used for normalization of the experimental data and calculation.

magnitude. So, one possible explanation of low polarization of the target during the experiment is mechanism of recombination of the atoms on the wall of injection tube coated by drifilm at room temperature. In a new design of the injection tube we plan to shorten the teflon part of the tube to 60 mm and introduce additional cooling.

Magnetic system of the ABS[4] consists of five superconducting sextupole magnets: first two magnets are tapered and three others are cylindrical. They are gathered in two groups and driven by different power suppliers. To understand the focusing properties of the magnet system detail studies of the intensity of the beam from ABS versus the currents through the coils of the magnets have been performed. Fig. 3 presents the results of the measured and calculated intensity. To get satisfactory agreement between the calculation and measurement some procedure was used in calculations. Thus the position and the size of the source of atoms was varied to get good description of ratio of the intensities when only one group of the magnet was ON at highest current. This led to put the source of atoms nearer to the magnet with enlarge size as that for the nozzle. When the intensity of the focused beam is not as high the experimental data and calculations agree well but at highest intensity available from the source the experimental result significantly low the predicted value. It seems that at high intensity some additional mechanism of the beam attenuation is began to operate. When the magnetic field in the magnets is increased the density of the atoms along the beam axis is changed significantly. Fig. 4 has shown the calculated density along the source near the beam axis.

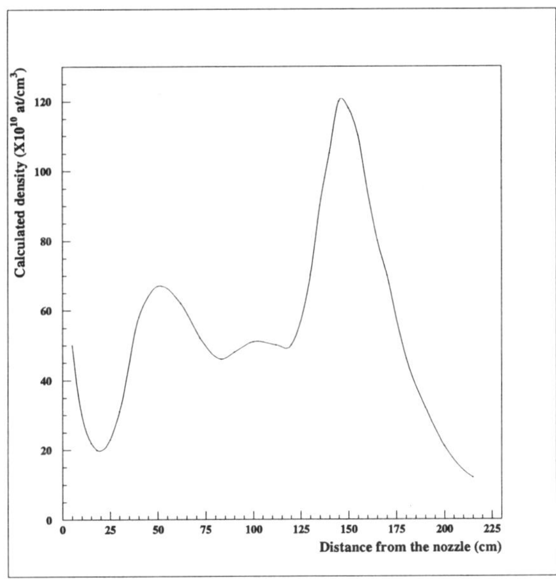

Figure 4. Calculated density near the beam axis along the atomic beam source.

The calculated density is normalized to the flux measured at full currents through the magnets. The integral of the density over the length along the beam axis up to the interaction point is about $10^{14} at/cm^2$. This area density is big enough that the intra beam scattering may be occurred. The mean free path of the molecule in the gas is given by expression:

$$\lambda_g = \frac{1}{\sqrt{2} \cdot \sigma \cdot n} \tag{1}$$

where σ - is gas kinetic effective cross section, n - is a gas density. The formula for the mean free path of the molecule in the supersonic beam may be approximately written as[5]:

$$\lambda_b = 3 \cdot \lambda_g \cdot \sqrt{2} \cdot \frac{V_d}{\delta V} \tag{2}$$

where the quantity $3 \cdot \lambda_g$ is approximately defined the mean free path of atoms or molecules in the beam having Maxwell velocity distribution and additional factor $\sqrt{2} \cdot V_d/\delta V$ is described the supersonic properties of the beam, here V_d is the drift velocity of the flow and δV is a temperature spread of the velocities around V_d. For estimation of the effect of attenuation of the beam intensity due to intra beam scattering let assumed $\lambda_b = 10 \cdot \lambda_g$, the length of the source $L_s = 1.5$ m, the effective cross section $\sigma = 1 \cdot 10^{-14} cm^2$ and the density of atomic beam n $\approx 1 \times 10^{12} at/cm^3$. If to put these numbers in the expression for the beam intensity at the exit of the source $I_{exit} = I_0 \cdot exp(-L_s/\lambda_b) = I_0 \cdot exp(-0.21)$, it's becomes obviously that the intra beam scattering is important. Morover the effective cross section may be even larger than it was used for the estimation because the temperature of the beam is very low, about 20 K.

The ABS was also tested to produce intensive hydrogen beam. Since the saturation vapour pressure of hydrogen is much higher (four or more order of magnitude) than for deuterium for given temperature the temperature of the cryo surface (inner bore of magnets) was reduced up to 2.5 K by pumping the helium gas in the cryostat. It allowed to increase the pumping speed for hydrogen atoms. Fig. 5 presented the results on the intensity measurements of the hydrogen and deuterium atomic beam versus the temperature of the liquid helium in cryostat. The intensity of the hydrogen beam is depended dramatically from the temperature whereas the intensity of the deuterium beam is a constant. The results show that for deuterium beam the vacuum conditions are good at 4.2 K temperature, but may be still limit the intensity of the hydrogen beam even at 2.5 K.

The highest intensity of the polarized deuterium and hydrogen atoms injected into the storage cell was measured to be 8.2×10^{16} and 7.8×10^{16} at/sec respectively.

This work was supported in part by the Russian Fund for Fundamental Research under Grants 01-02-16929, 01-02-17276 and State Scientific and Technical Program (Fundamental Nuclear Physics) 01-31/SC-FNP.

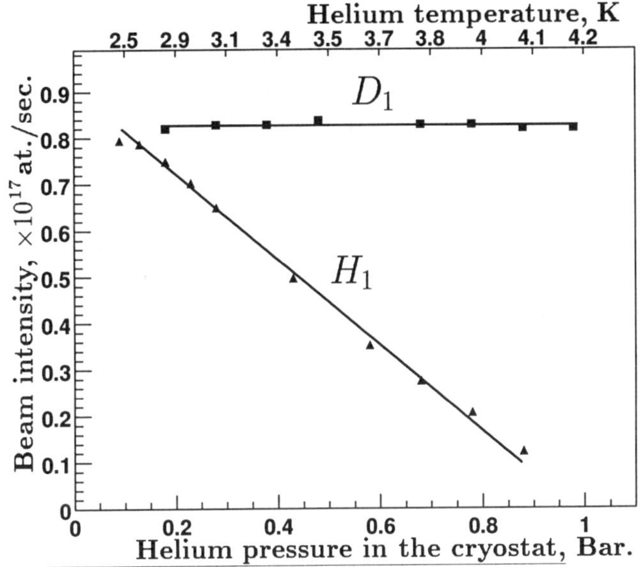

Figure 5. Measured intensity of the deuterium and hydrogen atomic beam versus the pressure of helium gas in cryostat.

References

1. D.M.Nikolenko *et al.*, Nuclear Physics **A684** (2001) 525.
2. M.Ferro-Luzzi *et al.*, Phys. Rev. Lett. **77**, (1996) 2630.
3. R.Gilman *et al.*, Nucl. Instr. and Meth. **A327** (1993) 277.
4. L.G.Isaeva *et al.*, Nucl. Instr. and Meth. **A414** (1998) 201.
5. V.S.Troitski. JETP, v.**41** (1961) 389. In Russian.

THE JEFFERSON LAB POLARIZED ^{3}He TARGET SYSTEM

KEVIN KRAMER[1] FOR THE JEFFERSON LAB POLARIZED ^{3}He
COLLABORATION

*(1) Department of Physics, College of William and Mary, P. O. Box 8795,
Williamsburg VA 23187, USA*

Jefferson Lab, a 5.7 GeV electron accelerator facility, has a broad program to study the structure of the neutron and the ^{3}He nucleus using polarized ^{3}He. The ^{3}He gas target system designed for this program allows a wide range of scattering angles and maintains high polarization in a continuous electron beam. The ^{3}He is contained in a high-pressure double-chambered glass cell centered in a 25 Gauss magnetic field. The ^{3}He is polarized by spin-exchange with optically polarized rubidium vapor. Two independent methods of measuring the polarization have been integrated into this system: NMR, a water calibrated measurement of the nuclear magnetic resonance and EPR, a measurement of the shift in the electron paramagnetic resonance of rubidium before and after a helium nuclei spin-flip. These methods agree with each other to high precision and agree with the extracted polarization from an elastic asymmetry measurement.

1 Polarized ^{3}He Physics in Hall A Jefferson Lab

Electron scattering experiments use polarized ^{3}He to study the structure of the neutron and the ^{3}He nucleus. Interest in these subjects has led to a broad program of physics using a polarized ^{3}He target in Hall A at Jefferson Lab.

Jefferson Lab is a 5.7 GeV electron accelerator capable of over 100 μAmps of continuous current with beam polarizations of over 80%. It has three halls where the beam is delivered to fixed targets. The polarized ^{3}He program has worked exclusively in Hall A to take advantage of the two high resolution spectrometers located there. The combination of high current, high beam polarization and high resolution spectrometers makes Jefferson Lab an ideal place to do precise measurements of the neutron and ^{3}He substructure.

There have been four completed experiments in Hall A using the polarized ^{3}He target system. The first, E94-010, was a measurement of the neutron spin structure functions in the Q^2 region connecting the Bjorken sum rule and the generalized GDH Sum Rule and it ran from September to December 1998. Immediately following this experiment was E95-001, the measurement of the inclusive spin-dependent quasi-elastic transverse asymmetry A_T from which the magnetic form factor of the neutron can be extracted. The two other completed experiments were run this summer. The first, E99-117, was the measurement of the neutron asymmetry A_1^n at large x_{Bj}. The second, E97-103 was a search for higher twist effects in the spin-structure function

E97-103 was a search for higher twist effects in the spin-structure function $g_2^n(x, Q^2)$.

There are two future experiments that have been approved for running in Hall A using the polarized ^{3}He target. The first, E97-110, is scheduled to run next summer and will measure the generalized GDH sum rule using nearly real photons. The second, E01-012, measurement of the neutron spin structure functions in the resonance region is waiting to be put on the schedule.

There have been several other proposals using the polarized ^{3}He target. They include a measurement of the electric form factor of the neutron, measurement of the ^{3}He wave function and several entries in the Jefferson Lab 12 GeV Upgrade white paper.

2 Polarizing ^{3}He using Optically Polarized Rubidium

The method used for polarizing ^{3}He in the Jefferson Lab polarized ^{3}He target is spin-exchange with optically polarized rubidium vapor. This procedure starts with putting the helium and rubidium in a 25 Gauss magnetic field. This field seperates the $m = 1/2$ and $m = -1/2$ states of the rubidium electrons through Zeeman splitting. The rubidium vapor is then exposed to circularly polarized 795 nm laser light that can only be absorbed by electrons in the $m = -1/2$ state. These electrons will be excited from an ^{5}S $m = -1/2$ state to a ^{5}P $m = 1/2$ state. Once excited, these electrons will eventually decay, through photon emission or collision, to either their original ^{5}S $m = -1/2$ state or the ^{5}S $m = 1/2$ state. Continual exposure to the circularly polarized laser light will eventually pump all the electrons out of the ^{5}S $m = -1/2$ state into the ^{5}S $m = 1/2$ state. Depending on the laser power, this process can polarize the rubidium vapor completely within a few minutes.

Once polarized, the rubidium can pass its polarization to the ^{3}He nuceli through the hyperfine interaction between the rubidium electrons and the ^{3}He nucleus. The cross-section for this interaction is small and therefore the process of polarizing a significant amount of ^{3}He takes many hours.

3 The ^{3}He Target Cell

The target cell, where the helium and rubidium are contained, is double-chambered and made of aluminosilicate glass. One chamber, called the pumping chamber, is spherical and contains the rubidium. This chamber is heated to 170C, to create rubidium vapor, and exposed to 90 W of laser light. This is where spin-exchange between the rubidium and helium occurs. The lower chamber, called the target chamber, is cylindrical, either 25 or 40 cm long, not heated and is where the electron beam scatters off the helium.

The characteristics of the cell strongly influence the maximum polarization of the helium it contains. Without optical pumping, the cell's polarization will decay exponentially. The coefficient in the exponential is called the lifetime of the cell and can be written:

$$P(t) = P_0 e^{-\Gamma t} \tag{1}$$

where $P(t)$ is the polarization as a function of time, P_0 is the initial polarization and Γ is the cell's lifetime. The lifetime can be further broken down into four signifcant sources of depolarization:

$$\Gamma = \Gamma_{He-He} + \Gamma_{wall} + \Gamma_{Beam} + \Gamma_{\Delta B} \tag{2}$$

where Γ_{He-He} is the contribution from helium-helium collisions, Γ_{wall} is the contribution from helium collisions with the cell walls, Γ_{Beam} is the depolarization due to electron beam effects and $\Gamma_{\Delta B}$ is depolarization due to field gradients[1]. Γ_{He-He}^{-1} is 744/[^{3}He] where the [^{3}He] is typically 7.5 to 10 atmospheres at room temperature[1]. Γ_{wall}^{-1} is for a good cell around 90 hours[1]. Γ_{Beam} has been roughly calculated to be 655 hours per μAmp of beam[2]. The program generally runs at between 10-15 μAmps. $\Gamma_{\Delta B}^{-1}$ is around 1700 hours and is not a significant contributor to target depolarization. The lifetime of the cell determines the maximuim polarization the cell will attain and therefore is the best indicator of the cell's quality. For the experiments this summer, the average lifetime of the cells we used was 54 hours out of beam.

4 Target Polarimetry

There are two methods of measuring the polarization of the ^{3}He integrated into our system. The first, NMR, uses the adiabatic fast passage (AFP) technique to drive the ^{3}He nuclei through their magnetic resonance. Adiabatic Fast Passage requires a high-frequency oscillating magnetic field perpendicular to the holding field to be on while the holding field linearly increases its strength. If the frequency of the oscillating field and the strength of the holding field is chosen carefully, at some point the nuclei will be able to absorb the energy from the oscillating field and the spins will flip in the opposite direction. At this point there will be strong magnetic field in the direction transverse to the holding field which will be proportional to the helium polarizatrion. This transverse field is measured by a set of pick-up coils.

To get a polarization from the signal from the pick-up coils, the signal height must be calibrated to a known polarization. A water cell, in the same shape as our helium cells but filled with deionized water, is used to calibrate the NMR system, since the polarization of water is well known. In addition to this calibration, geometric and density characteristics of the cell must be considered to get a meaningful calibration out of the NMR signal.

The other polarimetry method is EPR, the measurement of the rubidium's electron paramagnetic resonance. The frequency of the resonance is directly proportional to the external field. One component of the field is from the polarized ^{3}He nuclei. If the ^{3}He nuclei are flipped using AFP, the frequency shift of the resonance will be solely from the component coming from the ^{3}He and therefore proportional to the nuclei's polarization. To perform this measurement, the rubidium's resonance must be excited by a small RF coil and the photons resulting from the resonance must be detected by a photodiode. The AFP in the case of EPR has to be performed by sweeping the oscillating frequency through resonance and leaving the holding field steady.

The experiment E94-010 did a detailed target analysis and found that the EPR and NMR agreed with one another to 1.8%. This result also agreed with two elastic asymmetry measurements the experiment performed. Individual polarimetry measurements have an error of 3.6%[2] mostly due to uncertainty in the density and geometry of the target.

5   Target Setup

The target has two sets of helmholtz coils which provide the holding field. There are two sets so the field can be rotated in any direction in the horizontal plane. The target is polarized by 6 30 W 795 nm diode lasers, 3 lasers to polarize in the direction of the beamline and 3 to polarize in the perpendicular direction. A set of lenses, mirrors, quarter wave plates and polarizing beam splitters converge and circularly polarize the light from each laser.

6   Target Performance

The target performed extremely well during this summer's running with an average polarization of 42%. There were four cells used this summer. One cell ruptured, two were replaced after a month of running each and last one ran until the end of the experiment. This is an improvement on the previous set of experiments where the average polarization was only 35%[2] and there were four ruptured cells.

7 Conclusion

The Jefferson Lab target has performed well in four different experiments, running at high polarization in a high current polarized electron beam. The two polarimetry methods agree to 1.8%and provide an error on the polarization of individual measurements of $3.6\%^2$. The target has many experiments in front of it and hopefully the system will perform even better in the future.

References

1. A. Deur, Ph.D. Thesis, University of Blaise Pascal, 2000
2. I. Kominis, Ph.D. Thesis, Princeton University, 2001

THE STORAGE CELL FOR THE TRI-EXPERIMENT AT COSY-JÜLICH

O. FELDEN, R. GEBEL, M. GLENDE,
A. LEHRACH, R. MAIER, D .PRASUHN, P. VON ROSSEN

Forschungszentrum Jülich GmbH, Leo-Brandt-Strasse, D-52425 Jülich,
Germany
E-mail: m.glende@fz-juelich.de

J. BISPLINGHOFF, P.D. EVERSHEIM, F. HINTERBERGER

Institut für Strahlen- und Kernphysik, Universität Bonn, Nußallee 14-16, D-53115 Bonn
Germany

At the EDDA experiment in the cooler synchrotron COSY in Jülich an atomic beam target is used which provides the designed polarization and density distribution. To increase the target density significantly a storage cell has been developed and implemented. This will contribute to a higher accuracy for the test of Time Reversal Invariance (TRI) which will be performed at the EDDA target place.

To obtain the higher luminosity the target density and the transmission of the COSY beam through the cell were determined in their dependence on the cell aperture. Low storage cell apertures increase the target density in the cell but reduce the transmission of the circulating proton beam. To find the optimal cell design the transmission of the COSY beam was measured with movable scrapers and tested with an aperture at EDDA simulating the storage cell. The target density was calculated by Monte Carlo simulations for several cell geometries. An additional gain in target density is achieved by cooling the cell. A Teflon[1] coating of the cell reduces depolarization of the target gas.

First measurements with the EDDA detector have shown that the target density as well as the polarization arc within the range of the expected values.

1 Introduction

A polarized atomic beam target [1][2] is in operation at the EDDA experiment [3] to measure polarization observable as excitation functions [4]. A succeeding experiment at the EDDA target place aims to test Time Reversal Invariance (TRI) [5] which will be performed as a null test [6] to increase the experimental accuracy substantially. Another increase of the experimental accuracy will be achieved by a storage cell. It compresses the target gas of the atomic beam and deflects it in the direction of the accelerator beam, which results in a higher luminosity.

In this paper the implementation of the storage cell at the EDDA target and results of first measurements are presented.

---

[1] Registered trade mark of firm Du Pont

2   The Implementation of the Storage Cell at the EDDA Target Place

The storage cell at the EDDA target is a T-shaped pipe at the intersection point of the atomic beam and the COSY beam (figure 1). The cell is filled by the atomic beam part (left side) through the feeding tube. Most of the target gas is stored in the cell to interact with the COSY beam. About 1 % of the target gas passes the cell through a small hole at the backside of the cell without any wall collision for a polarization analysis in the beam dump (right side).

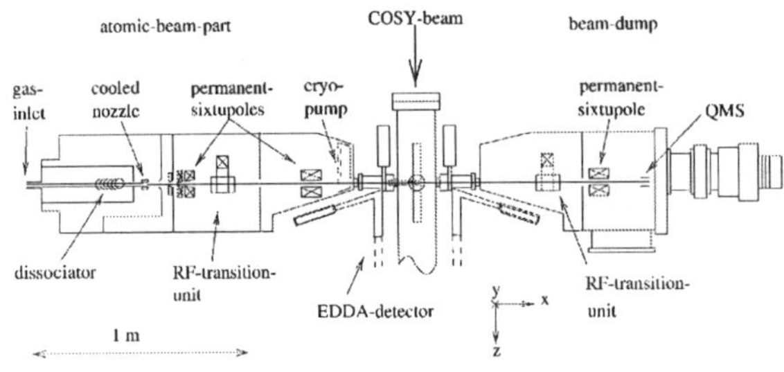

Figure 1. Scheme of the Target Region

The storage cell is mounted in the COSY beam pipe from above and connected to a cold head via a cold finger to operate the cell at a typical temperature of 80 K to 100 K. Lower temperatures can be achieved but will cause depolarization of the target gas. The heat of the cell is carried away by a cold finger of 58 cm and a proper turbo molecular pump is installed to take care for the vacuum. If the storage cell is not in use it can be lifted to the upper edge of the beam pipe for a free passing COSY beam. This is necessary because the aperture of the storage cell is smaller than the transversal beam size at injection energy and reduces the COSY beam intensity. But this reduction of the COSY beam is more than compensated by the high gain in target thickness resulting in much higher luminosities.

The storage cell (figure 2) is machined out of a single piece of aluminum by spark erosion to achieve a good heat conduction. The feeding tube can be fastened at the side of the cell. It must be removable if the storage cell is moved to the upper edge of the beam pipe. As the feeding tube is cooled by the storage cell a spring mounting cares for a continuous pressure between the cell and the feeding tube to achieve a high heat conductance.

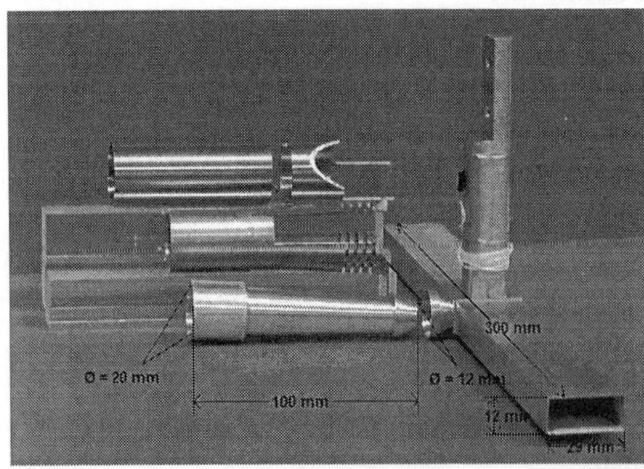

Figure 2. The Storage Cell with Feeding Tube

The cell aperture of 12 mm height and 29 mm width and the cell length of 300 mm is the result of examinations about its influence on the target density and the COSY beam transmission through the cell to reach high luminosities. The length of the feeding tube is limited to 100 mm by the valve to the atomic beam part.

3   The COSY Beam Tuning for the Storage Cell

In order to achieve a high luminosity the aperture of the storage cell must be small enough for a high target density and wide enough for a sufficient COSY beam transmission. For a study of the influence of different cell apertures on the COSY beam scrapers were moved in the coasting beam form the top, bottom, right and left to different end positions and the final beam intensity was measured. This was done at injection energy, when the beam envelope is at its largest size. By these measurements the acceptance of the COSY ring can be derived to be $a_x = 129$ π mm mrad and $a_y = 21$ π mm mrad in accordance to previous calculations. The measurement also allows to calculate the beam transmission for any aperture. Together with the results of the target density calculations for different cell geometries a cell aperture of 12 mm height and 29 mm width was found to deliver the highest luminosity.

With a test aperture of this size at the EDDA target place a COSY beam transmission of 50 % is achieved, which is higher than the expectation by the scraper test (40%). The transmission through the cell amounts to 43 %.

To reach for these beam transmissions several improvements on the beam properties were done. The implementation of the storage cell was the first time that such a high beam quality was needed at injection energy and for the beginning acceleration when the envelope is at its largest size. Experiments with an aperture at the EDDA target place show that a vertical beam displacement of only 0.1 mm results in a transmission reduction of 7 %. Therefore the orbit must be precisely held constant and well aligned to the cell especially at the beginning of acceleration.

4   Results and Discussion

The target density (figure 3) and the polarization (figure 4) was measured with the EDDA detector in a first test without feeding tube and without cooling.

The target thickness of a storage cell depends on the intensity of the injected atomic beam and the geometry of the cell. At an atomic beam intensity of $2.9 \cdot 10^{16}$ atoms/s in one state a target thickness of $8.5 \cdot 10^{12}$ $1/cm^2$ is expected for

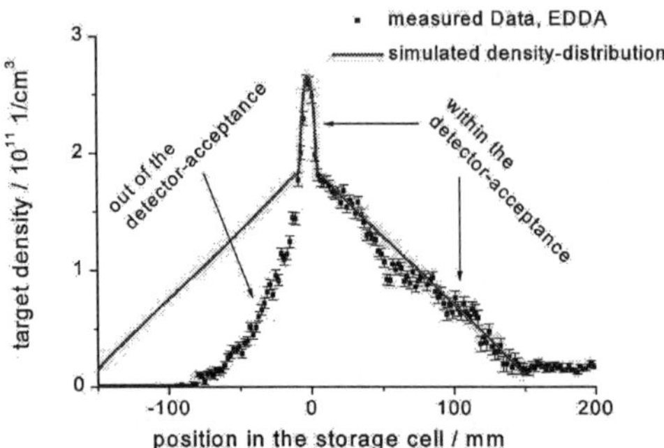

Figure 3. Distribution of the Target Density in the Storage Cell in COSY Beam Direction

the storage cell of the present design corresponding to an increase of a factor 43 compared to a pure atomic beam target. In the first test the target density was measured to be 55 % of the calculated value. To reach higher values more tests are necessary varying the cell position in respect to the atomic beam.

Comparing the measured density distribution in COSY beam direction to a density distribution delivered by a Monte Carlo simulation and reduced to 55 % the density profiles appear to be in a good accordance in the area where the storage cell is surrounded by the EDDA detector.

The polarization of the target gas in the storage cell depends on the dwell time of the atoms on the cell surface and therefore on their number of wall collisions.

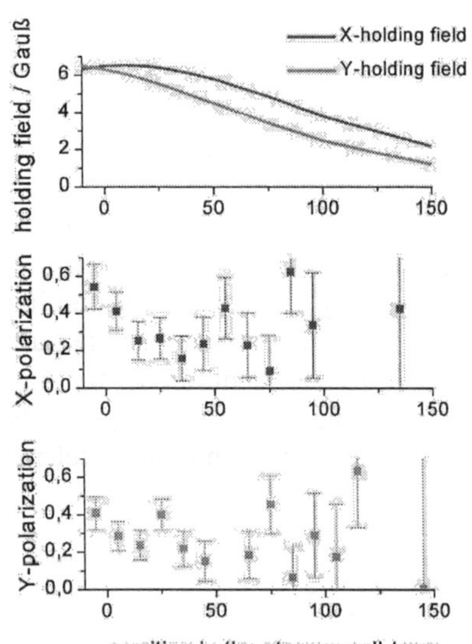

Figure 4. Distribution of the Target Polarization in the Storage Cell in COSY Beam Direction

With the results of the Monte Carlo simulation for the storage cell the polarization of the target gas in the cell can be estimated to be between 91% to 99% of the atomic beam polarization. Measurements for the Teflon coating by Price and Haeberli [7] yield 88 %. As the test measurement for the storage cell was performed with an atomic beam in two states the target gas polarization in the cell is expected to be 44 % for a holding field of 3 to 6 Gauss.

The measured distributions of the polarization in both transversal directions show that the polarization exceeds the expected value in the middle of the cell. But in order to decide if the polarization is constant in the rest of the cell, more measurements of a longer duration are necessary to reduce the error bars.

5 Acknowledgements

Many thanks to the EDDA collaboration and the COSY team for their effort and contributions to this work.

References

1. M. Altmeier, Die Entwicklung und der Bau des polarisierten Wasserstoffatomstrahlteils für das EDDA-Experiment an COSY-Jülich, Dissertation, Bonn 1998
2. O. Felden, Die Inbetriebnahme des polarisierten Atomstrahltargets für das EDDA-Experiment am Cooler Synchrotron COSY, Dissertation, Bonn 1998
3. J. Bisplinghoff, F. Hinterberger, AIP Conf. Proc. 221 (1991), 312
4. F. Hinterberger, Anregungsfunktionen im p-p-Stoß, Jülich Spezial 489 (1989)
5. P.D. Eversheim, AIP Conf. Proc. 421, Urbana, 501 (1997)
6. H.E. Conzett, 7th Intl. Conf. on "Pol. Phen. Nucl. Phys.", Paris (1990) 2D
7. J.S. Price & W. Haeberli, Nucl. Instr. Meth. A349, 321 (1994)

LASER DRIVEN H/D TARGET AT MIT-BATES

C. CRAWFORD[1], B. CLASIE[1], D. DUTTA[1], H. GAO[1], J. SEELY[1], F. XIONG[1], AND W. XU[1]

(1) Lab. for Nuclear Science, Massachusetts Institute of Technology, Cambridge, MA 02139, USA

A polarized hydrogen/deuterium internal target is being developed at MIT to be used in the South Hall Ring (SHR) at the MIT-Bates Linear Accelerator Center. In principle such a target can yield a higher figure of merit than that of the conventional atomic beam source (ABS) because of its greater thickness. Potassium vapor is optically pumped by circularly polarized laser light, tuned to the 770 nm D1 transition, in a spherical spincell which is optimized to reduce recombination and depolarization. The angular momentum is then transfered to the H/D atoms through spin exchange collisions. Even in the 100 mTesla holding field needed to overcome radiation trapping and efficiently pump potassium, the large number of hydrogen collisions in the spincell enhances the hyperfine interaction and drives the system to spin temperature equilibrium, yielding high nuclear polarization without RF transitions. We report preliminary measurements of hydrogen dissociation in operating conditions, and also atomic polarization using a Stern-Gerlach style polarimeter.

1 Introduction

In the last decade, several groups[1,2,3] have demonstrated the feasibility of polarized H/D targets by spin-exchange optical pumping as an alternative to the established atomic beam source (ABS) technology. Because it is actively pumped, the laser driven target (LDT) can be up to 25 times thicker, compensating for the smaller polarization and allowing for a potentially higher overall figure of merit.

A great candidate for the use of such a target is the new BLAST detector currently being installed in the South Hall Ring at the MIT-Bates Linear Accelerator Center. This large acceptance spectrometer is designed to study the spin-dependent electromagnetic responses of few-body nuclei using a longitudinally polarized electron beam at energies up to 1 GeV. We are developing a LDT for the experiment "A precision measurement of the charge radius of the proton" (RpEX)[4] at BLAST. Our goal is to achieve a 60% dissociation fraction and 50% polarization at a hydrogen flow rate of 2×10^{18} atoms/s, which corresponds to a thickness of 2.8×10^{15} cm^{-2}.

2   Spin Exchange Optical Pumping

The spin-exchange optical pumping technique involves the transfer of angular momentum from a circularly polarized laser beam to potassium and subsequently to atomic H/D via spin-exchange collisions. The potassium $4\,^2S_{1/2}$, $m_e = -1/2$ ground state is excited by a σ_+ photon to the $4\,^2P_{1/2}$, $m_e = +1/2$ state, which decays back to both ground state levels, populating the $m_e = +1/2$ state over time. A strong 100 mTesla holding field helps to overcome the radiation trapping effect in order to reach high optical pumping efficiency. The spin is transferred to the H/D atoms through spin exchange collisions. Even though the nuclear and atomic spin states decouple in the high magnetic field, the large number of H-H collisions enhance the hyperfine interaction allowing the atomic and nuclear states to reach spin temperature equilibrium (STE)[2,3,5]. At this point the nuclear polarization is the same as the atomic polarization for H, and even larger for D vector polarization. Such a target has been used for the first time in a nuclear experiment at IUCF[6].

3   Experimental Setup

The optical pumping occurs in a 5 cm diameter spherical Pyrex spincell, combined with a 5 cm transport tube leading to the target cell. The spincell is optimized to minimize the number of wall bounces, while having a large enough dwell time to reach STE, as shown in table 1. It is coated with drifilm to minimize recombination and depolarization effects of wall collisions. Hydrogen or deuterium gas is fed through a flow control unit into an RF dissociator connected to the spincell through a 1 mm restriction. Potassium vapor is also introduced to the spincell from a heated ampoule. The whole spincell and transport tube is kept above 180°C in a forced air oven to prevent potassium from condensing on the surface. The spincell is driven with a 3 W Ti:Sapphire standing wave laser. Laser diagnostics include power meters before and after the spincell, a wave meter, and a spectrum analyzer. Because of geometry constraints, two polarization-preserving mirrors are needed to periscope the 3 cm expanded beam into the spincell. The spincell/transport tube sits on top of a 40 cm long 1.25 cm diameter storage cell, also coated with drifilm. The 6 mm entrance aperture ensures that the atoms reach STE.

The polarimeter is precisely aligned with metal bellows to either of 2 sampling holes placed at 90° to the target cell entrance. These bends in the atomic path verify that polarization and degree of dissociation are being measured from the target cell and not directly from the spincell. The first chamber has a 1 Tesla, 15 cm long permanent sextupole magnet, a shutter for

Table 1. Calculated parameters for a single H atom in the LDT at 1×10^{18} atoms/s.

	spincell	transport tube	storage cell
wall collisions	640	96	768
K-H collisions	11.1	0.4	3.0
H-H collisions	443.9	25.0	119.9
dwell time (ms)	6.8	0.38	3.9
spin temp const (s)	7.5×10^{-5}		

background measurement, and 3 alignment apertures (blank conflat gaskets drilled with 3 mm holes) for beam collimation and differential pumping. The second chamber has a 20 Hz chopper and a Balzers Prizma quadrupole mass analyzer (QMA) which was modified to output a pure analog signal. A 300 l/s ion pump in each chamber and a NEG pump in the final chamber keep the pressure at 5×10^{-9} Torr with 1 sccm of gas flow.

The polarimeter is used to measure both the dissociation and polarization. The degree of dissociation is obtained by measuring the percentage drop in the mass two signal as the RF dissociator is turned on. The subtracted background is measured with the polarimeter misaligned instead of the shutter closed, to include background from the target chamber (10^{-5} Torr), where the contribution is the greatest. Similarly, the polarization is measured by the fractional rise or fall in the mass one signal when the laser shutter is opened to allow optical pumping. The laser helicity is switched by flipping the waveplate and tuning the laser frequency.

4   Results

A large effort was put into improving the dissociation, by varying the pressure in the dissociator. Fig. 1 shows the results of different dissociators as a function of flow rate. The maximum efficiency (up to 97%) was achieved at pressures near 200 mTorr in the dissociator. In temperature dependence tests (Fig. 2), the dissociation stayed at 57% after heating up the oven to 180°C and then introducing potassium. Fig. 3 shows a preliminary measurement of the polarization at 20-23%. However these measurements were made using polarization preserving mirrors with manufacture defects. The laser circular polarization was only 88% after reflection off both mirrors. We expect a substantial improvement after installing the correct mirrors.

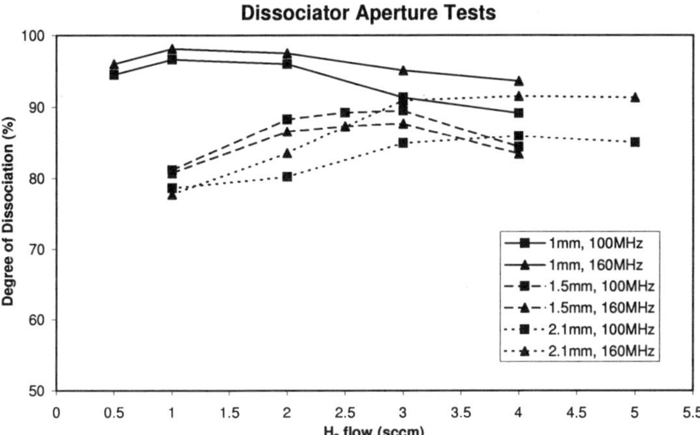

Figure 1. Dissociation vs. flow rate for the dissociator only, with three exit aperture sizes and operated at two RF frequencies.

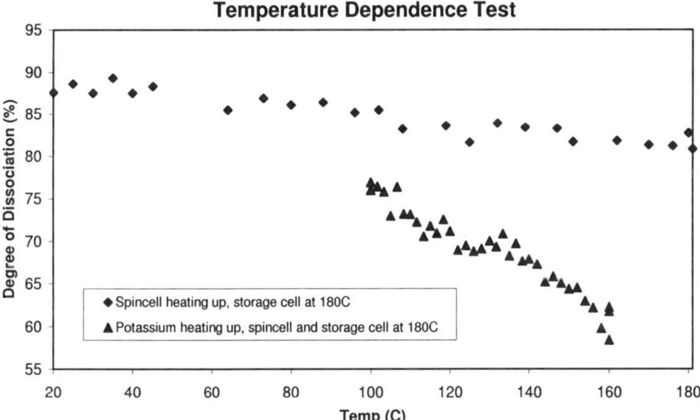

Figure 2. Temperature dependence of dissociation in the target cell. Diamonds represent the spincell being heated without potassium. Triangles show the dependence on the potassium ampoule temperature, with the spincell at $180°C$.

5   Conclusion and Outlook

We have finished the design and construction phase of a Laser Driven Target and have had promising preliminary results, especially in terms of dissocia-

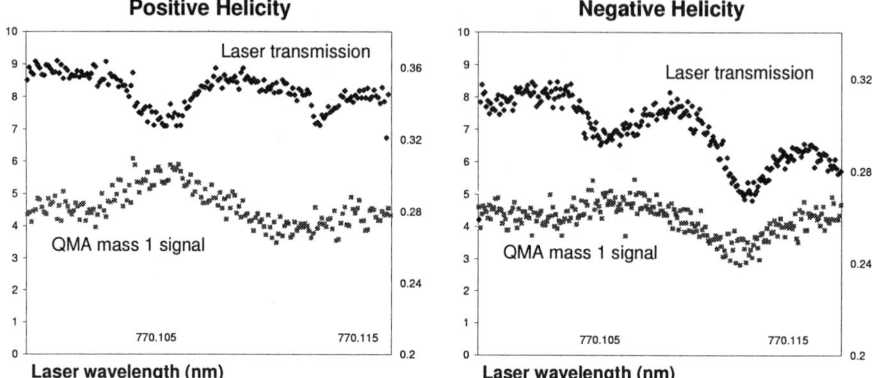

Figure 3. Preliminary polarization results with defective mirrors. The Ti:Saph laser is scanned over the potassium D1 transition, first right circular polarized, and then left-circular polarized.

tion. Future work will include replacing the polarization preserving mirrors, examining the efficiency of the sextupole magnet, and optimizing the operating parameters. In the end we will have a working target which will be used at MIT-Bates for fundamental few-nuclei experiments such as RpEX.

Acknowledgments

We thank Tom Wise and Willy Haeberli for the construction of the storage cells. We also thank Michael Grossman and George Sechen for their technical support, and Tom Hession for the fabrication of the spincells. This work is supported in part by the U.S. Department of Energy under contract number DE-FC02-94ER40818. H.G. acknowledges the support of an Outstanding Junior Faculty Investigator Award from the DOE.

References

1. M. Poelker et al., *Nucl. Instrum. Methods* A **364**, 58 (1995).
2. J. Stenger et al., *Phys. Rev. Lett.* **78**, 4177 (1997).
3. J. A. Fedchak et al., *Nucl. Instrum. Methods* A **417**, 182 (1998).
4. MIT-Bates PAC Pr. 00-02, spokespersons: H. Gao and J. R. Calarco.
5. T. Walker, L. W. Anderson, *Nucl. Instrum. Methods* A **334**, 313 (1993).
6. R. V. Cadman et al., *Phys. Rev. Lett.* **86**, 967 (2001).

THE ERLANGEN LASER DRIVEN SOURCE FOR POLARIZED HYDROGEN AND DEUTERIUM

J. WILBERT, W. HAUPT-NAGENGAST, K. RITH, F. SCHMIDT*

Physics Department, University of Erlangen-Nürnberg, 91058 Erlangen, Germany

The Erlangen laser driven source is a high intensity source for nuclear spin polarized hydrogen and deuterium atomic beams. The polarization is achieved by spin-exchange collisions between optically pumped alkali atoms and the H or D atoms. With several diagnostic tools it is possible to control most of the operating parameters. Together with an online readout of all important parameters we are able to make detailed studies of the source, for example of the optical pumping or the dissociation fraction.

1 Introduction

High intensity sources for nuclear spin polarized hydrogen and deuterium atoms are needed to investigate spin dependent phenomena in high and intermediate energy physics using internal gas targets in storage rings. In the laser driven source (LDS) the nuclear spin polarization is achieved by spin-exchange collisions between optically pumped alkali atoms and the H or D atoms. In the Erlangen LDS the alkali atoms potassium and rubidium can be used. They are optically pumped via circular polarized light from a Titanium-Sapphire laser. For the optical pumping of rubidium high power laser diodes are also available. The polarized alkali is mixed with dissociated H or D. Via spin-exchange collisions between the alkali atoms and the H/D-atoms the electron spin polarization is transferred from the alkali to the H/D. The nuclei become polarized mainly due to spin-exchange collisions among the H/D-atoms themselves. It was shown that the LDS is in spin temperature equilibrium for both hydrogen[1] and deuterium. In this state the polarization is maximum and the nuclear spin polarization can be calculated from the electron spin polarization.

2 Diagnostics

2.1 Breit-Rabi Polarimeter

The Breit-Rabi polarimeter[2] (BRP) consists of two high frequency transitions to interchange the population of two hyperfine substates and two permanent

---

*NOW AT LUCENT TECHNOLOGIES GERMANY, 90411 NÜRNBERG, GERMANY

sixtupole magnets to separate the atoms according to their electron spin state. With the BRP the electron and nuclear spin polarization and the dissociation fraction in the atomic beam can be measured.

2.2 Faraday Polarimeter

The Faraday polarimeter[3] uses a linear polarized laser beam from a tunable laser diode and determines the rotation angle of the plane of polarization (Faraday rotation) after passing the alkali vapor. For that the intensities of two orthogonal polarized components of the light are measured. The Faraday polarimeter allows a precise measurement of the alkali density, which is an important parameter because on the one hand it is important for an efficient spin transfer to the H/D-atoms and on the other hand it is critical for the lifetime of the drifilm coating of the pumping cell, which is necessary to reduce recombination on the surface. Measurements have shown that the temperature of the alkali reservoir is not sufficient to determine the density as it also depends on the surface of the liquid alkali and the conductance from the reservoir to the pumping cell.

2.3 Optical Dissociation Monitor

The optical dissociation monitor[4] analyzes the light emitted from the plasma. Therefore the areas in the emission spectrum according to one of the atomic lines and the continuous UV spectrum are selected with special filters, and the intensities are measured with photo diodes and photo multipliers. Once calibrated to the value of dissociation fraction measured with the BRP, the optical monitor allows to control the dissociation in the dissociator independently from effects in the pumping cell.

3 Studies on optical pumping

For high polarization of the H/D-atoms a high polarization of the alkali atoms is needed. Hence efficient optical pumping of the alkali atoms is very important.

Under LDS conditions the absorption profiles are Doppler broadened to several GHz. So it is necessary that the laser emission is adapted to the absorption profile. This is done by an electro optical modulator (EOM). Comparing the alkali polarization measured with EOM on or off, a big difference can be seen both for σ^+ and for σ^- polarized light. (Fig. (1)). In the case of σ^+ for example, the maximum value of the potassium polarization increases from

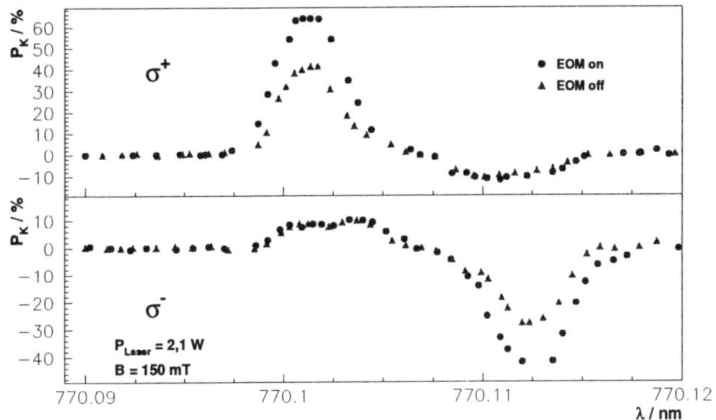

Figure 1. Polarization P_K of potassium with EOM on and off by variation of the pump laser wavelength λ.

40% with EOM off to over 60% when the EOM is switched on.

Another interesting point is the influence of the pump laser power on the polarization. Fig. (2) shows the dependence of the potassium polarization on the pump laser power for two different potassium densities. In both cases the

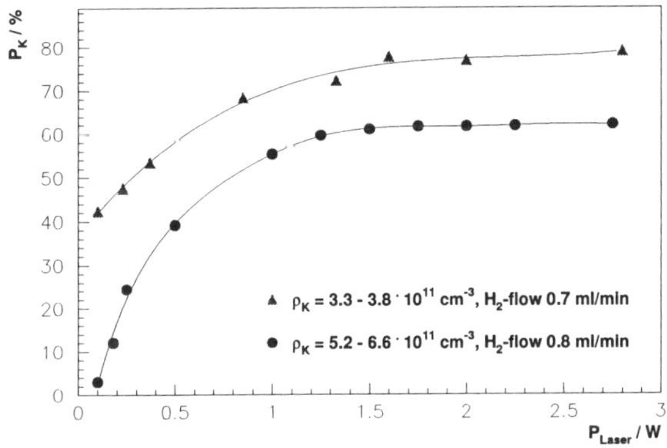

Figure 2. Polarization P_K of potassium versus pump laser power P_{Laser} for two differnt potassium denseties ρ_K.

maximum polarization is reached at a laser power of 1.5 W.

From these two aspects it is clear that the polarization is more limited by the spectral adaption than by the laser power. Therefore for efficient optical pumping the laser light has to cover up most of the absorption spectrum of the alkali either by modulation with an EOM or with several lasers detuned to each other. In this case relative low laser power is sufficient for high polarization.

4 Studies on the dissociation fraction

For detailed studies on the recombination it is important to know where the effects take place. To distinguish between effects in the dissociator and the pumping cell or exit tube the optical dissociation monitor is an important tool. Once calibrated to the value measured by the BRP it allows to control the dissociation in the dissociator itself while the BRP gives the dissociation fraction of the atomic beam after passing the pumping cell and the exit tube.

Fig. (3) shows the dissociation fraction versus the hydrogen flow. Both curves, the fraction measured with the BRP and the one measured with the optical monitor, show the same behaviour. This indicates an effect in the

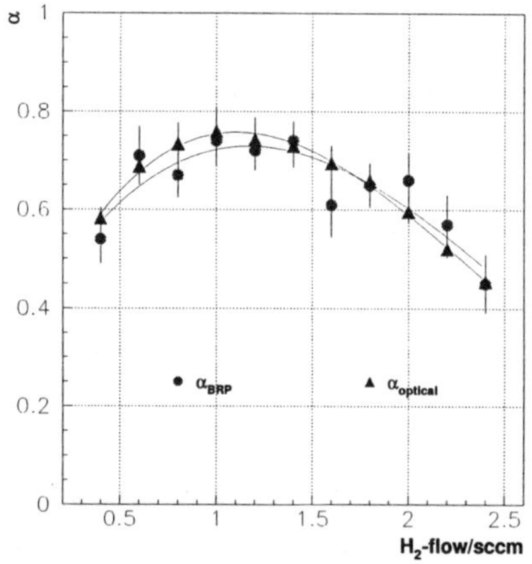

Figure 3. Dissociation fraction α versus hydrogen flow measured with BRP and optical dissociation monitor.

dissociator, not in the pumping cell.

It is different when the pumping cell is heated up. Fig. (4) shows a measurement where the pumping cell, which was uncoated, is heated up in steps and then cooled down again. While this has only small influence on the dissociation fraction by the optical monitor, which is mainly caused by an

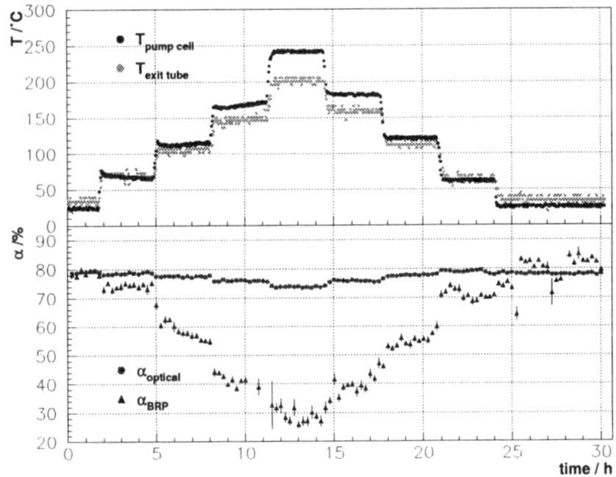

Figure 4. Comparison of the dissociation fraction α measured with the BRP and the optical dissociation monitor while heating up the pumping cell.

increase of the pressure in the dissociator, the dissociation fraction measured with the BRP is decreasing from 80% to 25% when the pumping cell is at the maximum temperature of 240°C. It recovers to the starting point when the pumping cell is cooled down again. This indicates that the recombination is extremely temperature dependent. Additionally it shows the need for a coating which reduces this effect. The improvement of these drifilm coatings should be the main focus in future investigations.

Acknowledgments

This work was supported by the Deutsche Forschungsgemeinschaft DFG.

References

1. J. Stenger *et al*, *Phys. Rev. Lett.* **78**, 4177 (1997).
2. C. Grosshauser *et al*, 12th International Symposium on High-Energy Spin Physics Amsterdam 1996, Eds. C. W. de Jager, T. J. Mulders, J. E. J. Oberski, M. Oskam-Tamboezer (1996) 340.
3. J. Stenger *et al*, *Nucl. Instrum. Methods* **384**, 333 (1997).
4. F. Schmidt *et al*, 8th International Workshop on Polarized Sources and Targets Erlangen 1999, Eds. A. Gute, S. Lorenz, E. Steffens (1999) 212.

II. Polarized Solid Targets

DEVELOPMENT ON DYNAMIC NUCLEAR POLARIZED TARGETS

S. I. PENTTILÄ

Los Alamos National Laboratory, Los Alamos, New Mexico 87545, USA

Our interest in understanding the spin content of the nucleon has left its marks on the recent development of the dynamic nuclear polarized (DNP) targets. This can be seen from the targets developed at CERN and SLAC for the measurement of the polarized spin structure functions in deep inelastic scattering. The results of the experiments indicated that less than 30% of the nucleon spin is carried by the quarks. This unpredicted small value initiated planning of new polarized target experiments to determine the gluon polarization on the nucleon using polarized real photons and polarized ^{6}LiD targets. In several facilities very intense polarized photon beams are available at a wide energy range. During the next few years these photon beams with DNP targets will be used to test the fundamental GDH sum rule. Other DNP target developments are also discussed.

1 Introduction

The development of the dynamic polarized nuclear (DNP) targets is strongly dictated by the large and complicated experiments with these targets. The effort has focused on the target technology and equipment that allow more precise experiments. This includes development of target materials that not only provide an accurate and high degree of polarization with low scattering backgrounds but also could handle high beam intensities that in many cases are required for the precision experiments.

In recent years, the question of the spin distribution of the nucleon among its constituents has received major attention. The simple presentation for the total nucleon spin of 1/2 is given by

$$\frac{1}{2} = \frac{1}{2}\Delta\Sigma + \Delta G + L_z, \tag{1}$$

where $\Delta\Sigma(= \Delta u + \Delta d + \Delta s + ...)$ determines the quark contributions, ΔG represents the gluon spin contribution, and L_z is the sum of the orbital angular momentum contributions coming from the quarks and gluons. During last decade, using the polarized deep inelastic scattering (PDIS) on polarized targets – mainly dynamic nuclear polarized (DNP) targets – the spin contribution on the nucleon carried by quarks has been extensively studied at CERN by the SMC collaboration [1] and at SLAC by E154 [2] using a polarized ^{3}He gas target, E143 [3,4], and E155 [5,6] with DNP targets. Before these experiments were possible, a major DNP target development has to have taken place. All the experiments to date, including the polarized internal gas target experiments at HERMES [7], have extracted the value of $\Delta\Sigma$ in the order of 0.3 [6]. A good discussion on the spin physics and polarized structure functions can be found in Ref. [8].

The PDIS experiments, up to date, do not distinguish spin distributions between the various polarized quarks. At present, there is also no direct measurements of ΔG and L_z available. Although, experiments that will attempt to measure ΔG are underway or planned.

2 Dynamic Nuclear Polarized Targets for Nuclear and Particle Physics Experiments

A dynamic nuclear polarized (DNP) target is a solid-state target where free nucleons of interest are polarized by the dynamic nuclear polarization (DNP) technique, that requires a low temperature, 1–0.3 K, and a high magnetic field, 2.5–7 T. A microwave range oscillating magnetic field is used to drive the dynamic nuclear polarization process *via* paramagnetic centers in the target material. A strength of the DNP target is that the target polarization can be reversed without changing any magnetic fields by changing only the microwave frequency. This polarization reversal does not cause any false asymmetry to the experiment. The nuclear polarization is monitored typically by using the CW nuclear magnetic resonance (NMR). The theory for the DNP process can be found from Refs [9,10]. A thorough review of the DNP targets is given by Crabb and Meyer [11] and references herein.

The development of DNP targets includes the development of suitable target materials that have a high degree of nucleon polarization, high density of polarizable free nucleons of interest and low density of bound nuclei i.e. a large dilution factor f, ratio between free and bound nucleons [12]. In a number of today's experiments, the target material has to function in a high intensity charged particle beam without the loss of polarization through radiation damage. In the latest targets paramagnetic centers for the DNP process were created by a charged particle irradiation [13]. To date, the most used target material is hydrogenous and deuterated ammonia. Also, chemically doped hydrocarbons like butanol are used in present targets [14,15].

The DNP process requires the temperature range of 0.3-1 K. These temperatures can be achieved with a ^{4}He or ^{3}He evaporator or with a dilution refrigerator. Commercially suitable ^{4}He and ^{3}He evaporators are available, but dilution refrigerators still have to be designed and constructed by DNP target groups. The main problem is to have a refrigerator with large enough cooling power to remove heat created by the microwave irradiation during the

DNP process and in some cases, by an intense beam. Depending on the target material, magnetic field, and operating temperature, the heat load varies from 1–20 mW per gram of material. A more complicated mode of operation of a DNP target is a frozen-spin mode where the DNP process is turned off and the target is cooled to 50 mK or less. At these temperatures and at a 0.3 T holding field the polarization decay time constants of days have been achieved [16].

An uncertainty in the target polarization affects directly the overall uncertainty of an experiment. The measurement of the absolute value of target polarization using CW NMR is a technique that has been developed over the whole period that DNP targets have been in use. The NMR system for these experiments has been largely standardized and is well understood [17]. The systematic uncertainties in a polarization measurement in a proton (deuteron) target are typically in the range of 2 to 7%. This uncertainty consists of errors in the polarization calibration measurement about 1% (0.8%), in the temperature measurement during the polarization calibration about 1% (0.4%), and then the 1.2% (1.8%) systematic error in the NMR measurement of the enhanced signal [14]. In the latest DNP experiments, the target polarization uncertainty was a significant part in the overall experiment error, and, therefore, more development work is needed to reduce the target uncertainty [17].

3 Polarized Deep Inelastic Scattering Experiments with DNP Targets

The PDIS experiments of the spin structure function measurements used longitudinally (transversely) polarized DNP targets and longitudinally polarized muons at CERN and electrons at SLAC. The experiments measured the lepton-nucleon asymmetries

$$A_{\parallel} = \frac{\sigma_{\uparrow\downarrow} - \sigma_{\uparrow\uparrow}}{\sigma_{\uparrow\downarrow} + \sigma_{\uparrow\uparrow}} = \frac{1}{f_{RC}} \frac{C_N}{f P_b P_t} \frac{(N_- - N_+)}{(N_- + N_+)} + A_{RC}, \qquad (2)$$

where the arrows indicate the beam and target polarizations, f_{RC} and A_{RC} take into account radiative corrections [6], C_N is a correction factor for the polarized nucleons other than the nucleons of interest, like ^{15}N in a NH_3 target, f is the dilution factor representing the fraction of events originating from polarized free hydrogen within the target, and N_- (N_+) is the number of scattered electrons per incident beam charge for the negative (positive) beam helicity. Also, transverse spin asymmetry $A_\perp$ data has to be taken in order to determine the longitudinal spin structure functions [5]. $A_\perp$ corresponds to transverse nucleon spin orientation with respect to the beam direction. From measured $A_{\parallel}$ and $A_{\perp}$ the deep inelastic spin structure functions $g_1^{p,d}(x, Q^2)$ and $g_2^{p,d}(x, Q^2)$ were extracted from the proton and deuteron data in a wide range of the Bjorken x-space. The $g_1^n(x, Q^2)$ can be deduced from the proton and deuteron data. The Bjorken scaling variable is defined as a fraction of the nucleon momentum carried by the struck parton, $x = Q^2/2M\nu$, where $-Q^2$ is squared four momentum, M is the nucleon rest mass and ν is the energy of the virtual photon. The experiments were run at small x and high Q^2 so that pQCD could be applied reliably.

From the measured spin-dependent structure functions the net quark helicity $\Delta\Sigma$ in nucleon was obtained to be $\Delta\Sigma = 0.23 \pm 0.04 \pm 0.06$ at $Q^2 = 5$ $(\text{GeV}/c)^2$ [6]. This result can be compared to the Bjorken sum rule, a fundamental QCD prediction that connects the nucleon spin structure to the weak axial charge g_A [18];

$$\int_0^1 [g_1^p(x, Q^2) - g_1^n(x, Q^2)]\mathrm{d}x = \frac{1}{6}\frac{g_A}{g_V}C_{NS} = 0.171 \pm 0.006 \qquad (3)$$

at $Q^2 = 5$ $(\text{GeV}/c)^2$. The data from the neutron beta decay experiments provide the ratio of the weak axial coupling to the weak vector coupling constant $g_A/g_V = 1.2573 \pm 0.0028$. The constant C_{NS} represents the non-singlet QCD correction. The result validates the Bjorken sum rule in the experimental accuracy of 5–10%. The uncertainty is mainly due to the extrapolation to the unmeasured region of x.

Although, the precision spin structure function experiments at CERN and SLAC were run some time ago, they are briefly described here firstly, because they represent two different type of DNP targets, and secondly, because of their influence on the present development in the field.

The SMC experiment at CERN covered the kinematic range of $0.0008 < x < 0.7$ and $0.2 < Q^2 < 100$ $(\text{GeV}/c)^2$ [1]. The experiment scattered longitudinally polarized 100 GeV/c and 190 GeV/c muons on the polarized DNP target that consisted of two cells that were polarized in opposite directions. The total length of the target was 1.5 m and the total volume was 2.5 liters. The large target thickness was necessary to compensate for the small muon scattering cross sections. The target was operated at 0.05–0.5 K and 2.5 T field by the world's most powerful dilution refrigerator built by the collaboration [14,19]. Since the direction of the muon beam polarization could not be flipped – a normal approach to control systematic errors in a polarized asymmetry experiment – they instead rotated adiabatically the target polarization by a magnetic field system of 0.5 T at a temperature of less than 100 mK. The rotation operation required about 35 min [14]. Other polarized target experiments that will use target materials with long polarizing times are planning to adapt this target polarization rotation method to optimize their beam-on-target time. The SMC used NH_3, p-butanol and d-butanol as target materials. The experiment quoted typical proton and deuteron polarizations of $\pm 94\%$ and $\pm 60\%$, respectively. Considerable attention was paid to the accuracies of the NMR polarization measurements and their analysis [20,21,22,23,14]. The target related systematic uncertainty for the both nucleons was $2 - 3\%$ [14].

The SLAC spin structure function experiments with DNP targets E143 and E155 [3,4,5,6] had to take a different approach because of the energetic $(30 - 50$ GeV$/c)$ and intense electron beam. An advantage of the SLAC beam was that the systematic errors of the experiments could be controlled by flipping the helicity state of the beam in every pulse at the rate of 120 Hz. On the other hand, the intense beam damaged rapidly the targets and, therefore, the targets had to be periodically annealed to restore the polarization. The targets typically were 2.5 cm in diameter and 3 cm long. The small size of the target made it possible to install two DNP targets and one background target to a single target ladder. This arrangement minimized the beam-off time used otherwise for a target annealing process or a target change. The target was cooled by a ^{4}He evaporator to 1 K and polarized in a 5 T magnetic field. This 1K/5T DNP target greatly simplified cryogenics and made the operation of the target system very reliable [24]. At present, a number of 1K/5T DNP targets are used in different experiments [24,25,26,27,28]. In the SLAC experiments NH_3 was used for the proton data and ND_3 and ^{6}LiD for the deuteron data [5]. This was the first use of ^{6}LiD in a high-energy electron beam. Polarizations of 55–90% and 20–40% were reported for proton and deuteron, respectively. The experiments covered the kinematic range of $0.014 < x < 0.9$ and $0.7 < Q^2 < 40$ (GeV$/c)^2$.

The feasibility of the polarized nuclear targets like ^{3}He or ^{6}LiD for the PDIS experiments has been discussed widely in the literature [29,30,31,32,33]. The experiments did not observe any differences in the neutron data when polarized ^{3}He, ND_3, or ^{6}LiD were used. The use of ^{6}LiD as a polarized target in spin structure experiments requires that the nuclear properties of both lithium and deuteron are understood and also the kinematic x dependence of the nucleon polarization. The ^{6}Li structure, to first order, is well described by the alpha+deuteron picture. Thus half of the nucleons in ^{6}LiD are polarized resulting the dilution factor of f=0.50 that can be compared to f=0.16 of ammonia. Lithium deuteride is also five times as radiation resistant as ND_3 [29].

For the design of a DNP target, it is useful to know the contributions of typical DNP target components such as windows, NMR coil, liquid helium, etc. to the experimental asymmetry of equation (2). For example, the SLAC NH_3 target contained in the detector acceptance about 13% free polarized protons, 66% ^{15}N (slightly polarized), 10% ^4He liquid, 6% Al, and 5% Cu-Ni (NMR coil) by weight [6]. These nuclei have an effect on the dilution factor that has a x dependence [6,14].

In addition to the SMC and SLAC experiments the spin structure functions have been measured by the HERMES experiment with polarized internal gas targets [7]. The world's data on the proton and the neutron g_1^p and g_1^n are shown in Ref. [6].

4 Towards Gluon Polarization Measurements

The PDIS experiments access only the quark spin distribution on the nucleon. To learn the gluon spin distribution, two new DNP target experiments, specifically designed for determining of the gluon spin density $\Delta g(x)$ within the nucleon, are under construction at CERN and SLAC. Both experiments, COMPASS at CERN [34] and E161 at SLAC [35], will measure the asymmetry of polarized photoproduction of charmed quarks from polarized targets. Photoproduction of open charm via the photon-gluon fusion process, $\vec{\gamma}g \to c\bar{c}$, will be tagged by decays of D mesons. Contributions from the quark distributions can be neglected in this process because there is no or only a small intrinsic charm quark content in the nucleon.

The experiments will measure the polarized photon-nucleon cross-section asymmetry

$$A^{\vec{\gamma}\vec{N}\to c\bar{c}}(k) = \frac{\Delta\sigma^{\vec{\gamma}\vec{N}\to c\bar{c}X}(k)}{\sigma^{\vec{\gamma}\vec{N}\to c\bar{c}X}(k)} = \frac{1}{P_t P_b f} \frac{N^{\uparrow\uparrow} - N^{\downarrow\uparrow}}{N^{\uparrow\uparrow} + N^{\downarrow\uparrow}}, \tag{4}$$

where k is the photon energy, P_t is the target polarization, P_b is the photon beam polarization, and f is the target dilution factor.

In these experiments the quark content of the target is not important except that the target has to have as high an average polarization per nucleon as possible, therefore, both experiments are planning to use ^6LiD that has the best dilution factor, $f = 0.5$, compared to other practical DNP target materials. The polarization of ^6Li is measured to be equal to the deuteron polarization. But the polarizing times are long, up to 20–40 h, depending on the polarizing temperature and magnetic field. At 300 mK temperature and at a magnetic field of 6.5 T 70% ^6Li polarization has been measured [36,37]. Other studies have confirmed the polarizability of ^6LiD [38,39,40].

The experiments will detect open charm production through various D meson decays. The decay products have large production angles with respect to the incoming beam thus requiring a large polarized target magnet opening to match a large solid angle spectrometer.

The COMPASS experiment (common muon and proton apparatus for structure and spectroscopy) will determine the gluon polarization from the cross section asymmetry for polarized open charm muonproduction at $x = 0.09$. With the 160 GeV/c muon beam photons at the energy range of 35 to 85 GeV will be produced. The experiment is planning to use the SMC refrigerator and its 2.5 T solenoid magnet with two targets 3 cm in diameter and total 60 cm long. Depending on physics goals ^{6}LiD, d-butanol or NH$_3$ will be used. Apart from the gluon polarization measurement, COMPASS plans to study additionally at high Q^2 the transversity structure function h_1 [34,8], spin-flavour decomposition of the structure functions, and lambda polarization [34,41].

The E161 at SLAC will measure the gluon spin distribution within the nucleon using polarized open charm photoproduction. The 35–45 GeV photon beam will be produced by polarized 45–50 GeV/c electrons hitting an oriented diamond crystal. The experiment will cover the x-range of 0.1 to 0.5. E161 will use a horizontal dilution refrigerator [42,43] and a 6.5 T warm bore solenoid magnet to obtain a high degree of ^{6}Li polarization. The diameter of the ^{6}LiD target will be 1 cm and length 8 cm.

5 Test of the GDH sum rule

The Gerasimov-Drell-Hearn (GDH) sum rule [44] is one of the most fundamental relations in hadronic physics. A high precision measurement of the sum rule will test physics beyond standard model. The GDH sum rule relates the difference in the total hadronic photo-absorption cross section for left- $(\sigma_L^{\lambda N})$ and right-handed $(\sigma_R^{\lambda N})$ circularly polarized photons interacting with longitudinally polarized nucleons to the nucleon's anomalous magnetic moment κ_N;

$$\int_{k_\pi}^{\infty} \frac{dk}{k} \Delta\sigma^{\gamma N}(k) = \frac{2\pi^2 \alpha \kappa_N^2}{M^2}, \tag{5}$$

were k is the photon energy, $\Delta\sigma^{\gamma N}(k) = \sigma_L^{\gamma N}(k) - \sigma_R^{\gamma N}(k)$, the threshold energy $k_\pi = 0.15$ GeV is needed to produce at least one pion, α is the electromagnetic coupling constant, and M is the mass of the nucleon. $\mu_p = (1 + \kappa_p)\mu_B$ and $\mu_n = \kappa_n \mu_B$. From the experiment we have $\kappa_p = 1.79$ and $\kappa_n = -1.9$.

The experimental test of the integral is one of the major challenges for photoproduction experiments over the coming years. The prediction of the GDH sum rule is 204 μb for the proton, 232 μb for the neutron, and 219 μb $(-15 \mu$b) for the average isoscalar (isovector) combinations. The fundamental meaning of the sum rule is that any particle with a non-zero anomalous magnetic moment must have an excitation spectrum and internal structure.

A worldwide program with polarized targets is already underway to test the GDH sum rule at the large photon energy range. The first result is reported by the experiment at MAMI where the value of $226 \pm 5 \pm 12 \mu$b was obtained for the proton at the photon energy range of 200–800 MeV [45]. This experiment is now relocated to the Bonn electron accelerator facility ELSA where higher photon energies are available [46]. Most of the GDH experiments require a frozen-spin DNP target that has to allow the detection of emitted particles in a very large angular acceptance [16,47].

The following list indicates facilities and the available photon energy range in GeV for GDH experiments: SLAC; E159 ($4 < E_\gamma < 45$) [48], ELSA ($0.14 < E_\gamma < 3$), JLab ($E_\gamma \to 4$), SPring-8 ($1.5 < E_\gamma < 3.5$), GRAAL ($0.5 < E_\gamma < 1.5$; with HD target) [49], LEGS ($E_\gamma \to 0.47$) [50], and HIγS ($0.002 < E_\gamma < 0.225$) [47].

6 Other DNP Target Developments

6.1 DNP target experiments at the Thomas Jefferson National Accelerator Facility (JLab)

Since the start of the JLab polarized target program in 1998, experiments have been run in Hall C with the University of Virgina 1K/5T target and in Hall B with the JLab 1K/5T polarized target [28] specifically designed for experiments with the CLAS spectrometer. The JLab polarized target program consists of polarized structure function measurements on the proton and deuteron at the wide x range including the nucleon resonance region, a precision measurement of G_e^n on polarized ND_3, tests of the GDH sum rule, and a study of the helicity structure of single pion electroproduction.

6.2 Single-spin asymmetry experiment

The results from the measurements of the analyzing power A_N in proton-proton elastic scattering from 24–28 GeV/c at the PS at CERN [51] and at AGS at BNL [52] indicate that A_N increases as a function of momentum transfer squared, $p_\perp^2$. This behavior is not explained by pQCD. In order to measure A_N at higher proton energies, the updated University of Michigan 1K/5T polarized target system will be transported to IHEP, Protvino, for use in the SPIN@U-70 experiment [27].

6.3 Low-energy $N - N$ polarization experiments

The Triangle Universities Nuclear Laboratory TUNL continues to produce precise data on low energy spin-dependent $N - N$ interactions. The latest experiment measured the $N - N$ tensor force from the $\vec{n} - \vec{p}$ scattering [53,15]. The group is in process to convert their target system to a dilution refrigerator based target. Furthermore, they are building a frozen-spin deuteron target for the GDH experiment at the HIγS Facility at the Duke Free-Electron Laser Laboratory (DFELL) [47].

6.4 Other DNP target developments

6.4.1 New type of DNP targets

The production of highly polarized protons in a low magnetic field and at high temperatures had been a DNP target dream for a long time. Protons in a crystal of naphthalene or p-terphenyl doped with pentacene have been polarized to 32% and 18%, respectively, at liquid nitrogen temperature and in a magnetic field of 0.3 T by means of microwave-induced optical nuclear polarization (MIONP) [54]. A MIONP proton target is under development at RIKEN for experiments with radio-isotope beams [55].

At PSI the protons in an organic scintillator doped with TEMPO, were polarized to 80% [56]. The scintillating DNP target material offers new possibilities to perform high precision polarization experiments, allowing the particle detection in coincidence and thus an improvement in the S/N ratio.

6.4.2 DNP target applications

The DNP proton targets have been used to polarize low-energy neutron beams since 1964 [57,58,26]. A polarized proton target is the most efficient way to filter the spin of epithermal neutrons (see Ref. [59]).

In recent years, polarized low-energy neutron scattering on DNP targets has been used to study structures of proteins and other biological molecules [60,61].

7 Discussion

A complete description of the spin structure of the nucleon requires the determination of three structure functions: the momentum distribution $f_1(x, Q^2)$, the longitudinal helicity distribution $g_1(x, Q^2)$, and the transverse helicity distribution $h_1(x, Q^2)$. The gluons contribute to $g_1(x, Q^2)$ but not to $h_1(x, Q^2)$. Using the DNP targets the quark part of the longitudinal helicity distribution functions have been determined and the experiments under construction, COMPASS at CERN and E161 at SLAC, will determine the gluon spin distribution. In addition, COMPASS will also attempt to measure $h_1(x, Q^2)$. After the preliminary HERMES gluon polarization result, it seems that we need also to measure the orbital angular momentum, L_z, part of the nucleon spin. Unfortunately, there is, not yet a good experiment to measure it.

The DNP targets will be needed to continue the measurement of the quark flavour distributions on the nucleon spin and their x dependence. Some of these measurements will take place at JLab.

The big task of the coming years will be the precision test of the GDH sum rule. This will require construction of frozen-spin DNP proton and deuteron targets.

In the few years we will see interesting results with some surprises from the experiments with DNP targets but before that we have to overcome some challenges.

References

1. B. Adeva *et al.*, Phys. Lett. **B 412** (1997) 414; D. Adams *et al.*, Phys. Rev. D. **56** (1997) 5330; B. Adeva *et al.*, Phys. Lett. **B 420** (1998) 180; B. Adeva *et al.*, Phys. Rev. D **60** (1999) 2004; B. Adeva *et al.*, Phys. Rev. D **58** (1998) 112001; B. Adeva *et al.*, Phys. Rev. D **58** (1998) 112002.
2. K. Abe *et al.*, Phys. Rev. Lett. **79** (1996) 6620.
3. K. Abe *et al.*, Phys. Rev. Lett. **74** (1995) 346.
4. K. Abe *et al.*, Phys. Rev. Lett. **75** (1995) 25; K. Abe *et al.*, Phys. Rev. Lett. **76** (1996) 587; K. Abe *et al.*, Phys. Rev. Lett. **78** (1997) 815; K. Abe *et al.*, Phys. Rev. D **58** (1998) 112003.
5. P. L. Anthony *et al.*, Phys. Lett. **B 458** (1999) 529; P. L. Anthony *et al.*, Phys. Lett. **B 458** (1999) 536; P. L. Anthony *et al.*, Phys. Lett. **B 463** (1999) 339.
6. P. L. Anthony *et al.*, Phys. Lett. **B 493** (2000) 19.
7. K. Ackerstaff *et al.*, Phys. Lett. **B 404** (1997) 383; A. Airapetian *et al.*, Phys. Lett. **B 442** (1998) 484.
8. B. Lampe and E. Reya, Phys. Rep. **332** (2000) 1.
9. A. Abragam and M. Goldman, Rep. Prog. Phys. **41** (1978) 395.
10. A. Abragam and M. Goldman, *Nuclear Magnetism: Order and Disorder*

(Oxford University Press, Oxford, 1982).

11. D. G. Crabb and W. Meyer, Ann. Rev. Nucl. Part. Sci. **47** (1997) 67.

12. St. Goertz *et al.*, in these proceedings.

13. D. G. Crabb *et al.*, Phys. Rev. Lett. **64** (1990) 2627; W. Meyer *et al.*, Nucl. Instr. and Meth. **A 215** (1983) 65.

14. D. Adams *et al.*, Nucl. Instr. and Meth. **A 437** (1999) 23.

15. D. Markoff *et al.*, In these proceedings.

16. Ch. Bradtke *et al.*, Nucl. Instr. and Meth. **A 436** (1999) 430.

17. G. R. Court *et al.*, In these proceedings; G. R. Court *et al.*, Nucl. Instr. and Meth. **A 177** (1980) 218.

18. J. D. Bjorken, Phys. Rev. D **1** (1970) 1376.

19. J. Kyynäräinen, Nucl. Instr. and Meth. **A 356** (1995) 47.

20. C. M. Dulya, Nucl. Instr. and Meth. **A 356** (1995) 88.

21. D. Krämer, Nucl. Instr. and Meth. **A 356** (1995) 79.

22. T. Niinikoski, Nucl. Instr. and Meth. **A 356** (1995) 62.

23. Y. Semertzidis, Nucl. Instr. and Meth. **A 356** (1995) 83.

24. D. G. Crabb *et al.*, Nucl. Instr. and Meth. **A 356** (1995) 9.

25. D. G. Crabb *et al.*, Phys. Rev. Lett. **64** (1990) 2627.

26. S. I. Penttilä *et al.*, Proc. 11th Int. Symp. on High-Energy Spin Phys., (AIP Conf. Proc. **343**, 1995) p. 532.

27. D. G. Crabb *et al.*, in these proceedings.

28. M. Anghinolfi, Proc. Int. Workshop on Polarized Sources and Targets, eds. A. Gute *et al.*, Erlangen 1999 p. 442.

29. S. Bültmann *et al.*, Nucl. Instr. and Meth. **A 425** (1999) 23.

30. N. W. Schellingerhout *et al.*, Phys. Rev. C **41** (1993) 2714.

31. F. M. Steffens *et al.*, Phys. Lett. **B 447** (1999) 233.

32. O. A. Rondon, Phys. Rev. D **60** (1999) 035201.

33. V. Guzey, Phys. Rev. D **64** (2000) 045201.

34. The COMPASS Collaboration, COMPASS proposal, CERN/SPSLC 96-14, SPSC/P297 (Geneva, 1996); CERN/SPLC 96-30 (Geneva, 1996).

35. The E161 Collaboration, SLAC-PROPOSAL-E161, (September 2000).

36. A. Abragam *et al.*, J. Physique-Letts. **41** (1980) L-309.

37. V. Bouffard *et al.*, J. Physique **41** (1980) 1447.

38. B. van den Brandt *et al.*, Proc. 9th Int. Symp. on High Energy Spin Phys., eds. W. Meyer, E. Steffens, and W. Thiel (Springer-Verlag Berlin Heidelberg, 1991), Vol. 2, p. 320.

39. G. Durand *et al.*, Proc. Int. Symp. on High Energy Spin Phys., Nagoya, Japan 1992, eds. T. Hasegawa *et al.*, p. 355.

40. S. Goertz *et al.*, Nucl. Inst. and Meth. **A 356** (1995) 20.

41. R. L. Jaffe, Phys. Rev. D **54** (1996) R6581.

42. T. O. Niinikoski and J.-M. Rieubland, Proc. 9th Int. Cryo. Eng. Conf. (Butterworth, Guildford, 1982) p. 580.

43. C. V. Gaulard *et al.*, Phys. Rev. C **60** (1999) 024604.

44. S. D. Drell and A. C. Hearn, Phys. Rev. Lett. **16** (1966) 908; S. B. Gerasimov, Yad. Fiz. **2** (1966) 598; S. J. Brodsky and J. R. Primack, Ann. Phys. **52** (1969) 315.

45. J. Ahrens *et al.*, Phys. Rev. Lett **87** (2001) 022003.

46. G. Reicherz, in these proceedings.

47. D. G. Haase *et al.*, In these proceedings.

48. The E159 Collaboration, SLAC-PROPOSAL-E159, (September 2000).

49. M. Bassan *et al.*, in these proceedings.

50. M. M. Lowry *et al.*, in these proceedings.

51. J. Antille *et al.*, Nucl. Phys. **B 185** (1981) 1.

52. D. G. Crabb *et al.*, Phys. Rev. Lett. **65** (1990) 3241.

53. J. R. Walston *et al.*, Phys. Rev. C **63** (2000) 014004.

54. M. Iinuma *et al.*, Phys. Rev. Lett. **84** (2000) 171.

55. T. Wakui *et al.*, in these proceedings.

56. B. van den Brandt *et al.*, Nucl. Instr. and Meth. **A 446** (2000) 592.

57. D. Draghicescu *et al.*, Phys. Lett. **12** (1964) 334.

58. S. Ishimoto *et al.*, Japanese J. Appl. Phys. **25** (1986) L246.

59. O. Zimmer, in these proceedings; W. M. Snow *et al.*, in these proceedings; A. Gorzel *et al.*, in these proceedings; T. R. Gentile *et al.*, in these proceedings.

60. H. B. Stuhrmann *et al.*, Nucl. Instr. and Meth. **A 356** (1995) 124.

61. B. van den Brandt *et al.*, in these proceedings.

PROGRESS ON THE LEGS POLARIZED HD TARGET

M. M. LOWRY[1], F. LINCOLN[1], L. MICELI[1], T. SAITOH[2], A. M. SANDORFI[1], X. WEI[1], AND C. S. WHISNANT[3]

(1) Dept. of Physics, Brookhaven Natl. Lab., Upton, NY 11973, USA.

(2) Dept. of Physics, Virginia Tech., Blacksburg, VA 24061, USA.

(3) Dept. of Physics and Astronomy, U. of South Carolina, Columbia, SC 29208, USA.

A deuterium-hydride ice target with frozen spin polarization is under development for use at LEGS, TJNAF, and elsewhere. This target features favorable dilution factors and an open geometry. HD gas is isotopically purified by distillation, doped with a small amount of ortho-H_2 and frozen into a mesh of aluminum wires which conducts away the heat from the ortho to para conversion of the H_2. The wires constitute only about 20% of the target by mass and their contribution is readily determined by empty target measurements. The 2.5 cm diameter by 5 cm long cylinder of ice is created inside a dilution refrigerator-superconducting magnet system, where it is subsequently cooled to 15 mK in a 17 T magnetic field. The H reaches its equilibrium polarization of 80% in a few days. Transfers of spin from the H to the D, and subsequent re-polarizations of the H, allow a D polarization of 50% to be reached. After 6 to 8 weeks of aging, the ortho-H_2 decays sufficiently to allow removal of the target and storage in a separate dewar. The target is transferred into an inbeam dewar for bombardment. After an NMR calibration, the target is vaporized, ready to begin a new cycle. The current status and future program are discussed.

1 Introduction

The LEGS facility produces highly polarized and energy tagged gamma ray beams by Compton backscattering 263 nm laser photons off 2.8 Gev electrons at the National Synchrotron Light Source located on the Brookhaven National Laboratory site. The gammas have an average polarization of 90% at energies from pion threshold to the maximum of 470 MeV. A highly polarized, frozen spin mode, HD ice target, ideally suited for experimental tests of nucleon sum rules with that beam, is being developed by the LEGS group under the acronym SPHICE (Strongly Polarized Hydrogen deuteride ICE).

Figure 1 shows a schematic diagram of the target in use. The target itself is 3 grams of highly purified hydrogen deuteride ice, forming a cylinder 5 cm long and 2.5 cm in diameter. The cylinder is held within a shell of 0.5 mm poly chlorotrifluoro ethylene (PCTFE). It is threaded by 2000 aluminum wires, 51 micron diameter, with 99.99% purity. They occupy 1% of the volume and are 20% of the target mass. The shell and wires are attached at the upstream end to a copper ring which, in turn, is screwed into the nose of a pumped

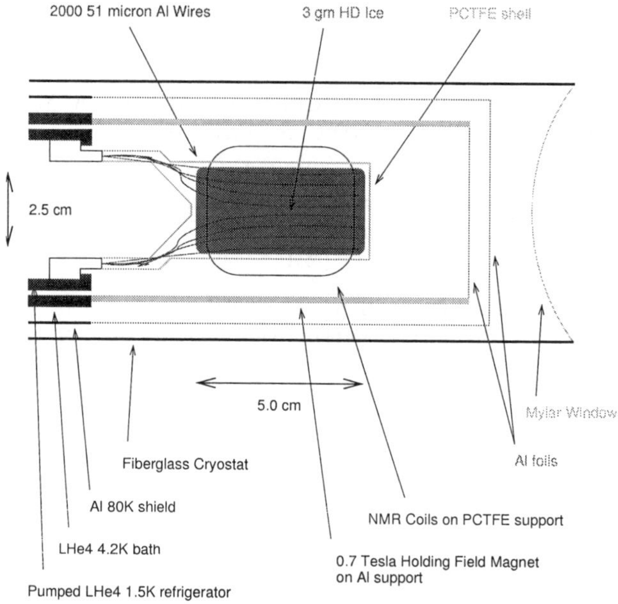

Figure 1. Schematic diagram of the HD ice target in use.

liquid He4 refrigerator running at 1.3 K. Just ouside the target shell are NMR coils on a PCTFE form that monitor the polarization. A 0.7 T holding field is provided by a NbTi solenoid, 0.6 mm thick on an Al mandrel, 0.5 mm thick, mounted on the liquid He4 shield. This is surrounded by an 80K radiation shield followed by a 2 mm thick fiberglass vacuum chamber wall.

This target system possesses a number of salient characteristics. The frozen spin polarization of the H should be 80% and that of the D 50%. The dilution factors associated with the proton and neutron are very good. The aluminum wires are only 20% by mass and their contribution is easily removed by empty target subtraction. The modest field and temperature required for the frozen spin mode means that the inbeam cryostat can give the detectors an excellent view of the reaction products. Of course, this depends on achieving long polarization retention times. Our goals are 35 days for H and 100 days for D.

We will discuss the basic concepts underlying this target system, the development history at BNL, and the future steps to be taken in that development. A parallel development effort is underway at Orsay, under the

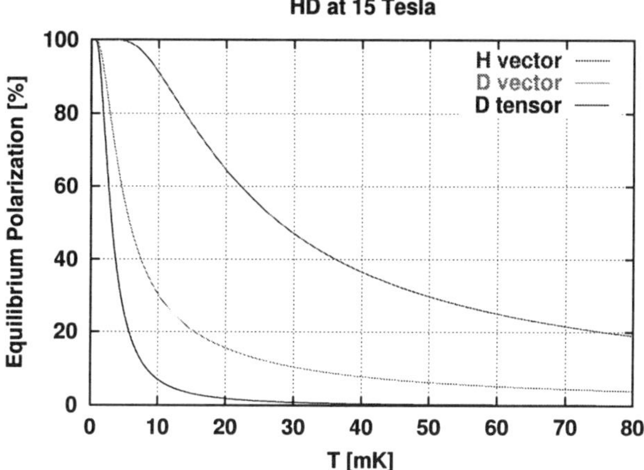

Figure 2. Equilibrium polarizations of H and D atoms in a 15 T field as a function of temperature. From top to bottom the curves are for H vector polarization, D vector polarization and D tensor polarization.

acronym HYDILE, and those interested are directed to their presentation at this workshop. The target system was proposed and initial development occurred at Syracuse University under the leadership of Arny Honig. Please refer to reference 1 and others cited therein for an account of the work there.

2 Concepts

Figure 2 shows the polarization reachable with brute force polarization of H and D atoms at high field and low temperature. With the 10 mK base temperature of our dilution refrigerator, an H polarization of over 80% is predicted. However, this value cannot be attained in a finite time by absolutely pure HD because the orbital angular momentum zero ground state of the molecule couples very poorly to the lattice, giving a very long T1. Of course, this long T1 is essential for the frozen spin mode of the target but must be defeated if we are to have any spin to freeze. The solution to this paradox is to add a small amount of ortho-H_2, a few parts in 10^4. The ortho-H_2 polarizes readily because its orbital angular momentum is constrained by Fermi statistics to be odd, with $L = 1$ being the ground state. By spin exchange, it polarizes the H in the HD molecules. But ortho-H_2 is metastable, decaying into the magnetically inert para-H_2. Thus, after six to eight weeks of waiting at po-

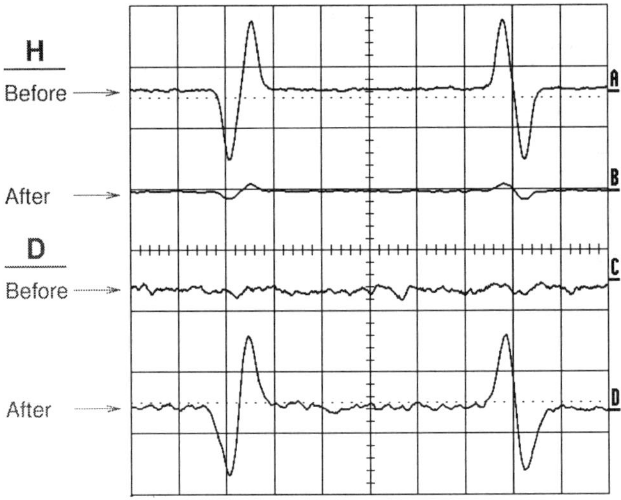

Figure 3. Transfer of polarization from H to D with forbidden adiabatic fast passage (FAFP). From top to bottom, the NMR traces are H before, H after, D before, and D after the FAFP. Efficiency of transfer was 67%.

larizing conditions, the H in the HD molecules will once again be de-coupled from the lattice. The target can then be removed by our cold transfer cryostat and placed in the inbeam conditions shown earlier or placed in a storage cryostat in order to free up the dilution refrigerator and high field magnet for polarizing the next batch of targets.

The question of D polarization is more complex. As Figure 2 shows, the equilibrium D polarization is only around 25%. So shortening T1D by adding para-D_2 does not lead to a high polarization and the 3 times slower decay rate of the para-D_2 makes this option unattractive anyway. Instead we will use a forbidden RF transition to adiabatic fast passage H polarization to the D. A recent test of this process is given in Figure 3. The NMR signals are from crossed coils using triangular magnetic field sweep so the resonance appears once on the way up and once on the way down. The RF is frequency moduated and a lockin amplifier is used to detect the resultant amplitude modulation so the signals are the derivative of the absorption. The signals demonstrate that polarization can be transferred by an RF field operating at the difference of the H and D frequency. In this case, 67% of the maximum possible transfer took place. The H can be subsequently re-polarize and the process repeated. We hope to achieve a final D polarization of 50% with this method.

3 History

The final development effort at Syracuse was a triple target production run carried out in the spring of 1999. This run served as a shakedown of the various tools and procedures required to condense 3 targets and start to polarize them. One target was successfully extracted from the dilution refrigerator and placed in the storage cryostat. The cryostat kept the target at 4.2 K and 2.0 T during an 8 hour journey by truck, demonstrating that transport between laboratories is possible. Following this, the dilution refrigerator, superconducting magnet, transfer cryostat, distillation apparatus, etc. were packed up and shipped to BNL for installation in a new facility there.

Construction on the new room was finished in Feb. 2000 and commissioning of the dilution refrigerator occurred in March. Following an initial run, which was aborted by a power failure, the first BNL triple target run was performed in the summer of 2000. This revealed several problems. One, the background in the NMR signal from H not in the target was quite severe. Second, the HD gas isotopic purity was inadequate and the attempt to circumvent the problem by pre-aging for several months was only partially successful. Finally, the H polarization achieved was only 27%. However, the two previous problems prevented a full appreciation of the significance of this fact. The attempt to utilize these targets was frustrated by a vacuum failure of the transfer cryostat and a magnet lead failure in the inbeam cryostat.

In order to fix the H background problem, the material of the target shell was changed from mylar to PCTFE and the NMR coils were affixed to their PCTFE forms by teflon threads instead of GE7031 varnish. The HD distillation apparatus was setup at the University of South Carolina and a double distillation on an initial 24 moles of HD was carried out to yield 6 moles of much higher purity gas. A new, larger diameter, bellows was installed in the transfer cryostat and the inbeam cryostat was shipped back to Orsay, France for repair of the magnet and upgrading of the cooling system. Extensive measurements of the T1H and T1D for the old gas, the single distilled South Carolina gas and the double distilled gas were carried out at 2 K and 4 K as a function of aging time.

In June of 2001, we began another target production cycle by condensing one target and implanting it in the bottom position of the dilution refrigerator. Subsequent NMR measurements confirmed our success in removing the H background. We also obtained T1 measurements that proved this double distilled gas had appropriate values for reaching equilibrium at the required temperatures and field. However, the target would not initially cool below 120 mK and its temperature slowly decayed with a time constant close to that

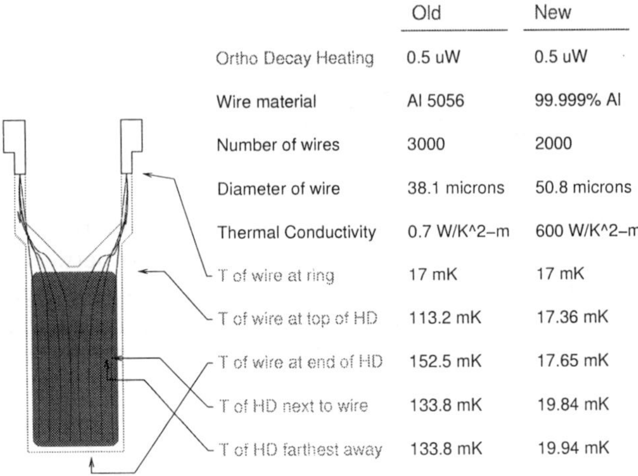

	Old	New
Ortho Decay Heating	0.5 uW	0.5 uW
Wire material	Al 5056	99.999% Al
Number of wires	3000	2000
Diameter of wire	38.1 microns	50.8 microns
Thermal Conductivity	0.7 W/K^2–m	600 W/K^2–m
T of wire at ring	17 mK	17 mK
T of wire at top of HD	113.2 mK	17.36 mK
T of wire at end of HD	152.5 mK	17.65 mK
T of HD next to wire	133.8 mK	19.84 mK
T of HD farthest away	133.8 mK	19.94 mK

Figure 4. Calculated temperature rises from ortho-H_2 decay heating with two types of cooling wires. HD Kapitza and bulk conductivities are adapted from Constable and Gaines. [2]

of the ortho-H_2 decay. As the calculations shown in Figure 4 demonstrate, the material chosen for the cooling wires, an aluminum alloy, was the problem. The mechanically stronger alloy was required for the gold coating that in turn was necessary for an easy solder joint between the wires and the copper target ring. Faced with this problem, we turned to a pure aluminum wire and developed a soldering technique with a solder specifically for aluminum. Unfortunately, we were only able to procure on a short timetable sufficent wire for one target so the third target retained the aluminum alloy wires.

Both these targets were injected on July 27, 2001 and NMR measurements confirmed the cooling wire diagnosis. The top target failed to cool just as the bottom had done while the middle pure-aluminum-wire target cooled properly. The measurements at 500 mK, 250 mK and 125 mK also served to calibrate the NMR measurements. On August 1, we went to 18 mK and 15 T to begin the polarization procedure.

After 45 days of aging at polarizing conditions, the field was lowered for NMR measurements. The measurements revealed an unexpected and as yet unexplained enhancement of the H NMR signal. As Figure 2 indicates, we expected an H polarization for the middle target of 65-70%. The top target should have cooled somewhat before the T1 grew too long for the polarization to respond to the dropping temperature so we expected a polarization in the 20-30% range, similar to the 27% obtained the year before. Instead we obtained signal sizes corresponding to 36% for the top and an impossible 450% for the middle. In contrast, the D signal for the middle target corresponds to 15% polarization as expected from Figure 2.

4 Future

The future steps to be taken are clear. Our first priority is to deal with our polarization determination problem. We need to understand the source of our H signal enhancement. Alternative polarization determination methods must be examined. One possibilty is to measure the total magnetization due to the H from the shift in magnetic field of the NMR line when the polarization direction is flipped by an allowed adiabatic fast passage. For a 100% H polarization, this will produce a 10 Gauss shift, about the same as the line width.

We need to determine the T1H and T1D of these aged targets under the conditions of the inbeam cryostat, 1.3 K and 0.7 T. Assuming these are adequate, we will transfer the targets into the inbeam cryostat and begin data taking. While the lower polarization of the top target is not as efficient for data taking, it is a valuable test case for the transfer and usage of the middle target. At the time of this workshop, the inbeam is running at 1.25 K and 0.7 T making ncutron detection efficiency determinations with a target cell filled with H_2 instead of HD and which was successfully loaded in with the transfer cryostat.

The next step for the dilution refrigerator is to transfer the results of the FAFP studies referred to earlier. The internal resonant transformer coils must be re-installed without introducing any H background. We can then investigate the amount of RF power that can be applied without unacceptable loading of the dilution refrigerator. We must then study the T1H and T1D aging dependence at the required fields and temperatures. That will establish the optimum time periods for H polarization, transition to low field for FAFP and then backup to high field for H re-polarization, etc., in order to achieve maximum D polarization. We can then produce targets with a high D polarization.

At that point we will have succeeded in developing a unique new polarized target system ideally suited for experimental tests of nucleon spin sum rules.

References

1. A. Honig, X. Wei, F. Lincoln, M. Lowry, A.M. Sandorfi, A. Lewis, S. Whisnant, C. Commeaux, J.P. Didelez and C. Schaerf in **Proceedings PST99, Erlangen, Germany, Sep. 27 - Oct. 2, 1999**, ed A.Gute, S. Lorenz, and E. Steffens [1999, University Erlangen-Nürnburg] 392-9.
2. J. H. Constable and J. R. Gaines, *Phys. Rev.* B **8**, 3966 (1973).

HIGH PRECISION MEASUREMENT OF TARGET POLARIZATION IN SOLID STATE TARGETS WITH NMR

G.R.COURT AND M.A.HOULDEN

Physics Department, Liverpool University, Liverpool L69 7ZE, UK

And

D. G. CRABB

Physics Department, University of Virginia, Charlottesville, VA, USA

The circuit related factors that can affect the systematic uncertainties in CW NMR measurements of the polarization in solid state targets are reviewed. The main problem with the large signals typically obtained with protons is non-linearity arising from failure of the constant RF drive current assumption., which can be reduced by operation at lower than normal signal levels. With the much smaller deuteron signals, instability of the background signal generated by the resonant length cable, used in the standard system, may be the limiting factor. A new system is proposed using non-resonant length cables, which can potentially reduce uncertainties arising from this problem to a low level.

1   Introduction

Continuous wave (CW) NMR has been the technique generally used to measure the polarization in solid state polarized targets since they were first developed over thirty years ago. The systematic uncertainty in this measurement currently obtainable is typically around 5%, however there are experiments underway, or being planned, in which a better precision is required. How this might be achieved is the subject of this paper.

CW NMR is observed when a constant RF field is applied to the target material at a frequency at or close to the spin transition resonance frequency (ω_0) of the relevant nucleon in the target applied static field (B_0). This RF field, normally produced by a coil wound around the material, must be directed perpendicular to B_0 and scanned over a frequency range, which covers the width of the nucleon spin resonance line. The spin transition(s) are driven by the RF field and absorb or emit energy in the process. An analysis by Abragam [1] shows that, for a thermal equilibrium (TE) NMR signal at temperature T, the RF energy absorbed per unit time (W) by a sample containing N spins I of magnetic moment γhI is given by

$$W = N \frac{h^2 \omega \omega_0}{kT} \frac{\gamma^2 H_1^2}{12\pi} I(I+1) f(\omega)$$

112

where H_1^2 is the amplitude of the RF field, ω is its frequency and $f(\omega)$ is the resonance line shape function. This energy appears as a change in the effective Q or resistance of the coil (ΔR), hence the name Q-meter. This change in resistance is measured by measuring the change in voltage (ΔV) at constant RF current (I_c). Then the power in the signal is

$$\Delta V I_C \, f(\omega) = K\,P\,N\,H_1^2 f(\omega)$$

from the previous equation; where P contains all the terms specifying the TE state populations and hence the polarization and K includes all the constant terms. H_1 is proportional to I_c so it follows that

1. The polarization is proportional to the integral (over ω) of the voltage signal generated by the coil resistance change.
2. The integrated signal size at fixed values of P and N is proportional to I_c.
3. Observation of an energy emitting (negative polarization) signal is only possible if the effective coil intrinsic resistance is made greater than the maximum value of ΔR
4. The resonance signal is complex so also has an imaginary part which can influence the measurement.

Two very important assumptions are made in this analysis

- H_1 and hence I are maintained constant during the measurement
- The value of H_1 used does not significantly disturb the spin state populations (no saturation).

2 Standard circuit arrangement

There are a number of practical problems that need to be overcome in a real system. Firstly the coil will normally have a reactive impedance (Z_{coil}) which is much larger than the $\Delta R(\omega)$ generated by the signal. This effect is normally minimised by operating the coil in resonance at ω_0. Z_{coil} is then purely resistive when $\omega = \omega_0$ but is reactive in the wings of the frequency scan so a curved background to the resonance signal is generated. Using a phase sensitive voltmeter to measure the real part of the signal can eliminate this background. A second practical problem arises because the target material must be maintained at a low temperature and therefore the coil cannot be conveniently mounted in close proximity to the remainder of the Q-meter circuit. This problem has historically been overcome by making the connection with a $n\lambda/2$ length cable, which to first order images the Z_{coil} into the rest of the circuit. However this cable introduces a large asymmetric reactive impedance in the wings of the signal generating a large background signal which is only partially removed by measuring the real part of the signal. Further problems can arise from the use of a resonant length cable because the complex impedance of the cable – coil combination modifies both the real and the imaginary parts of the imaged Z_{coil}, which can produce signal distortion. The standard circuit is shown in figure 1.

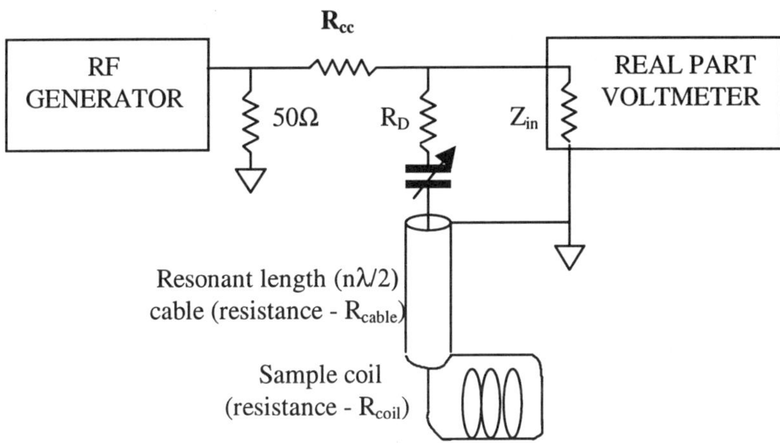

Figure 1. Standard circuit arrangement for the NMR Q-meter

The normal operating conditions can be characterized by the two extreme situations, corresponding to protons in 5 T static field where the frequency is 213 MHz and the scan width is around 500 KHz , and deuterons in 2.5 T at a frequency of 16 MHz with a similar scan width (with conventional materials). The capacitor C is used to tune the circuit to series resonance at the operating frequency. The cable wavelength integer n, typically has value three to eight with proton systems and unity with deuterons. The phase sensitive voltmeter system in general use, operates with an industry standard ($Zin = 50\Omega$) input stage. R_D is the level setting resistor which must have a value such that the sum of R_D, R_{coil} and R_{cable} is large compared with ΔR the maximum change in circuit resistance generated by the signal (see section 1. condition 3) . The constant current resistor (R_{cc}) sets the value of I_c and must be very much larger than ΔR to maintain a constant current. This particular condition can normally be easily satisfied.

Highly polarized proton signals obtained with typical target configurations are generally rather large and can have peak ΔR values as high as $\pm 5\Omega$. In this situation there can be significant circuit non-linearity because of coil current changes arising from the relatively low voltmeter Z_{in} in parallel with the resonant circuit. This can be a major source of systematic uncertainty in proton measurements. There are a number of possible solutions to this problem. Firstly, non-linearity corrections can be calculated if the effect is small and the circuit parameters are well known. This is a complex and tedious process but has been successfully carried out in a number of experiments. Secondly, the absolute size of the signal can be reduced by reducing the value of I_c. This leads to reduced S/N ratio in the TE signal, which although a problem in the past, is now acceptable because of the development of very sophisticated noise reduction systems using signal-averaging techniques. Finally the voltmeter Z_{en} can be increased, which is possible but technically difficult if a large increase is required.

With deuteron style signals the situation is completely different. They are always very much smaller than the proton signals and circuit non-linearity is not an issue. The most serious problem in this case is instability in the background signal which is generally very much larger than the resonance signal and which can lead to large systematic uncertainties in the measurement of the TE calibration signal. These instabilities arise because the background signal size and shape is critically dependent on the resonant cable parameters which are temperature dependent. Circuit operation can be optimised, in this case, by operating with the maximum possible value of I_c consistent with not violating the assumption that there is no significant disturbance of the spin state populations. However, a much larger improvement can in principle be obtained by eliminating the resonant length cable.

It should be noted that there are possible sources of systematic uncertainty which are not directly related to the circuit arrangement used. One of these is overall linearity and stability of the phase sensitive voltmeter system which typically better than 1% and therefore not normally significant. A second possible source is the uncertainty in the calibration parameters used to calculate the polarisation at TE, in particular the temperature measurement. Finally, systematic uncertainties can also arise because the coil may not spatially sample the target material uniformly. Both the latter problems are strongly dependent on the target configuration or operating conditions and will not be discussed further here

3   Improved circuit arrangement for the NMR Q-meter

The resonant length cable can be eliminated if the passive circuit components are mounted in the cold region of the target, in close proximity to the sampling coil.. The resonant length cable is then replaced by two matched cables, as shown in figure 2, for which the lengths are not critical. With this arrangement there is no cable associated background signal. However, there is still a small background signal generated by the interaction between the 'off resonance' complex impedance of the coil and the finite value of Z_{in}. There is secondary advantage to this system that arises because the constant current and level setting resistors operate in a low temperature environment so giving a reduction in the thermal noise level and an increase in circuit stability.

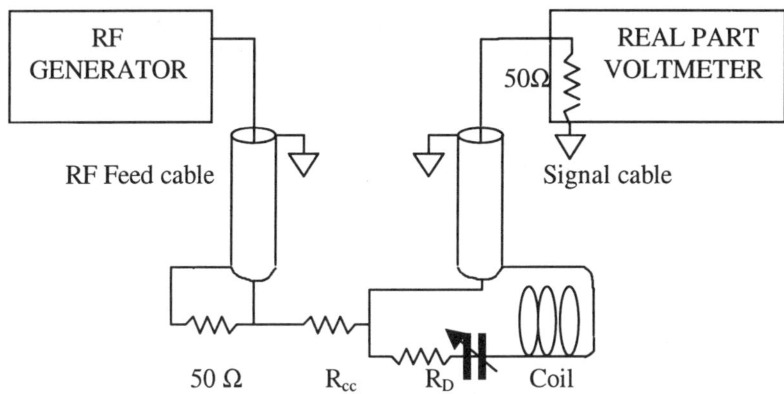

Figure 2 Non-resonant cable circuit arrangement for NMR Q-meter

Tests have been made with a deuteron target system operating at 5T static field [2] which show that a large improvement can be made in the stability and S/N ratio of a deuteron TE signal using this technique.

4   Conclusion

The dominant circuit related systematic uncertainty in the measurement of the polarization with proton style signals generally arises from non-linearity caused by failure of the constant current assumption. This can in principle be reduced to a very low level by operating with relatively small signals obtained by using a low coil current.   The use of the non-resonant cable circuit in this situation potentially gives a higher system stability and a reduction in signal distortion effects.

The main circuit related systematic uncertainty in deuteron polarization measurement is usually in the measurement of the TE calibration signal. The source of this problem is the instability of the large background signal generated by the resonant cable and potentially a very large reduction in this uncertainty is possible with the use of the non-resonant cable technique.

References

1. A.Abragam, Principles of Nuclear Magnetism, OUP (1983) p 41.
2. D.G.Crabb et al, Proc.14th Int. Spin Physics Symposium Osaka 2000, AIP proceedings 570 p 819.

ON THE HUNT FOR THE MOST EFFICIENT PARAMAGNETIC CENTER FOR THE DNP PROCESS

ST. GOERTZ, JÖRG HARMSEN, JÖRG HECKMANN, W. MEYER,
A. NÜRENBERG, E. RADTKE AND G. REICHERZ

*Institut für Experimentalphysik I AG, Ruhr-Universität Bochum, 44780 Bochum,
Germany*
E-mail: goertz@ep1.ruhr-uni-bochum.de

Achieving high dynamic polarizations using the Dynamical Nuclear Polarization (DNP) process in a paramagnetically doped solid is a much more difficult task for deuterons in comparison to protons, a fact which is mainly attributed to the small magnetic moment of the deuteron. But from the concept of spin temperature [1], which is the most successful theoretical approach to the DNP process, it may be seen that this problem can be circumvented by the use of a suitable paramagnetic center. If the centers are produced by irradiating the solid with ionizing particles, the DNP process even benefits from the weak magnetic environment of the free electron. Amorphously frozen beads of perdeuterated butanol has been irradiated with $20\,MeV$ electrons at liquid argon temperature and subsequently polarized in the Bochum dilution refrigerator. Deuteron polarizations up to -54 % at $2.5\,T$ and up to -71 % at $5.0\,T$ have been reached. A nuclear relaxation time of about 200 hours have been observed under low field and low temperature conditions showing the suitability of the material for *frozen spin* experiments.

The radiation induced center could be identified by EPR studies of different partially deuterated alcohols to be a hydrogen vacancy at a regular position in the molecule. These studies have been also extended to other corbonhydrate compounds like polyethylene as well as to alkanes of shorter molecular lengths. In particular in the case of polyethylene a paramegntic center could be produced, which is not only DNP active, but also possess a long term stability at room temperature.

1 Introduction

In the recent years polarized particle physics experiments became more and more important throughout the whole range of accessible energies. Among those, experiments with fixed solid state polarized targets play an essential role, because they provide a reasonable luminosity, even if beam fluxes as low as some MHz are used. Besides other parameters the efficiency of a polarized experiment is mainly determined by the achievable polarization of the colliding particles. In the case of a solid state target proton polarizations near 100 % are relatively easy to achieve more or less independently of the particular target technology in use. Maximum polarizations for deuterons so far range from about 30 % to 50 % depending on the particular material. Especially in experiments at energies close to or not far above the nucleon resonance region the extraction of the physics quantities often doesn't permitt the use of target materials, which contain polarizable nuclei in addition to the desired protons or deuterons. Additionally most of these experiments at intermediate energies require a large angular acceptance detector around the target, which has to be operated in the so-called *frozen spin mode*: The polarization once achieved during a polarization stage of several hours is maintained for a long time at very low temperatures $(50 - 70\,mK)$ at a small magnetic field $(0.4 - 0.5\,T)$, which is provided either by the fringe field of an external magnet or by a thin superconducting coil as a part of the refrigerator nowadays. In those cases one so far depends on the use of alcohols or carbonhydrate compounds, because of their spinless 'background nuclei'. The most common material for this purpose is n-butanol in its protonated or perdeuterated form with maximum polarizations of roughly 80 % and 30 %, at $2.5\,T$, respectively. Although the concept of the *frozen spin target* enables otherwise not realizable experiments, it envolves some inherent disadvantages:

- Due to relaxation during data taking, the average polarization is usually not larger the 80 % of the maximum achievable polarization.

- The materials free of polarized background provide the smallest ratio of protons or deuterons to the total amount of nucleons present (dilution factor) of all established target materials.

If these drawbacks combine with the relatively small achievable maximum polarization in the deuterated alcohols, the figure of merit of the corresponding experiment may be reduced below the limit of feasibility. From this point of view a further development of target materials suitable for frozen spin measurements seems to be very important.

2 Theoretical Predictions

The concept of spin temperature, the so-called Equal Spin Temperature (EST) theory, provides the most successfull approach to the DNP process in todays target materials. Although the underlying *Provotorov equations* can only be explicitly solved in the so-called *high temperature approximation*, in which an electronic polarization considerably lower than 1 is assumed, some predictions can also be extended to the low temperature high field case. In particular the dynamic nuclear polarization is given by an expression of the same form as in the case of the thermal equilibrium (TE) polarization [2], namely by the

Brillouin function $\mathcal{B}$.

$$P \;=\; \mathcal{B}\left(\frac{\hbar\gamma_k B}{2kT}\right) \tag{1}$$

γ_k is the gyromagnetic ratio of the corresponding nucleus. T denotes either the actual temperature of the lattice (TE) or the temperature of the so-called non-Zeeman reservoir of the electrons, which consists of all electronic non-Zeeman interactions, i.e. dipol-dipol interaction, but also electron-nucleus hyperfine interaction. Even different energies due to a spread of the g-factor may be included in this reservoir. By saturating microwave irradiation the *spin temperature* can be reduced by a factor of several hundred below the lattice temperature provided a frequency is chosen somewhat below or above the electron Larmor frequency ω_0. In the high temperature limit the *inverse spin temperature* $\beta = 1/T_S$ can be written as:

$$\beta(\Delta) \;=\; \beta_L \,\frac{\omega_0\Delta}{\Delta^2 + a\,D^2} \;. \tag{2}$$

$\beta_L = 1/T_L$ is the inverse lattice temperature, ω_0 the Larmor frequency and D a measure of the energy of the non-Zeeman reservoir. The factor a is defined as the ratio of the longitudinal relaxation times of the Zeeman and the non-Zeeman energies, a number usually of the order of a few units[2]. If the frequency offset Δ is varied through the resonance line, the absolut value of the inverse spin temperature reach their maxima

$$|\beta|_{max} \;=\; \beta_L \,\frac{1}{\sqrt{a}}\frac{\omega_0}{2D} \tag{3}$$

at positions $\Delta(\beta_{max}) = \pm\sqrt{a}D$. Assuming an equal spin temperature of $T_S \simeq 2\,mK$ for protons and deuterons it can be directly seen from Eq. 1 that the proton polaritation ($P_p \simeq 85\,\%$) is a factor 3.4 higher than the corresponding deuteron polarization ($P_d \simeq 25\,\%$). Actually this relation reflects relatively well the results obtained from nitroxyd doped h- and d- butanol polarized in a dilution refrigerator at $2.5\,T$ ($P_p \simeq 80\,\%$, $P_d \simeq 30\,\%$). In the case of the nitroxyd radicals the electronic energy spread D is mainly given by a large hyperfine interaction with the nitrogen nucleus as well as by an anisotropic g-factor. Therefore, D and thus also the maximum invers spin-temperature are practically unchanged, when going from a protonated to a deuterated environment. If on the other hand the doping is carried out via irradiation instead of admixture of a chemically stable radical, the created paramagnetic vacancy experiences directly the magnetical environment of the material. Thus, in comparison to the protonated case, the hyperfine induced width of the electron resonance line is reduced by a factor $6.5/2 = 3.25$ in the corresponding deuterated material. Consequently a three times higher inverse spin temperature may be reached in the deuterated compound, if the g-factor contribution to the overall electronic linewidth is small.

3 Radiation Doped Butanol

3.1 Preparation

Desoxygenated perdeuterated d-Butanol has been frozen to amorphous beads of about $2\,mm$ diameter by dropping into liquid nitrogen. The beads were then exposed to the $20\,MeV$ electron beam of the Bonn injection Linac for about $5\,min$ at liquid argon temperature. The total radiation dose was about $1 \cdot 10^{16}\,e^-/cm^2$. The density of the created paramagnetic impurities was measured to be about $2 \cdot 10^{19}\,e^-/g$ corresponding to the usual density in chemically doped butanol.

3.2 The Radical Structure

Figure 1 shows the EPR resonance line of irradiated perdeuterated n-butanol (thick line) and of n-butanol $2, 2, 3, 3, 4, 4, 4\text{-}d_7\text{-}oh$ (thin line) taken at $9.35\,GHz$ under liquid nitrogen. Whereas the spectrum of the partially

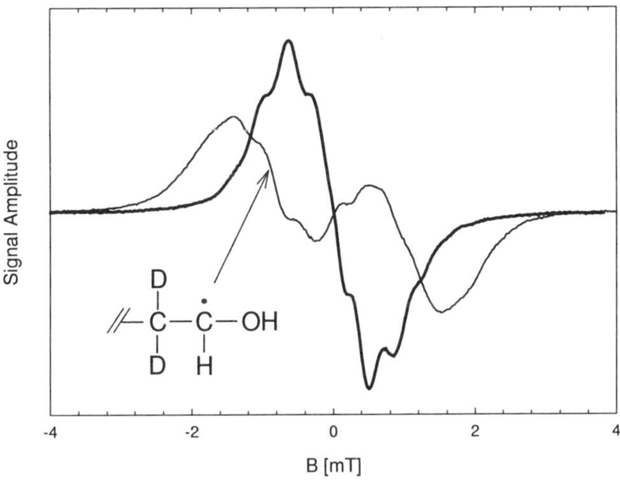

Figure 1. X-band EPR spectra of perdeuterated n-butanol (thick line) and n-butanol $2, 2, 3, 3, 4, 4, 4\text{-}d_7\text{-}oh$ (thin line). Shown are the derivatives of the absorption lines.

deuterated compound can be doubtlessly attributed to the defect structure shown in the figure, i.e. a hydrogen vacancy at the first carbon atom, the interpretation of the perdeuterated spectrum is more difficult due to the small deuteron HFS splitting, which is additionally washed out by intermolecular interactions. Nevertheless there is strong evidence that the paramagnetic center in perdeuterated n-butanol also consists of a hydrogen (deuterium) vacancy at some of the carbon atoms. The critical temperature, above which recombination of the centers sets in, was determined to be $120\,K$.

3.3 Polarization Results

The material prepared in the way described was dynamically polarized in the Bochum dilution refrigerator at magnetic fields of $2.5\,T$ and $5.0\,T$. The maximum negative polarization values observed were $-54\,\%$ and $-71\,\%$, respectively. Positive polarizations were not measured. Consequently, the maximum inverse spin temperature of radiation doped d-butanol is about twice as high as in the nitroxyd doped material at $2.5\,T$. The value taken at $5.0\,T$ can not be compared to nitroxyd doped butanol, because in this case it is impossible to give any final polarization value due to an extremely small build up rate[3]. The deuteron relaxation time was determined to be $200\,h$ at $0.4\,T$ and $70\,mK$, which is twice as long as the usual relaxation time in nitroxyd doped d-Butanol.

4 Conclusions

The unprecedentedly high deuteron polarization in radiation doped d-butanol in combination with its very long relaxation time make this material the best choice for future frozen spin measurements on polarized deuterons in all those cases, in which the temperature of the refridgerator can be kept considerably below $120\,K$ during the loading procedure of the material.
Similar investigations have also been started in the case of other carbonhydrate compounds like long chain polyethylene (CH_2) or short chain alkanes. In CH_2 two different kinds of radiation induced paramagnetic centers were observed. While both are able to drive the DNP process, one of the centers showed a very small sensitivity against exposure of the material to room temperature even for days. This property may open up the possibility of a 'room temperature stable' target material with all the benefits of the radiation doping method. These studies will be also extended to deuterated polyethylene.

References

1. A. Abragam and M. Goldman, *Rep. Prog. Phys.* **41**, 395 (1978)
2. M. Borghini, *Proc. 2nd Int. Conf. on Polarized Targets*, ed. G. Shapiro **1** (Berkeley, 1971)
3. J. Harmsen, PhD thesis in preparation

POLARIZED PROTONS DOMAINS IN MATTER

B. VAN DEN BRANDT[1], H. GLÄTTLI[2], I. GRILLO[3], P. HAUTLE[1], H. JOUVE[4],
J. KOHLBRECHER[1], J.A. KONTER[1], E. LEYMARIE[2], S. MANGO[1], R. MAY[3],
H.B. STUHRMANN[4] AND O. ZIMMER[5]

[1]*Paul Scherrer Insitute, CH - 5232 Villigen PSI, Switzerland*
[2]*CEA Saclay / SPEC, F - 91191 Gif-sur-Yvette, France*
[3]*Institut Laue Langevin, BP 156, F- 38042 Grenoble Cedex 9, France*
[4]*Institut de Biologie Structurale Jean-Pierre Ebel, F - 38027 Grenoble Cedex 1, France*
[5]*Technische Universität München, James-Franck-Strasse, D - 85748 Garching, Germany*

Domains of polarized protons have been created around paramagnetic centers in alcohols by dynamic nuclear polarization (DNP) methods, rf saturation and/or adiabatic fast passage polarization reversal. The decay of the polarization gradient has been indirectly observed by NMR methods.

We aim at taking advantage of the possibility to create such clusters of polarized protons in organic materials in order to investigate with higher resolution at will different parts of complex macromolecules by polarized neutron scattering.

1 Introduction

Dynamic nuclear polarization (DNP) is an efficient tool to create contrast in small angle neutron scattering (SANS) on hydrogenous samples by exploiting the strong spin dependence of neutron scattering from protons. The nuclear polarization arises from paramagnetic centers distributed in the samples, which transfer their spin ordering upon microwave irradiation to the nearby protons. In a second step this ordering spreads out into the bulk of the sample through spin diffusion. The DNP mechanism does not *a priori* lead to a homogeneous polarization distribution, as the two processes might proceed with different time constants. Situations can be imagined in which domains of highly polarized protons around the paramagnetic centers survive long enough to be observed by neutron scattering [1,2,3]. Such local contrast could highlight radicals, which play a key role in the biological activity of redox molecules, e.g. the tyrosyl radical in catalase [4].

This study explores the behaviour of the spin systems present in alcohols under rf irradiation in order to find optimum procedures to create and hold polarization gradients.

2 Apparatus

We investigated samples of propanediol with different isotopic substitutions (98%, 90%, 60%, 0% deuterated), doped with 5×10^{19} EHBA-Cr(V) paramagnetic centers per cm^3. They were polarized by DNP in our polarized target system [5] in a temperature range of 100 mK to 1 K at magnetic field values of 2.5 T and 3.5 T and their polarization was monitored with a Liverpool Q-meter. In order to observe fast changes in the NMR signal upon rf irradiation of the sample a rf system has been set up that allows a rapid execution of arbitrary irradiation / NMR measurement sequences.

3 Experiments: creation and decay of proton polarization domains

The cw-NMR measurement performed with a Q-meter can only monitor changes in the average polarization of the protons located outside the sphere of influence of the paramagnetic centers, i.e. of protons in the bulk. The protons close to a paramagnetic center experience a strong shift of their Larmor frequencies thus are not directly observable.

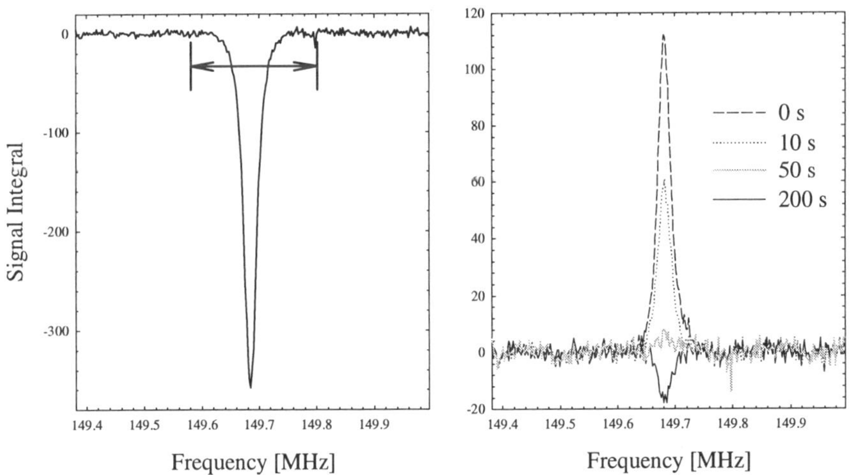

Figure 1. Left: The AFP polarization reversal (negative to positive) is performed by sweeping over the indicated frequency range, thus affecting mostly the bulk nuclei. Right: evolution of the NMR signal after the reversal, the signal decreases and becomes negative again.

This separation in frequency makes possible to selectively irradiate one "type of protons": *bulk protons* on the NMR signal, *close protons* on the wings of the signal. A polarization gradient might therefore be created by a specific rf

124

irradiation, e.g. destroying the bulk proton polarization, reversing it by AFP (figure 1.) or saturating the protons near the centers. The coupling times between the "two sorts of protons" could then be deduced by observing the evolution of the bulk protons NMR signals (figure 2).

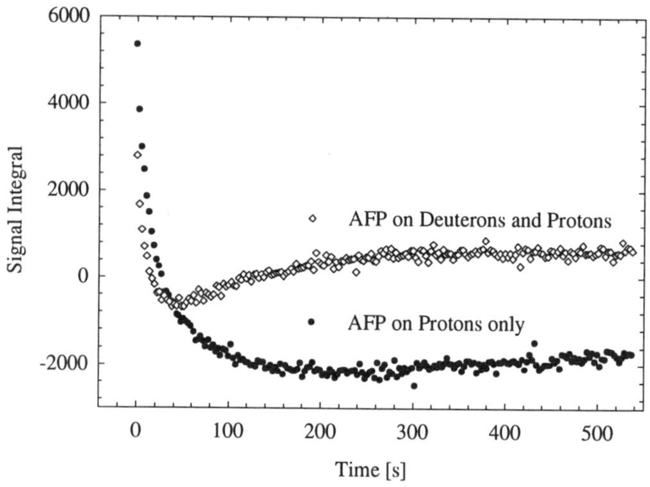

Figure 2. Evolution of the proton NMR signal of the 98% deuterated sample after an AFP polarization reversal (negative to positive) of the bulk nuclei (protons only or protons and deuterons). The close and bulk nuclei get into equilibrium after about 200 s. For a few seconds after the AFP, a strong polarization gradient between the two types of protons must exist.

The decay of the polarization gradient shows a quadratic dependence on the distance between neighbouring protons, given by the percentage of deuteration of the sample, as is expected for a spin diffusion dominated process (figure 3). It is slower at lower temperature and higher magnetic field, what hints at the involvement of a spin system in contact with the lattice, most likely the electron non-Zeeman system of the paramagnetic centers, which opens a path for the penetration of the diffusion barrier.

A much weaker influence of this system can be expected in biological samples, in which a much smaller number of *well-localised* paramagnetic centers are present. Preliminary NMR measurements on tyrosil radicals in catalase appear to support this assumption.

The clear indication of the existence of polarized protons domains and the timescale of their decay (a few seconds up to tens of seconds) suggested to study them more thoroughly by polarized SANS. In a first series of experiments we observed the expected strong change in neutron scattering intensity upon irradiating an initially unpolarized sample with microwaves as well as when reversing by an AFP sweep the polarization of the bulk proton spins only in a

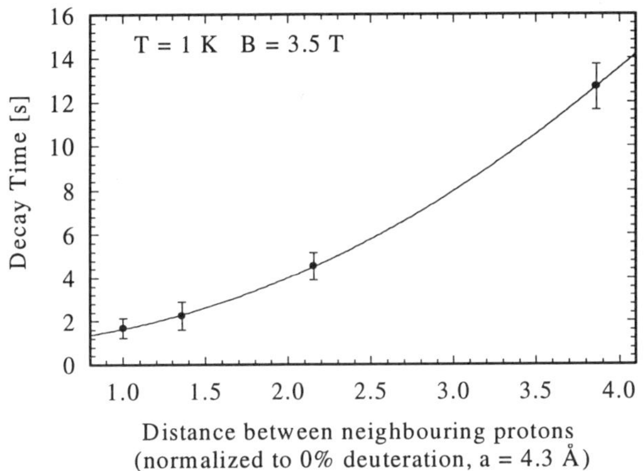

Figure 3. The decay time of the polarization gradient shows a quadratic dependence on the distance between neighbouring protons.

homogeneously polarized sample. In order to detect the most prominent changes in the local spin orientation, which take place in the first few seconds after the manipulation of the proton spin systems, the scattering data were taken in a stroboscopic manner using short time frames. To achieve the necessary statistics, a summation of several stroboscopic cycles was performed [6].

4   Conclusions

We created domains of polarized protons around paramagnetic centers in standard polarized target materials and observed their decay by NMR methods. We started to investigate the transient behaviour of such polarization domains by time resolved small-angle neutron scattering experiments.

References

1. J.B. Hayter, G.T. Jenkin, J.W. White, *Phys. Rev. Lett.* **33** (1974) 696.
2. M. Kohgi et al., *J. Phys. Soc. Japan* **56** (1987) 2681.
3. H.B. Stuhrmann *et al.*, *J. Appl. Cryst.* **30** (1997) 839.
4. D.C. Bicout, M.J. Field, P. Gouet, H.M. Jouve, *Biochemica et Biophysica Acta* **1252** (1995) 172
5. B. van den Brandt, J.A. Konter, S. Mango, *Nucl. Instr. and Meth.* **289** (1990) 526.
6. *to be published*

STATUS OF MICHIGAN POLARIZED PROTON TARGET*

D.G. CRABB[1], Z.B. ETIENNE[2], V.V. FIMUSHKIN[3], T. KAGEYA[2], A.M.T. LIN[2], V.G. LUPPOV[2], C.C. PETERS[2], A.F. PRUDKOGLYAD[4], R.S. RAYMOND[2], P.A. SEMENOV[4], K. YONEHARA

[1] Department of Physics, University of Virginia, Charlottesville VA 22901
[2] Spin Physics Center, University of Michigan, Ann Arbor MI 48109-1120
[3] JINR, Dubna, Russia
[4] IHEP, Protvino, Russia

The University of Michigan Polarized Proton Target is a 5 T, 1 K target built in the late 1980's for use, with radiation-doped NH_3, in an intense proton beam. It is now being updated in preparation for use in the SPIN@U-70 experiment, planned to begin in 2002, at IHEP, Protvino, Russia. Ongoing improvements to the magnet, refrigerator pumping, NMR and microwaves will be discussed. Tests of target materials will also be discussed..

The University of Michigan Polarized Proton Target [1], shown in Fig. 1, is a 5 T, 1 K, NH_3 target, built in the late 1980's, which was very successfully used in a p-p elastic scattering experiment [2] at the Brookhaven AGS in 1990. The target thickness is about 2×10^{23} polarized protons/cm^2. The AGS beam intensity of 10^{11} protons/s was limited by magnet quenches, so the luminosity was about 2×10^{34} /cm^2s; the Target's proton polarization averaged about 85% over the course of the experiment. The Target has been upgraded in many areas and will be used for the SPIN@U-70 [3] elastic scattering experiment at 70 GeV using the U-70 accelerator at IHEP, Protvino, Russia.

The Target magnet consists of a set of superconducting coils which produce a vertical 5 T field with a uniformity of 10^{-4} over a truncated 4 cm-diameter sphere. Its original main power supply is now unstable and a new power supply system was purchased. This uni-polar supply with an energy absorber was successfully used and tested with the magnet.

The 213 MHz NMR system was upgraded to a PC using LabView with National Instruments boards. A voltage ramp into an FM signal generator produces a frequency ramp which goes to Liverpool NMR Boxes. Signals from the Boxes are processed by a home-built manual amplifier/offset module and then fed back to the PC for analysis.

The Target is cooled by a ^{4}He evaporation refrigerator. A calibrated RuO resistor from Scientific Instruments was installed as the primary temperature standard, and a second temperature sensor (^{3}He bulb) is being considered. The holder for the NH_3 is made of Kel-F with aluminum end caps. The 70 GeV proton beam will be initially rastered only vertically; thus, the first material holder may be

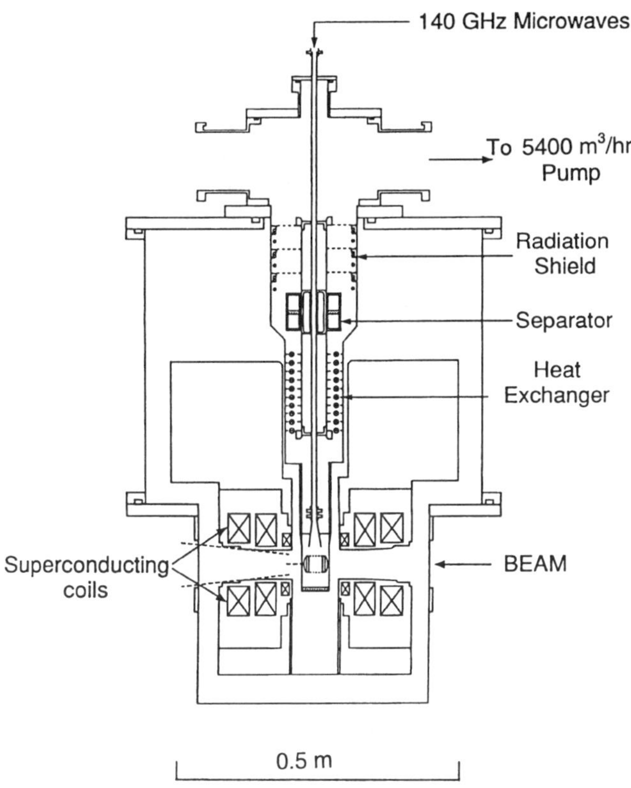

Fig.1 The Michigan Polarized Proton Target

rectangular, with inside dimensions of 5 mm wide by 20 mm high and 32 mm long.   Two NMR coils, each about 7 mm in diameter, seem small enough to avoid some earlier problems operating at 213 MHz.

For the earlier AGS experiment, the helium pumping system consisted of three Roots blowers (a 6000 m^3/hr, followed by a 3000 m^3/hr, followed by a 350 m^3/hr) followed by a mechanical pump. For the SPIN@U-70 experiment, we will use two 5400 m^3/hr Roots blowers in parallel, backed by three 227 m^3/hr mechanical pumps in parallel.   One blower and one mechanical pump were purchased and are being tested.   Two used 630 1 liquid helium transport dewars are being purchased.

In order to avoid air transportation difficulties we plan to irradiate the frozen ammonia with electrons at a Moscow State University accelerator. The 140 GHz microwave tube will be an EIO with a maximum output of 22 W.

First operation of the target with the 70 GeV proton beam is expected in late 2002. The SPIN@U-70 experiment is expected to continue until 2005.

The Target was recently tested using 1-butanol doped with TEMPO [4]. These tests included an effort to find the optimal dopant concentration for 1 K, 5 T. The results are summarized in Table 1.

Amount of TEMPO (grams)	Maximum Polarization (%)
0.05	35
0.10	67
0.15	68
0.20	61

Table 1. The measured proton polarization for different amounts of TEMPO dissolved into 19 ml of 1-butanol plus 1 ml of water.

* Supported by a Research Grant from the US Department of Energy.

References

1. D.G. Crabb., C.B. Higley, A.D. Krisch, R.S. Raymond, T. Roser, J.A. Stewart, G.R. Court, Observation of a 96% Proton Polarization in Irradiated Ammonia, Phys. Rev. Lett. **64**, 2627 (1990).
2. D.G. Crabb *et al.*, High-Precision Measurement of the Analyzing Power in Large-$P_\perp^2$ Spin-Polarized 24 Gev/c Proton-Proton Elastic Scattering, Phys. Rev. Lett. **65**, 3241 (1990).
3. SPIN@U-70 Proposal. See Michigan Spin Physics Center website: http://spinbud.physics.lsa.umich.edu.
4. 2,2,6,6-tetramethyl-piperidine-1-oxyl

CONSTRUCTION OF A FROZEN SPIN POLARIZED DEUTERON TARGET AT THE TUNL HIGS FACILITY FOR INVESTIGATIONS OF THE GERASIMOV-DRELL-HEARN SUM RULE

DAVID G. HAASE AND DIANE M. MARKOFF

*Department of Physics, North Carolina State University, Raleigh, NC, 27695-8202,
USA and the Triangle Universities Nuclear Laboratory, Durham, NC 27708, USA*

We describe the design of a frozen spin deuteron target for measurement of the Gerasimov-Drell-Hearn sum rule for the deuteron at the TUNL HIGS facility at the Duke Free Electron Laser Laboratory.

1   Introduction

A central motivation for the Triangle Universities Nuclear Laboratory (TUNL) High Intensity Gamma Source (HIGS) facility at the Duke Free-Electron Laser Laboratory (DFELL) is the investigation of the Gerasimov-Drell-Hearn sum rule for the deuteron. The GDH sum rule for a nucleon is

$$\int_{k_\pi}^{\infty} \left(\sigma^P(k) - \sigma^A(k) \right) \frac{dk}{k} = 2\pi^2 \alpha \left(\frac{\kappa_N \, hc}{2\pi M_N c^2} \right)^2$$

where k_π is the threshold energy for pion production from the nucleon, $\sigma_P(\sigma_A)$ is the total inelastic photon cross section when the nucleon and circularly polarized photon spins are parallel (anti-parallel) and κ_N is the anomalous magnetic moment of the nucleon. Significantly, this sum rule is based upon very general principles: causality, unitarity, gauge and Lorentz invariance. It has been shown that the deuteron can also be treated as the object of the sum rule, where k_2, the photodisintegration energy (2.2 MeV), is now substituted for k_π (~145 MeV).

$$\int_{k_2}^{\infty} \left(\sigma^P(k) - \sigma^A(k) \right) \frac{dk}{k} = 4\pi^2 \alpha \left(\frac{\kappa_d \, hc}{2\pi M_d c^2} \right)^2$$

The sum rule for the deuteron may be broken into two integrals as:

$$\int_{k_2}^{\infty} \left(\sigma^P(k) - \sigma^A(k) \right) \frac{dk}{k} = \int_{k_2}^{k_\pi} \left(\sigma^P(k) - \sigma^A(k) \right) \frac{dk}{k} + \int_{k_\pi}^{\infty} \left(\sigma^P(k) - \sigma^A(k) \right) \frac{dk}{k} = 0.6 \ \mu b$$

The second term on the right hand side will be measured elsewhere, e.g., LEGS and GRAAL. Its value may be estimated under the assumption that the impulse approximation is valid. Thus,

$$\int_{k_\pi}^{\infty} \left(\sigma^P(k) - \sigma^A(k) \right) \frac{dk}{k} \approx + 436 \ \mu b$$

It is proposed that the low energy integral, from k_2 to k_π,

$$\int_{k_2}^{k_\pi} \left(\sigma^P(k) - \sigma^A(k) \right) \frac{dk}{k} \approx -436 \ \mu b$$

be measured at HIGS.

If the assumptions underlying the GDH sum rules are generally valid, then asymmetries in the total cross sections dominated by high energy pion production and resonance excitation processes give a firm prediction for asymmetries in a very low-energy process, namely photodisintegration below pion threshold. The object of the TUNL HIGS GDH experiment is to test this important hypothesis.

2   Target Design

We have made design studies and begun construction of a polarized deuteron target that would be used at the TUNL HIGS Facility at the Duke Free-Electron Laser Laboratory to investigate the Gerasimov-Drell-Hearn sum rule for the deuteron. The experiment involves directing a beam of gamma rays onto a longitudinally polarized deuterium target and detecting the total flux of neutrons produced from the deuteron breakup. We expect a flux of circularly polarized γ's greater than 10^7/sec at energies of 2-225 MeV following implementation of the DFELL upgrade. The target will be a 1 cm diameter by 5.1 cm long cylinder of a deuterated organic material polarized to 30 - 40% via the dynamic nuclear orientation process.[1,2] The cylinder will be oriented with its axis parallel to the circularly polarized gamma beam from the DFELL Compton Backscattering facility. The target cryostat will be a high flow dilution refrigerator with a horizontal bore so that the gamma beam enters the cryostat along its main axis and encounters a minimum of extraneous material before reaching the deuterated target.[3]

Because the experiment requires neutron detectors located near the target in a low magnetic field the target will be held in the frozen spin mode.[4] A 0.5 tesla holding field will be provided by a small superconducting solenoid located in the target cryostat.[5] At this low temperature the nuclear spin lattice relaxation time becomes of the order of several days, so the 2.5 tesla magnet can be rolled away and replaced with a bank of neutron detectors that fit closely around the target cryostat.

It is anticipated that the design, construction and testing of the cryostat at TUNL will require 12 to 18 months before its final installation at the DFELL. The cryostat facility has several component sub-assemblies. The gas handling system has been designed and the major components acquired. We are now specifying the valves, gauges and layout for the pumping system for assembly at TUNL. The system requires two superconducting magnets, one 0.5 T holding coil mounted on the cryostat 4 Kelvin plate and a 2.5 T polarizing room temperature access solenoid in a separate dewar. This second magnet was designed by us and built by American Magnetics/Precision Cryogenics. The dilution refrigerator cryostat dewar is being designed.

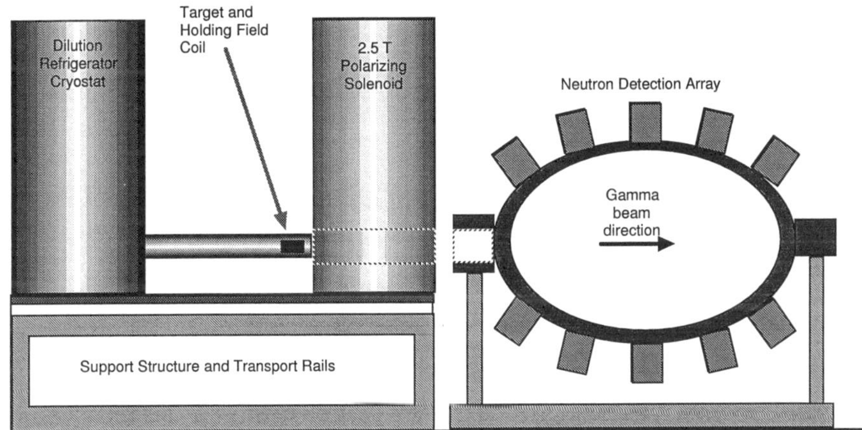

Figure 1. Conceptual diagram of the TUNL/DFELL target. The neutron detector array will be aligned and fixed in the gamma beam. The dilution refrigerator cryostat and the polarizing magnet will move on separate tracks. During polarization the polarizing solenoid will slide over the target cryostat and then move away when the target polarization is frozen. Then the target cryostat will slide into position inside the neutron detector array.

3 Acknowledgements

We acknowledge the collaboration of W. Tornow, H. Weller and B. Norum in the preparation of the GDH project. Work supported by the US Department of Energy, Office of High Energy and Nuclear Physics, Grants DE-FG02-97ER41033 and DE-FG02-97ER41042.

References

1. W. De Boer and T. O. Niinikoski, Nuclear Instrum. and Methods 114 (1974) 495.
2. W. de Boer, M. Borghini, K. Morimoto, T. O. Niinikoski, and F. Udo, J. Low Temp. Phys. 15 (1974) 249.
3. N. S. Borisov, B. B. Kulikov, A. B. Neganov and Yu. A. Usov, Cryogenics 33 (1993) 738.
4. T. O. Niinikoski, and F. Udo, Nuclear Instrum. and Methods 134 (1976) 219.
5. Ch. Bradkte, H. Dutz, M. Pluckthun, H. Peschel, R. Gehring, S. Goertz, W. Meyer and G. Reicherz, "Internal Superconducting Holding Coils: A New Technique for Frozen Spin Targets," presentation at Polarized Target Workshop, TRIUMF, May, 1996.

DEVELOPMENT OF POLARIZED SOLID PROTON TARGET FOR RI BEAM EXPERIMENT

T. WAKUI[1], M. HATANO[1,2], H. SAKAI[1,2], A. TAMII[2] AND T. UESAKA[3]

[1] *RIKEN, Saitama 351-0198, Japan*
[2] *Department of Physics, University of Tokyo, Tokyo 113-0033, Japan*
[3] *Department of Physics, Saitama University, Saitama 338-8570, Japan*

A polarized proton target that can be operated in a magnetic field of 0.3 T at a temperature of 77 K is being developed for experiments with RI beams. Protons in a crystal of naphthalene doped with pentacene have been polarized up to 18.4% by a pulsed DNP method.

1   Introduction

Development of radio-isotope (RI) beams extends the experimental nuclear physics to the nuclei far from the stability line and provides a rich field for the study of nuclear structure as well as the nuclear reaction. At RIKEN, the RI beam factory (RIBF) is currently being constructed.[1,2] The RIBF will be capable of providing more than 3000 species of unstable nuclei. A combination of the RI beams and a polarized proton target will provide us a lot of information on nuclear structure of unstable nuclei.

RI beam experiments utilize the inverse kinematics: one have to detect low momentum recoiled protons (typically less than 10 MeV). A conventional polarized proton target system is not suitable for such experiments because it requires a high magnetic field ($\geq$ 2.5 T) and a very low temperature ($\leq$ 1 K). The condition will create a difficulty in detection of recoiled protons. Therefore, a proton target that can be polarized in a low magnetic field ($\leq$ 0.3 T) at a high temperature ($\geq$77 K) is desirable for RI beam experiment.

Protons in a crystal of aromatic molecules can be polarized in such condition by means of microwave-induced optical nuclear polarization (MIONP) combined with the integrated solid effect (ISE) [3]. This method was proposed by Henstra *et al.* to enhance NMR signal amplitude of rare abundant spins in a material. Recently, Iinuma *et al.* was achieved a proton polarization of 32% by using this method in 0.3 T at 77 K.[4] We apply this method to produce a polarized proton target for RI beam experiments. The equipment for polarizing proton was assembled last year and the tuning of each element has been started.[5] The goal is to attain a polarization of more than 60% in a target with sizes of a diameter of 20 mm with a thickness of 2 mm.

134

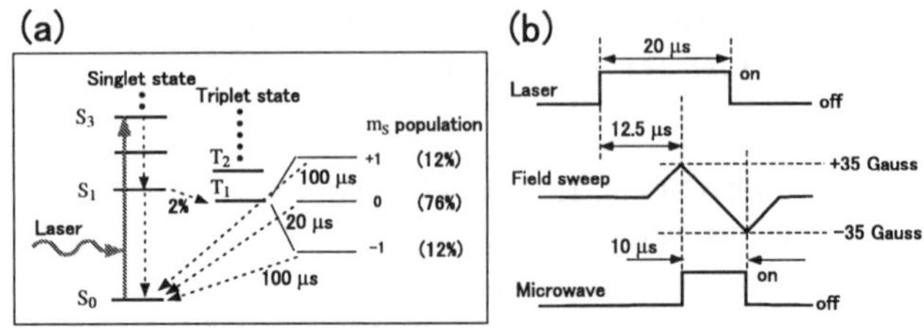

Figure 1. (a) Energy levels of pentacene. The electron alignment spontaneously occurs in the lowest triplet state. The population difference between two sublevels in the triplet state is 73%. (b) Time sequence of pulses for an application of the integrated solid effect, a polarization transfer method. In this method, magnetic field is adiabatically swept, simultaneously with microwave irradiation.

2 Proton polarization in a crystal of aromatic molecules

As a target material, we use a crystal of naphthalene or p-terphenyl doped with pentacene. The crystal is produced by the Bridgman technique after purification of materials by the zone-melting method.

In the crystal, pentacene molecules are excited to a higher singlet state by laser irradiation as shown in Fig. 1 (a). The excited pentacene has a probability of transition to the lowest triplet state owing to the intersystem crossing.[6] About 2% of excited pentacene decays to the lowest triplet state in which populations of sublevels are different from each other due to the angular momentum selection rule.[7,8] The electron alignment occurs spontaneously. Thus, the population difference between two sublevels can be obtained independent of the magnetic field strength and the temperature. The population difference is about 73%.[9]

Then, the population difference is transferred to proton polarization by the ISE method, before pentacene molecules in the triplet state decay to the ground state. In this technique, the magnetic field is adiabatically swept through an electron spin resonance (ESR) line in the triplet state, simultaneously with pulsed microwave irradiation as shown in Fig. 1 (b). During the adiabatic passage, the Hartmann-Hahn condition for cross polarization is fulfilled between the effective Larmor frequency of the electron spins in the rotating frame and the Larmor frequency of the proton spins in the laboratory frame. Then, the electron spins and the proton ones exchange their polarization via the dipolar interaction. Moreover, all electron spins in the ESR line make the contribution to the polarization transfer because of the magnetic field sweep. By using this method, the population difference can be efficiently transferred to proton polarization even in a magnetic field as low as 0.01 T.

3 Proton polarizing system

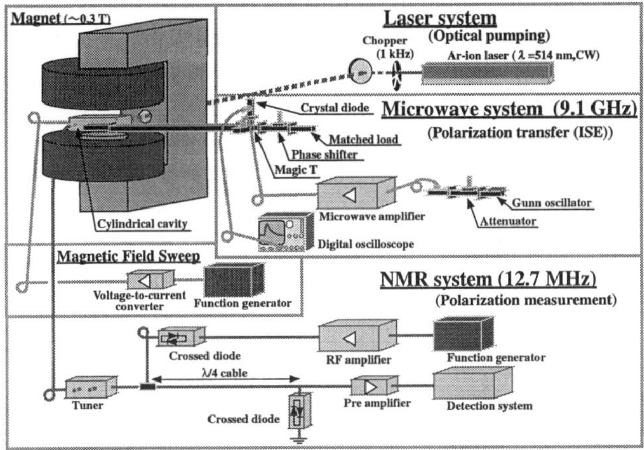

Figure 2. Schematic of the proton polarizing system. The system consists of a magnet, a laser system for optical pumping, a microwave system and a magnetic field sweep system for polarization transfer, and a pulsed NMR system for polarization measurement.

Figure 2 shows a schematic of the proton polarizing system. A target crystal, whose typical dimension is 4 mm $\times$ 5 mm $\times$ 2 mm, is placed in a microwave cavity mounted in the center of a C-type magnet. The magnetic field is about 0.3 T. The field inhomogeneity at 0.3 T is 3.5×10^{-4} over the target size. An Ar^+ laser is used to excite pentacene molecules. The laser beam is chopped by an optical chopper and irradiates the target crystal through a hole of the magnet. A microwave system and a field sweep system are used for the polarization transfer. The microwave cavity used is TE_{011} cylindrical one with a quality factor of 4000 at the room temperature. The resonance frequency is about 9.1 GHz. A microwave from a Gunn oscillator is pulsed and amplified by a microwave amplifier whose maximum output power is 20 W. Pulsed microwaves are split with a magic tee, one arm of which is connected to the cylindrical cavity. Typical microwave power for the polarization transfer is about 2.5 W and the pulse width is about 10 μs. During the pulsed microwave irradiation, the magnetic field is swept through the entire width of an ESR line. The field sweep is done by applying a triangular wave current of 53 A (peak-to-peak) to a set of coil placed in the microwave cavity.

This current produces the maximum magnetic filed of ±35 Gauss. A pulsed NMR system is used for measuring the proton polarization. The RF signal produced by a function generator is pulsed and amplified by an RF amplifier, whose maximum output is 500 W. The frequency and the pulse width are 12.7 MHz and 1.6 μs, respectively. The RF pulse is passed through a tuner for impedance matching and applied to the RF coil. The RF coil picks up the NMR signal. The proton resonance frequency component of the picked-up signal is obtained by using a phase sensitive detection technique.

4   Proton polarization

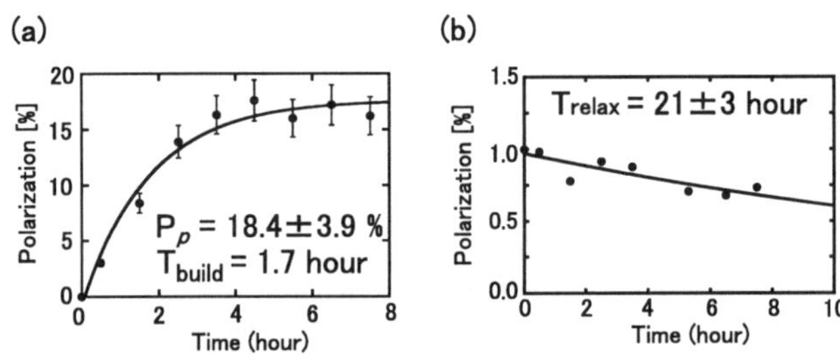

Figure 3. (a) Buildup of the proton polarization in 0.3 T at 100K. A crystal used was naphthalene doped with 0.01% pentacene. The buildup time was 1.7 hours and the proton polarization was 18.4%. (b) Relaxation time of the proton polarization in 0.3 T at 100 K. Relaxation time was 21 hours.

Recently, we have succeeded in polarizing protons up to 18.4 ± 3.9% in a magnetic field of 0.3 T at 100 K. Figure 3 (a) shows the buildup time of proton polarization. The crystal used was naphthalene doped with 0.01 mol% pentacene. The measured buildup time was about 1.7 hours. Figure 3 (b) shows the relaxation time of proton polarization. The measured relaxation time was about 21 hours. The proton polarization is smaller than that of reported value of 32% in Ref. 4, though the relaxation time is longer than that of reported value of 15 hours in Ref. 4. This smaller polarization value is due to the smaller laser power. The laser used in this study is a continuous wave Ar^+ laser having the maximum power of 5 W. However, to obtain high population difference in the triplet state, the laser beam have to be pulsed by an optical chopper and resulting power becomes 40 μJ/pulse. This power corresponds to the number of photons only 0.1% of pentacene can be excited to the lowest triplet state. When a powerful laser is used, for example, 3 mJ/pulse, a proton polarization higher than 50% can be expected.

5 Conclusion

We are developing a polarized proton target system that can be operated in 0.3 T at 77 K. The equipment for polarizing proton has been constructed and the tuning of each element has been started. With this system, we have succeeded in polarizing proton up to 18.4% in a crystal of naphthalene doped with pentacene. We are planning further improvement of the system, including the introduction of a powerful laser and an enlargement of the crystal size, toward the first experiment at the RIBF where high intensity RI beams will be available in 2005.

6 Acknowledgements

The authors would like to thank Dr. I. Tanihata at RIKEN for his constant support of this study. We also wish to thank Dr. M. Iinuma and Dr. K. Takeda for valuable advice. One of us (T.W.) would like to acknowledge the Special Postdoctoral Researchers Program.

References

1. I. Tanihata, *Nucl. Instr. and Meth.* **B 126**, 224 (1997).
2. I. Tanihata, *J. Phys.* **G 24**, 1311 (1998).
3. A. Henstra, P. Dirksen, and W. Th. Wenckebach, *Phys. Lett* **A 134**, 134 (1988).
4. M. Iinuma, Y. Takahashi, I. Shaké, M. Oda, A. Masaike, and T. Yabuzaki, *Phys. Rev. Lett.* **84**, 171 (2000).
5. T. Wakui, M. Hatano, H. Sakai, A. Tamii, and T. Uesaka, 14th Int. Spin Physics Symposium, Osaka. 2000, AIP Conference Proceedings **570**, 861 (2001).
6. W. H. Hesselink and D. A. Wiersma, *Phys. Rev. Lett.* **43**, 1991 (1979).
7. M. S. de Groot, I. A. M. Hesselmann, J. Schmidt, and J. H. van der Waals, *Mol. Phys.* **15**, 17 (1968).
8. W. S. Veeman and J. H. van der Waals, *Mol. Phys.* **18**, 63 (1970).
9. S. S. Kim and S. I. Weissman, *Rev. Chem. Intermed.* **3**, 107 (1979).

III. Polarized Electron Sources

OVERVIEW OF POLARIZED ELECTRON SOURCE PROGRESS

KURT AULENBACHER[1]

*(1) Institut für Kernphysik der Universität Mainz,
J.J. Becherweg 45, D-55099 Mainz, Germany*

This report addresses progress that has been achieved in recent time concerning three major aspects of polarized electron source operation. As a first issue the restriction to $P_e = 80$ % is discussed. About 10% of the polarization loss can be explained by transport depolarization and inelastic processes in the surface region. Second, photocathode lifetime has seen a large improvement, so that average beam currents in the several hundred μA range are possible today. The extraction of high current densities -several A/cm^2 in pulse trains of hundreds of nanoseconds- from highly polarized structures is necessary for the next generation of e^+/e^--colliders. Here the improvements have lead to a reduced recombination time for the trapped surface charge, which should allow to achieve the desired current densities in the future.

1 Introduction

Electron scattering experiments with polarized beam require the optimization of the quality factor

$$Q = P^2 \cdot I$$

in order to get to an optimum statistical accuracy of the result. Usually the required intensity I on target will depend on the specific aspects of the desired experiment, whereas the need of high polarization P is inherent to all experiments. The phase space (transverse emittance, bunch length and energy spread) of the electrons produced by a polarized source must be matched to the needs of the beam transport system. These requirements range from high average current d.c.-operation to nanosecond pulse operation and consequently high peak current. Picosecond pulses may be required for the next generation of electron/positron colliders. It may seem surprising that all this demands can be met with one principle of a polarized electron source, namely the laser driven semiconductor source where the photocathode is a suitably chosen semiconductor heterostructure.

The semiconductor source was introduced around 1975 with photocathodes made from GaAs which had an inherent polarization limit to 50%. The major breakthrough towards higher P was the introduction of the strained layer photocathode[1,2] which was first used for an experiment at

SLAC[3]. Nowadays nearly all accelerators are equipped with a cathode of this (or related) type, the polarizations achieved are usually in the 70 to 80% range[6,13,14,15]. This talk tries to summarize some of the progress that has been achieved in polarized electron source technology in recent time. It will address the following aspects:

- Optimization of polarization

- Extractable charge on long timescale ('lifetime problem')

- Extractable charge on short timescale ('charge limit problem')

2 Polarization-optimization of Photocathodes

The achieved $P^2 \approx 0.64$ is high, but of course far from optimum. One should keep in mind that accelerator facilities have very high running cost and that for many experiments the required runtime scales like P^2.

To understand the limiting factors it is necessary to look closer at the photoemission process of a strained layer photocathode. The basic idea is to prepare a pure 'Zeeman' sublevel as the energetically highest occupied state in the semiconductor valence band. This done by an additional splitting that removes the degeneracy of the valence band electrons. It may be achieved by reducing the symmetry of the elementary cell of GaAs, what is done by growing GaAs (or related compounds) on a substrate of different lattice constant. In a σ^+ photoabsorption this should lead to a pure $m_s = -1/2$ state, hence to a completely spin polarized ensemble in the conduction band of the semiconductor. The electrons diffuse to the surface of the photocathode and may then be extracted into the vacuum. To make the emission energetically possible, a mono-atomic Cs:O layer has to be applied to the surface, in order to achieve a situation where the potential difference between the vacuum state and the conduction band is negative (Negative Electron Affinity, NEA). From this 'three step model of photoemission' one immediately can infer that there are also three possible stages of depolarization:

1. Impure preparation of the electronic state in the semiconductor.

2. Interaction during diffusion may depolarize the ensemble.

3. Possible interactions in the surface zone.

In the following section the two latter subjects will be addressed, because new experimental data have been achieved recently.

2.1 Influence of diffusive transport on depolarization

The diffusive transport of the electrons towards the surface takes some time during which depolarizing interactions will modify the polarization according to an equation:

$$P = \frac{P_0}{1 + t/\tau}$$

Where τ is the depolarization time which represents the sum of all depolarizing effects and t is the average time that an electron spends in the conduction band. The depolarization time τ can be measured - e.g. by time resolved observation of the polarization of the photoemitted electrons[17] - to be of the order of 100 ps for structures similar to the ones presently in use. The parameter t can be extracted by measuring the response time. Figure 1 shows the behavior of the pulse response as a function of the layer thickness. The lines in the figure show the results obtained from fitting the pulse response by a diffusion model[16]. The data points follow the model calculation down to a layer thickness of 200 nm below which the experimental resolution of our apparatus does not allow to check the model prediction.

The result shows that the restricted thickness of the strained layer cathode gives an advantage: The response-time of the cathodes is proportional to the square of the layer thickness. For the strained layer cathodes with typically 150 nm layer thickness the response-time is found to be less than 3 ps or -after deconvolution of the experimental resolution- even below 2 ps. Applying this to the equation above we see that the depolarization during the transport is less than 2%. Furthermore does the response-time of 2 ps fulfill even demanding requirements of r.f.-sources towards the pulse length.

2.2 Influence of surface effects

The surface escape process is a complex part of the photoemission, because its behavior is governed by the complicated structure of the NEA surface. Usually the activation of a photocathode is done in a way to generate a surface with a maximized quantum efficiency (q.e.=number of electrons per incident photon).

Figure 2 shows the observation[4] of the dependence of electron spin polarization on the quantum efficiency. The decrease was obtained by intentionally poisoning the surface with an introduced gas species. The result is that there is an increase of polarization with decreasing efficiency. This may be explained by a filtering effect, because inelastic processes (phonon emission) can occur in the band bending region. Electrons with an energy loss may still escape if

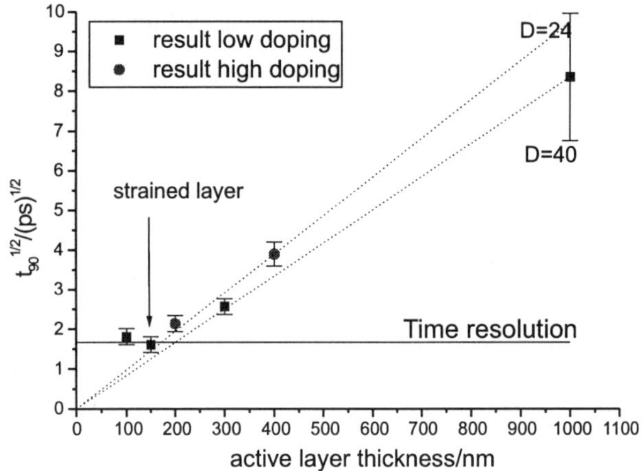

Figure 1. Measured square-root of response time vs. layer thickness. Lines are model calculations which are used to fit the responses for different doping concentrations

the NEA of the photocathode is large enough. During the aging of the photocathode the inelastically scattered electrons seem to be filtered off: It may be guessed that either the NEA gets to small -so that emission is energetically not possible- or the surface tunnel mechanism is no longer effective for those electrons because of a modification of the surface barrier. The effect leads to an relative increase of the value of P of about 9-10%. It seems that surface effects carry a large fraction of the depolarizing effects.

In operation at the accelerator surface deterioration will also lead to decreasing q.e. and hence to an increased polarization. Though this gives a certain advantage, it also requires polarization-monitoring during the experiment. During the last years experiments have been performed with aged photocathodes of the kind described above. Polarizations in excess of 85% have been achieved for example at MAMI. An aged photocathode tends to be less stable. So far it is not completely clear if it is possible to do an artificial pre-aging while simultaneously achieving the best possible stability.

3 Charge emission: life time and extractable charge

3.1 Transmission losses

In real-life operation the continuous operation time is limited by the parallel action of several effects, which deteriorate the NEA surface of the cathode

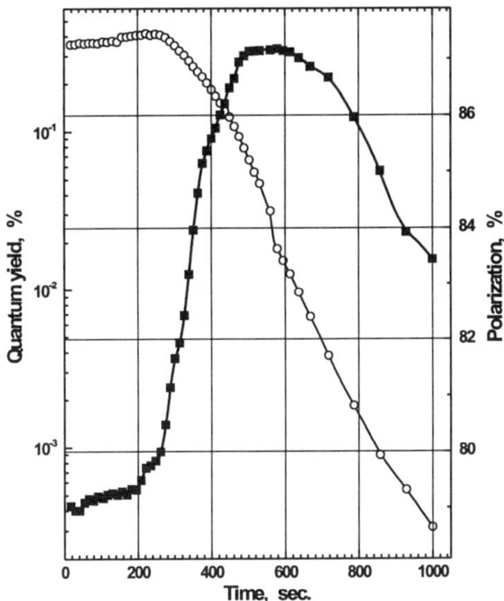

Figure 2. Increase of polarization as function of decreasing q.e., from: Mamaev et. al.[4]

and therefore limit the lifetime. The 'lifetime' τ is usually defined as the time interval until the q.e. of the photocathode has decreased to $1/e$ of its initial value. Each effect will generate it's own lifetime τ_i and the smallest of these (i.e. the dominant effect) will limit the performance of a source. Some effects are independent of the beam current like the interaction with the gas-species present in the system at base pressure, or with gas desorbed by field emission. For others the decay constant $1/\tau_i$ should be proportional to the current. Examples are the ionization of residual gas by the electron beam, or gas desorbed by beam loss. Nonlinear effects have also been observed, but seem to be not critical for well activated photocathodes. The current independent effects may be suppressed by careful vacuum and high voltage design to values where they do not interfere with the operation of a source on a timescale of several 1000 hours. In this situation it is the most urgent problem to fight against the proportional effects. If the performance is limited by such an effect the extractable charge during the 'lifetime' of a photocathode is limited to a constant value. Investigations during the last years have improved this limitation to charges of more than 100 Coulomb:

A possible candidate for a lifetime limiting effect are beam-losses which are generated by photoemission from the edge areas of the cathode (cathode holder). This effect is avoided by the anodized mask technique that was pioneered at TJNAF[6]: The photocathode is covered with an oxide which has practically no photoemission and only a small part of the photocathode (a few mm in diameter) is free of oxide (see figure 3). The beam is generated from this area. Anodized photocathodes have transmission losses which are at least two orders of magnitude smaller than nude cathodes.

At Mainz we have performed the so-called 'mechanical mask activation': The activation is done through a small hole, so that only a small active area of the cathode is covered with Cs:O, the rest of the cathode and its holder stay free from Cs. A comparison of the different transmission losses was done on an insulated tube 10 cm downstream from the anode which was 20 mm in diameter. The relative losses with a nude cathode were about $5 \cdot 10^{-4}$ whereas anodized and masked cathodes resulted in relative losses of 10^{-6} or below.

The advantage of the masked activation is that one can omit the mask in a later activation in order to do a comparison between a nude and a masked activation on the same photocathode. The result is shown in figure 3. The lifetime of the mask-activated cathode is obviously much longer. With such an activation we achieved a continuous extraction of 120 Coulomb charge during one photocathode lifetime without changing the laser-beam spot and without recesiation. We therefore believe that the elimination of beam losses in the vicinity of the anode of the source is a necessary condition for the extraction of high average beam current.

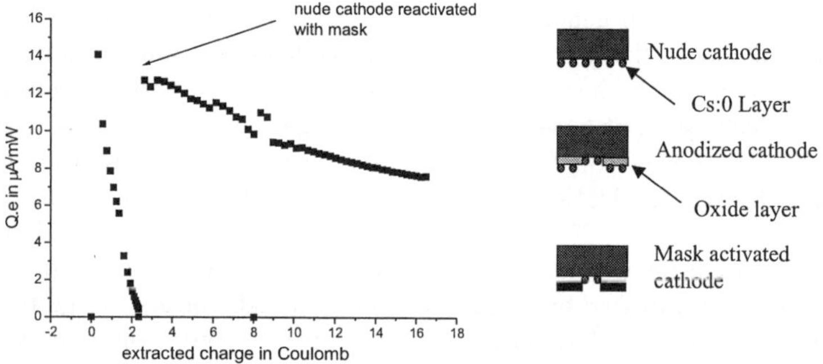

Figure 3. Left side: Lifetime after unmasked and masked activation. Right side: Different methods of cathode activation. (The mask is removed in operation.)

Recently the extracted charge from a strained layer photocathode at

TJNAF[5] was increased even to 300-600 Coulomb. This progress may be caused by other technical improvements like large apertures/low desorption and optimized effective pumping speed[6].

3.2 Beam current potential of the semiconductor source

With the aforementioned techniques one is now in a position to produce average currents in the several hundred microampere range for reasonable time intervals; beam currents in the milliampere range seem possible, even though appropriate reduction of thermal load may be a problem for strained layer cathodes.

Operation of high energy accelerators at this current levels would be difficult, because of thermal limitations of the targets, backgrounds etc. However the available current could be used in nuclear physics experiments in the MeV range where the beam power is smaller. Finally, studies are carried out for a $\vec{e}$-p collider, which would need beam currents in the several hundred mA region. This project would make use of the concept of power recycling, to avoid the beam power related problems.

4 Peak current densities from a polarized source

This topic is of concern for the operation of linear e^+/e^--colliders of the next generation. This machines have stringent requirements on the bunch trains that have to be produced from the source. The NLC for example requires the production of 90 bunches with 0.7ns length with a bunch separation of 2.8ns. The charge per bunch is 4.5nC, resulting in a peak current of about 6A[7]. Because of emittance requirements the beam-diameter on the cathode should not exceed about 1-2 centimeters. Therefore a high current density of several A/cm^2 must be produced. Similar requirements must be met for the JLC design.

As a first issue this requires high accelerating potentials and high gradients at the cathode. Field emission from the HV-electrodes acts in the same way as the transmission losses described above, therefore they must not exceed several nanoamperes. This behavior has been achieved at SLAC with a d.c.-high voltage of 120 kV and 1.8 MV/m gradient[3] and operation of activated GaAs at TJNAF and Nagoya was achieved at even higher Potentials and gradients[8,6,a] With parameters in this range the requested current densities can be achieved.

---

[a]Considerably higher gradients (60MV/m) for operation with activated GaAs have been demonstrated in an r.f.-source[18].

As a second problem there exists the so-called 'charge limit' which is caused by the 'surface photovoltage effect': Charge is accumulated in the surface zone of the photocathode, because a fraction of the electrons gets trapped instead of being emitted. This space charge modifies the NEA-state. Photoemission probability will therefore be strongly reduced, if the charge accumulation is faster than the recombination of the electrons with holes in the surface zone. For standard strained layer structures one finds that the steady state regime (i.e. current density for a long pulse) is of the order $1A/cm^2$. The achievable optimum is very sensitive to the surface conditions because a low escape probability increases the trapping of surface charges. Another important parameter is the time constant of charge decay in the surface zone. It has been determined on a conventional cathode to be about $70\,\mathrm{ns}$[10]. This sojourn time of the electrons is the reason for the limited current density. Whereas the desired parameters could be achieved in a single pulse at SLAC[3], it was not possible to reproduce the pulse again before the characteristic decay time had passed. An alternative collider project, TESLA, would find more favorable conditions: Here the single bunch charge is similar but the bunch spacing is $330\,\mathrm{ns}$[7].

Activities in the last years have concentrated on the optimization of electron escape probability and on increasing the recombination rate. The Nagoya group explores a superlattice cathode which has an electronic level scheme that can simultaneously fulfill the above requirements and gives a polarization of about 80%[9,14]. Another attempt followed at SLAC and Nagoya is to increase the recombination rate by high surface doping. Surface doping of an about $7\,\mathrm{nm}$ thick layer minimizes the depolarization which would be present in a homogeneously heavy-doped photoactive layer[11]. This method decreases the recombination time from typically 70 to $10\,\mathrm{ns}$. From the dynamics of the process[12] it can be guessed that this improvement will increase the available steady state current density by at least an order of magnitude. In recent years several experiments have been done with photocathodes of this type which gave promising results because the characteristic decrease of quantum efficiency with laser pulse energy could not be observed. Multibunch extraction of several highly charged bunches has been demonstrated[9] as well as the production more than $100\,\mathrm{ns}$ long bunches with high peak current[11]. However the desired pulse trains for NLC have not yet been demonstrated. All experiments were limited by their laser systems and/or by space charge problems of the electron source used. Therefore the full range of the photocathode improvement that has been achieved with the aforementioned methods has not yet been explored.

5 Conclusion

Typical beam polarizations for a photocathode operating with maximum q.e. $(5 * 10^{-3})$ at an accelerator are close to 80%. Operation at a maximum polarization of $P > 85\%$ has also been demonstrated at reduced q.e. $(1*10^{-3})$. The responsible mechanisms for this behavior are thought to be scattering processes in the band bending region at the surface of the photocathode. The response time of strained layer cathodes are very short, therefore the transport depolarization seems to be smaller than the surface effects. The reasons for the remaining difference of about 10% towards 100% are not yet completely understood.

The life-time of photocathodes can be improved by avoiding beam losses in the vicinity of the anode of the source. It was demonstrated, that a reduction from $5 * 10^{-4}$ to 10^{-6} relative losses increases the possible charge output during one photocathode lifetime from a few to 120 Coulomb. Even better values which seem to allow the extraction of 600 Coulomb have already been observed at TJNAF.

The investigations concerning the generation of bunch trains needed for the next linear colliders have improved the charge-limit effects especially by heavy doping in an ultrathin surface layer. Under this conditions the built up of surface charge has been minimized, so there is hope that a step forward to the collider requirements can be demonstrated as soon as appropriate laser-systems are available.

Acknowledgments

This work was supported by the Deutsche Forschungsgemeinschaft within the framework of the Sonderforschungsbereich 443 and by the Eurpean Union through INTAS grant 99-00125. The author wants to thank the following persons for the additional information and support in order to produce this overview: M. Farkondeh, W. Hillert, Y. Mamaev, T. Maruyama, T. Nakanishi, D. Orlov, M. Poelker, A. Terekhov, V. Tioukine.

References

1. T. Nakanishi et.al: Large Enhancement of Spin Polarization Observed by Photoelectrons from a Strained GaAs Layer, Phys. Lett. **A 158**, (1991) p.345-349
2. T. Maruyama et. al.: Observation of Strain–Enhanced Electron–Spin Polarization in the Photoemission from InGaAs, Phys. Rev. Lett., **66**

1991 p. 2376-2379

3. R. Alley et. al :, The Stanford Linear Accelerator Polarized Electron Source, NIM **A365**, 1995, p.1-27

4. Y. Mamaev et.al : Temperature Dependence of Spin Dynamics, in: Spin 2000, AIP conference proceedings, Vol 570, Melville, New York, 2001, p.920-925

5. M. Poelker: Private communication

6. C. Sinclair: Performance of the Jefferson Laboratory Polarized Electron Source at High Average Current, International Workshop on Polarized Sources and Targets, (PST99), Erlangen 1999, p.222-230

7. F. Richard, et. al. (eds.): TESLA technical design report, Hamburg 2001, p.II-110

8. T. Nakanishi: private communication

9. K. Togawa et. al.: Polarized Electron Source for Japan Linear Collider, in: Spin 2000, AIP conference proceedings, Vol. 570, Melville New York, 2001, p.982-987

10. T. Maruyama et. al.: Investigations of the Charge Limit Phenomenon in GaAs Photocathodes, in: Spin 2000, AIP conference proceedings, Vol. 570, Melville New York, 2001, p.976-981

11. T. Maruyama: private communication

12. A. Subashev et. al.: Emission from Semiconductor Photocathodes at High Intensity Inhomogeneous Excitation (This conference)

13. E. Tsantalovich et.al: Photoemission from Strained GaAsP with High Power Diode Laser at MIT-Bates Accelerator Center (This conference)

14. W. Hillert et. al. The 50 kV Inverted Source of Polarized Electrons at ELSA, in: Spin 2000, AIP conference proceedings, Vol. 570, Melville New York, 2001, p.961-964

15. K. Aulenbacher and Ch. Nachtigall et. al.: The MAMI–Source of Polarized Electrons, NIM **A391**,(1997) p.498-506

16. P. Hartmann et. al.:, A Diffusion Model for Picosecond Electron Bunches from Negative Electron Affinity Photo-Cathodes, Journal of Applied Physics, **86**, p. 2245 ff. (1999)

17. P. Hartmann: Aufbau einer gepulsten Quelle polarisierter Elektronen, PhD–thesis, Institut für Kernphysik der Joh. Gutenberg Universität Mainz, p.99 ff. (1997)

18. N.S. Dikansky et. al.: Present status of experimental S-band GaAs-photogun driven by the Solid State Pulse Laser, in: Spin 2000, AIP conference proceedings, Vol. 570, Melville New York, 2001, p.988-991

COLD ELECTRONS FROM GaAs(Cs,O)

D. A. ORLOV[1], U. WEIGEL[1], M. HOPPE[1], D. SCHWALM[1],
A. S. TEREKHOV[2], A. WOLF[1]

(1) Max–Planck–Institut für Kernphysik, 69029 Heidelberg, Germany
(2) Institute of Semiconductor Physics, 630090 Novosibirsk, Russia

A unique facility for preparing GaAs photocathodes and measuring energy spreads of electrons has been developed over recent years and is now used to analyze two-dimensional energy distributions of photoemitted electrons at different surface conditions. The possibility to obtain an intense cold electron beam is demonstrated.

1 Introduction

Semiconductor NEA photocathodes are in widespread use as effective sources of spin-polarized and monoenergetic electrons. [1] However, sensitive diagnostic methods which could help optimizing their properties for specific applications only started to become available. For example, the requirement of a low energy spread of electron beams at high dc current, as set by collision experiments at the ion test storage ring in Heidelberg, can be met only on the basis of detailed knowledge of the photoelectron energy distribution at the surface. To study complete energy distributions $N(E_{\parallel}, E_{\perp})$ as a function of transverse and longitudinal energies we developed a method [2] which proceeds by "marking" electrons with fixed longitudinal energy and a subsequent measurement of the associated differential transverse energy distribution, applying an adiabatic magnetic compression technique.

In this paper two-dimensional energy distributions of photoelectrons from GaAs(Cs,O) were studied at different thicknesses of the (Cs,O) activation layer with the aim to reduce electron scattering during the escape process and to increase the transmission of the interface for cold electrons. The photocathode performance data promise electron beams of several 10 mA/cm^2 with longitudinal and transverse energies ≤ 10 meV.

2 Experimental

We used transmission mode GaAs photocathodes consisting of a double layer heterostructure GaAs/AlGaAs bonded to a transparent glass carrier. The emitting p^+-GaAs(100) layer was epitaxially grown to a thickness of $\approx 1.5\,\mu$m and doped with Zn to a concentration of holes of about $5 \times 10^{18}\,\mathrm{cm}^{-3}$. The cleaning of GaAs surface started with removing oxides in a solution of HCL and isopropyl alcohol. This process was performed in a glove box filled with pure nitrogen. Then the photocathode was installed inside a hermetic all-metal vessel, which protected the semiconductor surface against oxidation and contamination, and was transferred to the loading chamber of our UHV setup without contact with the laboratory atmosphere. In a preparation chamber with a base pressure below 10^{-12} mbar the photocathode was subsequently heat cleaned and activated with (Cs,O) to maximum quantum efficiencies of about 25 % (in the reflection mode at 670 nm). For studying photoemission from GaAs with thin (Cs,O) layers an activation process was interrupted before reaching the maximum QE. The thickness of the (Cs,O) layer was calculated from the time of activation based on the assumption that the first Cs-peak corresponds to 0.5 ML and that the sticking coefficient of Cs to GaAs(Cs,O) is closed to unity during the activation. The details of the photocathode preparation setup are described elsewhere. [3] After the activation the photocathode was transferred into the electron gun of the measurement setup, where it could be cooled to 90 K. All measurements were performed at low temperature.

Illumination of the cathode with a 800 nm diode laser yields photoelectrons which are subsequently accelerated to $\approx$ 20 eV and guided by a longitudinal magnetic field to a retarding field analyzer, capable of measuring the longitudinal energy distribution of electrons. While the magnetic field B_0 at the cathode position was kept at a constant value of 0.05 T, the magnetic field $B_a = \alpha B_0$ at the analyzer was adjustable between $\alpha = 1.0$ and 2.5. Since $E_\perp/B$ is an adiabatic invariant [4] the energy components of the electron at the cathode $(E_\perp, E_\parallel)$ and at the analyzer $(E_\perp{'}, E_\parallel{'})$ are related by the following way: $E_\perp{'} = E_\perp + \Delta\alpha E_\perp$ and $E_\parallel{'} = E_\parallel - \Delta\alpha E_\perp$ with $\Delta\alpha = \alpha - 1$. A potential barrier in front of the photocathode, generated by a ring-shape electrode, was used to block electrons with longitudinal energies $E_\parallel < E_\parallel{}^f$ before the acceleration. To measure differential transverse energy distributions $N(E_\perp, E_\parallel{}^f)$ the cutting energy $E_\parallel{}^f$ was modulated in time with a small amplitude $\Delta E_\parallel{}^f$. The resulting modulation of the analyzer current originates only from electrons with initial longitudinal energies $E_\parallel$ near $E_\parallel{}^f$. The related ac - component of this marked analyzer current I^m was measured sensitively by synchronous lock-in detection. Recording the derivative dI^m/dU_{bias}, a signal proportional to the transverse energy distribution for electrons with $E_\parallel = E_\parallel{}^b$ could be obtained. [2] A series of measurements for different $E_\parallel{}^f$ samples directly the two-dimensional distribution $N(E_\parallel, E_\perp)$.

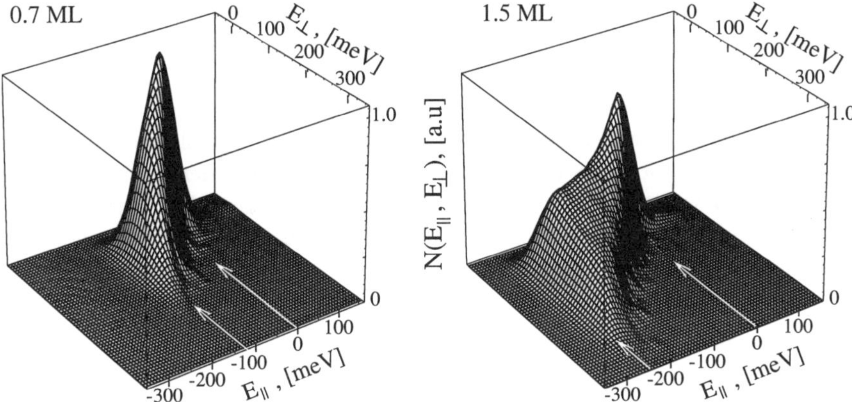

Figure 1. Two-dimensional energy distributions at 90 K measured at different thickness of (Cs,O) activation layer. The positions of the vacuum level and of the conduction band minimum ($E_\parallel = 0$) are shown by arrows.

3 Results

Figure 1 shows two-dimensional energy distributions of photoelectrons at 90K from GaAs activated with a thin ($\approx$0.7 ML) and a thick ($\approx$1.5 ML) activation layer. The thick layer corresponds to the maximum QE which was about 23% at room temperature and was increased to 29% with cooling. For the thin layer the rise of the QE was from 8% to 16%. The two-dimensional distributions are found to be dominated by a peak near E_c , which is by a factor of 1.3 higher for the thin layer. Both peaks are located at $E_\perp \approx 6$ meV and $E_\parallel \approx$ -25 meV, i.e. slightly below the initial longitudinal energy of photoelectrons in the bulk (E_c), where they are distributed in a narrow energy spread of about $kT \approx 8$ meV at 90 K. The high-energy side of the peak originates from the Boltzmann distribution of electrons in the bulk of the semiconductor and falls off exponentially with a slope of about 12 meV. Looking down from the peak to the vacuum level the distributions are limited with line of zero transverse energy to the right, and of constant total energy $E_\perp + E_\parallel \approx E_c$ to the left. For the thick layer the low energy side of $N(E_\parallel, E_\perp)$, after an initial steep decrease, gradually evolves into a plateau extending down to the vacuum level (245 meV below E_c) with a width continuously growing with the decrease of $E_\parallel$. For the thin (Cs,O) layer the plateau is almost not visible because of the cutting by the vacuum level, which is only 120 meV below E_c . This cutting not only decreases the QE but also "removes" electrons with a high average transverse energy. Indeed, the mean transverse energy is 37 meV and 66 meV for the thin and thick (Cs,O) layer, respectively. Adjusting the vacuum level with activation by thinner (Cs,O) layers or applying an external potential barrier, [3] which cuts electrons with $E_\parallel \leq E_c$, a cold electron beam with a spread of about kT can be obtained. The measurements with different samples show that the effective QE for the cold part of the distribution is about one per cent for a cathode activated to the maximum QE and by a factor of 1.3-2.0 higher for thin layers.

4   Discussion and outlook

In the bulk of the semiconductor photoelectrons are distributed near E_c with a narrow energy spread of about kT. Although almost all electrons reach the surface, only a small part of them leaves to vacuum, where they occupy all available momentum space. Indeed, at the right-hand limit of the distributions (looking down towards lower $E_\parallel$ from the peak, see Fig.1) the electrons lost a part of their total energy but kept a low transverse energy. At the left-hand limit the photoelectrons passed through the band bending region without energy loss, but were scattered elastically and emitted at a large angle to the surface normal. The region between those limits corresponds to electrons which were scattered and suffered energy loss as well. Although the mechanisms of the electron elastic and inelastic scattering are still controversial, the observation [5] of phonon replicas in some longitudinal energy distribution curves proves that phonon emission is at least one of the relevant mechanisms of electron energy loss near the interface. The dominant peak in the distribution might be explained [3,5] by an efficient capture of electrons to the upper two-dimensional quantized subband in the near-surface potential well, from where a large part of electrons escape to the vacuum by tunneling through the thin potential barrier (conserving the transverse momentum) or by scattering on local fluctuations of the potential near the surface. The possibility to suppress electron scattering in the near surface region and to enhance the escape probability of cold electrons has to be studied additionally.

One way of increasing the effective QE is to use thin activation layers. An activation curve, which is a result of competition between increasing the |NEA| and decreasing the transmission, is adjusted to reach the maximum QE, but not optimized under the point of view of the effective QE and of the energy spread of the photoemitted electrons. It is evident that at very thick layers the transmission drops due to electron scattering inside the bulk of the (Cs,O) layer. Even in not extremely thick layers this effect can still take place. [6] Indeed, it is seen (Fig.1) that the activation of GaAs to a 1.5 ML (Cs,O) layer causes not only a lowering of the vacuum level but also a decrease of the amplitude of $N(E_\parallel, E_\perp)$.

Basing on our data a cold electron beam of considerable dc intensity could be obtained. To make the most use of a GaAs(Cs,O) photocathode it has to be activated with thin (Cs,O) layers of about 1 ML and operated at low temperature. Under these conditions one can reach high transmission of the interface for cold electrons together with a NEA of about -100 meV, where the later finite rise of the NEA could serve as a "protection" against degradation. An additional potential barrier is required to adjust the cutting level for electrons to longitudinal energies $\leq E_c$. Altogether, this will provide an effective QE of about 1.5-2% and longitudinal and transverse energy spreads of ≤ 10 meV at 90 K.

In conclusion, the two-dimensional energy distributions of electrons emitted from GaAs(Cs,O) were studied at different thicknesses of the activation layer. It was shown that elastic as well as inelastic electron scattering are of crucial importance in the electron transfer through the GaAs(Cs,O)-vacuum interface. It was found that thin activation layers provide a higher effective quantum efficiency for cold electrons. Methods to obtain cold electron beams of several 10 mA/cm^2 with longitudinal and transverse energy spreads ≤ 10 meV can be based on these measurements.

References

1. H.-J. Drouhin, C. Hermann, G. Lampel, Phys. Rev. B **31**, 3859 (1985).
2. D.A. Orlov, M. Hoppe, U. Weigel, D. Schwalm, A.S. Terekhov, and A. Wolf, Appl. Phys. Lett. **78**, 2721 (2001).
3. S. Pastuszka, M. Hoppe, D. Kratzmann, D. Schwalm, A. Wolf, A. S. Jaroshevich, S. N. Kosolobov, D. A. Orlov, A. S. Terekhov, J. Appl. Phys. **88**, 6788 (2000).
4. T. M. O'Neil and P. G. Hjorth, Phys. Fluids **28**, 3241 (1985).
5. D. A. Orlov, V. E. Andreev and A. S. Terekhov, JETP Lett. **71**, 151 (2000).
6. D.A. Orlov, M. Hoppe, D. Schwalm, A.S. Terekhov, U. Weigel, A. Wolf, *Proceedings of the International Workshop on Polarized Sources and Targets*, Erlangen, Germany, 280 (1999).

PHOTOEMISSION FROM STRAINED GaAsP WITH A HIGH POWER DIODE LASER AT MIT-BATES LINEAR ACCELERATOR CENTER

M.FARKHONDEH, W.FRANKLIN, E.TSENTALOVICH, T.ZWART

MIT-Bates Linear Accelerator Center, 21 Manning Rd, Middleton, MA, 01949,
USA
E-mail: evgeni@mit.edu

First successful photoemission tests with high-polarization photocathodes at high peak currents have been conducted at Bates. Strained GaAsP crystals have been manufactured at the Laboratory of Spin-Polarized Electron Spectroscopy in St.Petersburg, Russia. A high power diode array laser (Spectra Physics model BFA1500-808-60-01) coupled to a fiber was used for photoemission. The laser operates at a wavelength of 808 nm and produces peak power up to 150 W at duty factor of 1.5%. Quantum efficiency (QE) and polarization have been measured. The effects of surface photocurrent saturation and crystal heating by laser beam have been investigated.

1 Introduction

MIT-Bates Linear Accelerator Center has conducted a number of experiments with polarized electron beam [1-3]. Bates linac properties define a short duty cycle (of order 1%) and a time structure very demanding for the polarized source operation: a repetition rate of 600 Hz with pulse duration of about 20 μsec. Peak current of up to 12 mA was required for the experiments. In the past, no commercial pulse laser systems with the required parameters were available, and we used a 7 W Ti:Sa cw laser driven by 25 W argon laser. Light pulses of needed duration have been chopped out of the cw beam.

All previous experiments have been conducted with bulk GaAs crystals. Although photocathodes with high polarization (strained or superlattice) have become available, they have significantly lower QE and could not be used with our Ti:Sa laser for practical purposes. Recently South Hall Ring has been built and commissioned. It has two modes of operation: stored beam mode and stretcher mode. While the first mode requires very low repetition rate (1 Hz), the operation in the stretcher mode still requires a rate of 600 Hz.

In the last few years diode array lasers of high power have become available. The advantages of these lasers are high peak power (150 W or more), high stability, no maintenance and convenient control of the intensity and time structure. The drawbacks are a fixed wavelength and a large divergence of the beam out of the fiber.

In this paper, we present the results of our first tests with photoemission from strained GaAsP crystals using high power diode lasers.

2 Experimental set up

The tests have been conducted on a dedicated test set up as well as on the polarized injector. The test set up consists of the 60 kV electron gun, transport line equipped with focusing solenoids and steering coils, Wien filter and Mott polarimeter.

At the injector the gun is mounted on a high voltage platform that is surrounded by a Faraday cage maintained at 300 kV relative to ground.

Bates linac is equipped with Moeller polarimeter. Additionally, we built a transmission polarimeter, which provides capabilities for rapid polarimetry at the front end of the accelerator. Both Mott and transmission polarimeters might have a systematic background which is rather difficult to extract. Since Moeller polarimeter provides the absolute measurements of polarization, for calibration purposes we measured asymmetries with the same photocathode with all three polarimeters, and used Moeller data to cross-calibrate Mott and transmission polarimeters.

2.1 Electron gun modifications

GaAs or GaAsP photocathode is mounted on the cathode stalk inside the electron gun. A tantalum cap holds it firmly in place. The diameter of the emitting surface is 11 mm, the gun however was originally designed to operate with a small (2-3 mm) laser beam spot. Since QE of high-polarization crystals is usually affected by surface current saturation effect, we have increased the size of the laser beam spot and illuminate the whole crystal. The geometry of the electrodes had to be changed in order to keep the size of the electron beam small enough to avoid any scrape-off. In particular, the cathode angle was reduced from 39° to 25° to reduce the focusing properties of the cathode. The first focusing solenoid has been moved closer to the gun to accommodate the new optics.

Since the diode laser beam has a very large divergence, it is no longer possible to transport the beam through our existing 15-m long transport line designed for the Ti:Sa laser system. An additional view port was built into the side of the gun. The laser beam is delivered through this port to the crystal with an angle of 37°. Due to high index of refraction of GaAs, the refracted ray is nearly perpendicular to the surface and thus the reduction of the degree of circular polarization is negligible.

2.2 Laser set up

The laser beam emerging from the fiber has a diameter of 1.5 mm and a divergence of 100 mrad. A set of lenses is used to control the beam size. The beam line also includes a polarizer, a Pockels cell for fast helicity reversal, and two $\lambda/2$ plates: one for a slow helicity reversal and one for a reduction of the Polarized

Induced Transport Asymmetry (PITA). Additionally, a second polarizer and a photodiode are used to measure light polarization.

The laser and the optical elements are mounted on the rail outside the Faraday cage. Only the last lens focusing the beam onto the crystal is left inside the cage. Laser beam ~50 mm diameter is transported 2 m in air to the Faraday cage.

The very large emittance of the diode laser beam requires a strong focusing through the Pockels cell. It leads to a large angular spread of the rays passing through the Pockels cell and reduces light polarization to as low as 90%. We considered to replace a helicity Pockels cell with a fast retractable $\lambda/2$ plate, which is much less sensitive to the angular spread. Unfortunately, the maximum frequency of the helicity reversal with a $\lambda/2$ plate doesn't exceed 1 Hz, while some experiments require this frequency to be 600 Hz.

In order to improve the beam polarization while keeping the ability of very fast helicity reversal we use a Pockels cell of very large diameter (70 mm). Using the lenses of large diameter, we increase the beam size to about 40 mm (the maximum beam size is limited by the aperture of available polarizers) and make it parallel. As a result, the circular polarization of better than 99% has been achieved.

A dedicated driver has been developed for the diode laser. It produces pulse of up to 200 A and 2-40 μsec duration. The rise/falling time of the pulses is about 200 nsec. The driver also provides a fast feedback if the improvement of the stability is needed and a fast (600 Hz) helicity-correlated intensity correction to compensate for the residual PITA asymmetry.

2.3   Transmission polarimeter

The transmission polarimeter provides a relative measure of the electron beam polarization at the front end of Bates linac, where the electron energy is 20 MeV. The electron beam passes through a thin BeO target and the subsequent Bremsstrahlung radiation retains a significant fraction of the incident electrons polarization. The beam of the polarized photons exits the accelerator and is then attenuated by a magnetized iron slug 15 cm in length. The asymmetry of the order of 1% is measured in the detector located downstream of the iron slug. The polarimeter provides an easy, inexpensive and fast polarization monitor. The statistical accuracy of 1% can be achieved in about 10 sec. The drawbacks of the polarimeter are the difficulties in the absolute calculation of the analyzing power and the inability to measure the transverse components of polarization.

3   Results

In our tests we used strained GaAsP crystals manufactured by SPIRE Co. and by the Laboratory of Spin-Polarized Electron Spectroscopy in St.Petersburg, Russia.

For the SPIRE sample the polarization as high as ~75 % at QE of ~0.05% has been measured. The results are shown in Fig.1. The wavelength dependence at low laser power was measured with a tunable Ti:Sa laser. The peak polarization of this sample is at a wavelength of 840-850 nm. Unfortunately, the manufacturer of the diode lasers at this wavelength discontinued production, and we limited our use of SPIRE crystals with these preliminary tests.

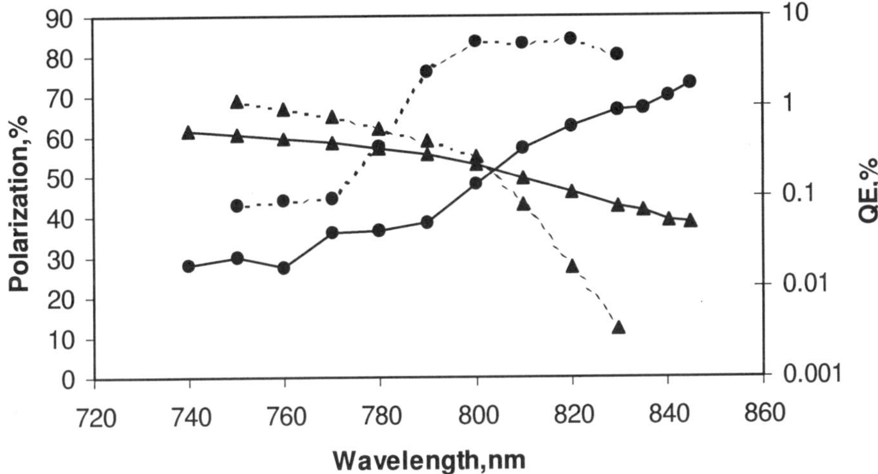

Figure 1. Wavelength dependence of the QE (triangles) and polarization (circles) for SPIRE (solid line) and St.Petersburg (dashed line) samples.

St.Petersburg Laboratory has the capability to adjust the band gap by varying the fraction of phosphor in the crystal, and thus placing a polarization peak in the desirable location. Commercially available diode laser has a wavelength of 808 nm, and we ordered crystals with a corresponding band gap. Two samples produced in slightly different conditions have been studied. The wavelength dependence of QE and polarization for the first sample is presented in Fig.1. This sample was also illuminated by the flashlamp Ti:Sa laser at λ=810 nm and peak power of 400 W. The photocurrent of about 100 mA was measured.

The second sample appears to have a polarization peak at longer wavelength. We measured the polarization of about 50% and QE > 0.3% at λ=810 nm.

Although the tantalum cap provides a very good electrical contact between the crystal and its holder, the thermal contact is poor. The laser delivers several Watts of average power to the crystal, and we noted that the parameters of the crystal (QE and polarization) drift rapidly under such conditions. We estimated that the temperature of the crystal might increase by several tens of degrees. In order to investigate the phenomena, we heated and cooled the crystal holder and measured

QE and polarization as a function of temperature (Fig.2). At higher temperatures, the polarization decreases, while QE increases. We attribute this phenomenon to the restoring current of holes affecting the surface photovoltage. The measured temperature dependence is a strong indication that the thermionic diffusion over the barrier makes a large contribution to the restoring current.

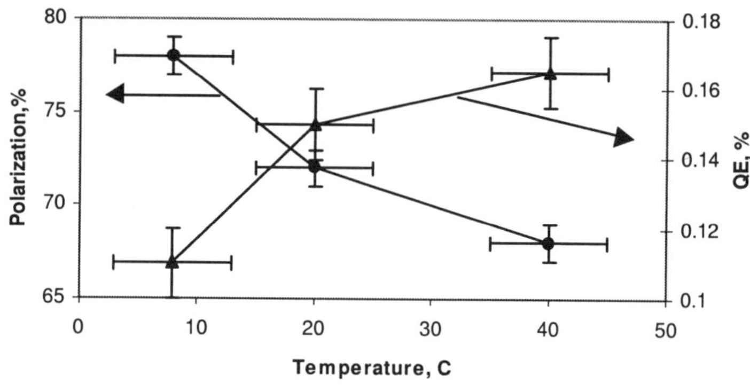

Figure 2.

We also noted that although QE increases at higher temperatures, the lifetime becomes rather short (15-20 hours). Adding Cs restores most of QE back. Most likely Cs evaporates faster from the surface of the crystal at elevated temperatures.

After these tests we changed our procedure for crystal installation. We inserted a thin (.002") indium foil between the crystal and molybdenum crystal holder and heat the assembly to about 200°C in vacuum to solder the crystal to the holder. That provides an adequate thermocontact and we don't observe any drift in photocathode parameters even at high average power of the laser.

The modified gun with a high-polarization photocathode and the new laser system are adequate for both stored beam and stretcher modes of South Hall Ring, as well as for parity violating experiments. Currently the gun and the laser system have been installed and tested on the main injector. The beam developments are nearly completed for a new SAMPLE experiment scheduled to take data this fall.

References

1. S.M.Dolfini et al, *Phys.Rev.* **C51** (1995) p. 3479.
2. D.T.Spayde, T.Averett et al, *PRL* **84** (2000) p.1106.
3. S.M.Dolfini et al, *Phys.Rev.* **C60** (1999) .

POLARIZED EMISSION FROM STRAINED SUPERLATTICES

A.D. ANDREEV

Ioffe Physical Technical Institute RAS, 194021 St Petersburg, Russia, E-mail:
andreev@aad.ioffe.rssi.ru

A.V. SUBASHIEV, YU.A. MAMAEV, YU.P. YASHIN,
A.N. AMBRAJEI, A.V. ROCHANSKY

Department of Experimental Physics,
State Technical University, 195251 St Petersburg, Russia
E-mail: arsen@spes.stu.neva.ru

Polarized electron emission from the barrier and well- strained short-period semiconductor superlattices is investigated. Strong features in the spin-polarization
dependence on the electron energy are identified with the Van-Hove singularities
in absorption spectra. Calculated spectra are close to the experimental spectra of
polarized electron emission.

Semiconductor superlattices are the most promising candidates for the
polarized electron sources of the next generation [1,2]. Further improvements
of the existing GaAs strained layer cathodes are limited by the controversial
character of the demands on the cathode heterostructure parameters. High
polarization requires high stress in the GaAs layer. Highly stressed layers have
small critical-layer thicknesses beyond which the strain gradually relaxes and
high concentration of structural defects is introduced in the active layer. On
the other hand, the active layer thickness should be much thicker than the
width of the band bending region. Therefore a thickness close to 100 nm,
which is 10 times larger than the critical thickness, is found to be optimal.
As a result the strained layer crystalline quality appeared to be rather poor
and not well controlled. It is even more difficult in this case to control the
properties of the band bending region and the activation layer which are
crucial for the electron emission in vacuum.

An alternative source of highly polarized electrons is a semiconductor
superlattice (SL) structure in which the 4-fold degenerate Γ_8 valence band
state is splitted into only 2-fold spin-degenerate states Γ_6 (heavy-hole) and
Γ_7 (light-hole miniband) by the hole confinement in the SL quantum wells
(QW), the expected maximum polarization value of the electrons excited by
the circularly polarized light being $|P| = 1$. The main advantage of SL-based
photoemitters is the possibility to vary the properties of the active layer over
a wide range by the appropriate choice of layer composition, thickness and

"

doping, and optimize the structure as a whole. While for the strained layer structures the polarization losses at all stages of photoemission were previously discussed [3,4], the analysis of the electron polarization at the excitation in superlattices is still lacking.

We report the results of the experimental and theoretical studies of the polarized electron emission from the strained InGaAs-AlGaAs and GaAs-AlInGaAs and GaAs-InGaP superlattices. The interpretation of the observed features in the electron polarization spectrum is obtained. We show that the potential for further improvements of SL sources is favorable.

The SL samples with different layer content were grown by the solid-source molecular beam epitaxy on GaAs(100)-oriented p-type substrates. The composition of the SL layers in the SLs was aimed to get large enough valence band splitting and small conduction band offset to provide high mobility of the excited electrons and low polarization losses in extraction to the surface in addition to high initial polarization.

Well-strained $Al_{0.15}Ga_{0.85}As$ – $In_{0}.15Ga_{0.85}As$ SL consisted of 20 pairs of 3 nm - thick layers homogeneously doped to p $= 1\times10^{18}$ cm^{-3}. The strained-barrier SL sample consisted of 17 pairs of GaAs (4nm) and $Al_{0.18}In_{0.16}Ga_{0.66}As$ (4nm) layers. The doping concentration with Be was p $= 1\times10^{17}$ cm^{-3} in the active layer, but p $= 6\times10^{18}$ cm^{-3} in the last 6 nm heavily-doped GaAs layer to increase polarization and quantum yield. The procedure of mole fraction and layer thickness estimation was based on knowledge of growth rates of AlAs and InAs binary compounds. The characterization of the samples was done using luminescence and X-ray diffraction techniques. The polarization of the emitted photoelectrons was measured at residual pressure 3×10^{-10} Torr by Mott scattering at 38 keV (effective Sherman function being equal to 0.224 ± 0.013).

We have calculated the miniband spectrum, optical absorption and spin polarization spectra of strained superlattices using the envelope-function approximation, in the framework of the Kane model including the conduction band Γ_6, the states of light and heavy holes of the valence band Γ_8 and also the states of the spin-orbit splitted band Γ_7.

The miniband spectrum of the strained $GaAs-Al_{0.18}In_{0.16}Ga_{0.66}As$ SL as a function of wave vector in the layer plane $k_{||}$ and along the growth direction, k_z parallel to [100] axis is presented in Fig.1 (a). The small conduction band offset shows itself in a very small splitting between the second and the third electronic minibands at the center of the Brilluin zone.

For the circularly polarized light absorption the spin polarization of electrons excited in all minibands of the conduction band at the excitation mo-

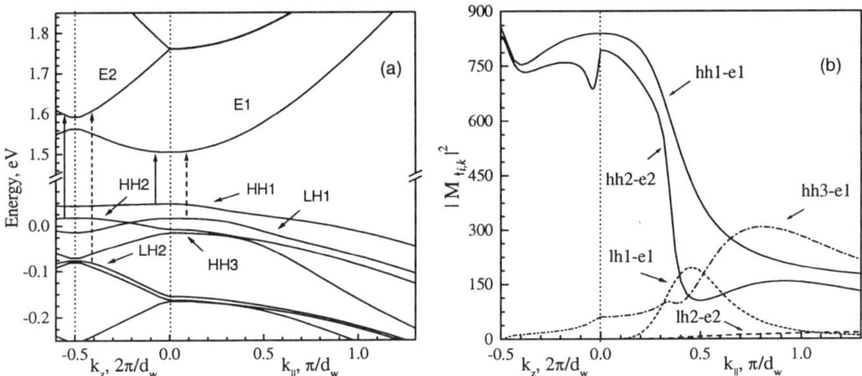

Figure 1. The electron miniband spectrum of GaAs-AlGaInAs supelattice (a), and matrix elements for the interband optical excitation by circularly polarized light in $|\uparrow\rangle$ state (b).

ment was calculated as

$$P = \frac{\sum_n (K^n_\uparrow - K^n_\downarrow)}{\sum_n (K^n_\uparrow + K^n_\downarrow)} \tag{1}$$

Here $K^n_\uparrow$ ($K^n_\downarrow$) is the absorption coefficient for the excitation in the $|\uparrow\rangle$ state with spin up ($|\downarrow\rangle$ for spin down) of the n-the electronic miniband from all the hole minibands. According to Eq. (1), the resulting electron polarization depends on the relative contribution of the light- and heavy-hole bands, the singularities in the polarization being originated from the Van-Hove singularities in the absorption spectra.

The calculated matrix elements $|M_{i,k}|^2$ for the main optical interband transitions in one $|\uparrow\rangle$ of the two electron spin states ($\uparrow$ and $\downarrow$) are depicted in Fig. 1 (b). Strong dependence on the lateral wave vector $k_\parallel$ is due to the form-factors of the profile of the band structure potential. Rapid decrease of hh1-e1 and lh-e1 matrix elements with $k_\parallel$ makes possible high polarization values far above the absorption edge at the onset of e2-hh2 transitions, since the absolute value of the e2-hh2 matrix element is found to be close to that for the e1-hh1 transition even in the case of small conduction band offset.

The smearing of the absorption edge is included as a broadening of the electronic spectra by replacing the δ-function in the absorption probability by a Lorentzian with the half width δ for all the interband transitions. The polarized emission spectral curve together with the calculated electron polarization at the excitation as a function of the excitation energy for GaAs-

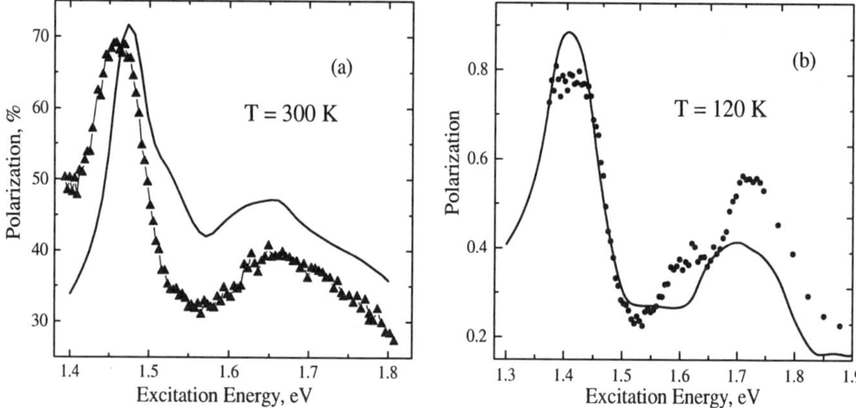

Figure 2. Polarization spectra of the emitted photoelectrons (dots) and calculated electron polarization at the excitation (solid lines) and for GaAs-AlGaInAs (a) and InGaAs-AlGaAs (b) superlattices as a function of optical excitation energy.

$Al_{0.18}In_{0.16}Ga_{0.66}As$ superlattice for T=300 K are presented in Fig.2 (a).

The dependence of the polarization and absorption spectra on the excitation energy follows the sequence of the miniband transitions. The polarization in the first maximum is very sensitive to ratio of the hh1-lh1 splitting to the broadening parameter which is associated with the smearing of the absorption edge by the hole scattering processes and by the band edge fluctuations. These processes lead to the non-zero contribution of the lh1 light-hole miniband to the absorption near the edge and populate the second spin state.

The estimated values of δ in analogous laser structures depend on the carrier concentration, doping and temperature and are varying in the range 5-30 meV[6]. For $E_{hh1} - E_{lh1}$=37 meV and δ=10 meV (typical for p-doped samples at T=300 K) the maximum polarization value does not exceed 72 %. The comparison with the experimental spectrum suggests lower δ since the polarization losses in transport to the surface and emission from the band bending region are in the range of 10 - 15 % [3]. To obtain $P \geq 90$ % the values of $\delta \leq 5$ meV are necessary which can be expected at room temperature in low doped SLs with weaker hole scattering.

Experimental polarization spectrum of $In_{0.15}Ga_{0.85}As$ - $Al_{0.15}Ga_{0.85}As$ SL for T=120 K is presented in Fig. 2 (b) in comparison with the calculation results. The splitting $E_{hh1} - E_{lh1}$=45 meV ensures maximum initial polarization $P_{ex} \geq 87$ % even with δ=10 meV. Note that the broadening estimated in Ref. [4] from the experimental data on polarized electron emission for GaAs/GaAsP strained layer (in the range of $\delta \approx 18$ -28 meV for differently doped samples at room temperature) did not lead to more than 6 % depolarization at the excitation. Higher sensitivity of polarization in the SL structures can be attributed to stronger mixing of the heavy- and light-hole states with the growth of the lateral hole momentum.

The second maximum in the polarization spectra is found to be due to the transitions at $k_z = \pi/d$ of the miniband edge. For both considered superlattices it corresponds to electronic states with the energy above the barriers in the conduction band. In the InGaAs-AlGaAs the maximum polarization in this peak can exceed 50 % since its position falls in the region of reduced contribution of the HH1-E1 transition to the absorption. Due to the differential character of polarization spectrum its details (especially, the polarization values in the maxima and minima) are highly sensitive to the SL structure and the band offset ratio.

To summarize, we have preformed the studies of the polarized electron emission from the barrier and well strained superlattices. Three features originating from hh1→e1, lh1→e1 at $k_z = 0$ and hh2→e2 at $k_z = \pi/d$ are shown to dominate in the polarization spectra. Taking into account the electron spin relaxation, the calculated spectra are in a good agreement with the observed excitation spectra of polarized electron photoemission both for InGaAs-AlGaAs and GaAs-AlInGaAs strained superlattices. The maximum value of the polarization in the excitation is found to be more sensitive to the smearing of the absorption edge than in case of the strained layer which can be compensated by the larger valence miniband splitting and lower doping. The calculated wave functions can be used to estimate polarization losses in the electron extraction to the surface and emission in vacuum, which makes band structure engineering for the photoemitting SL structures a close possibility.

This work was supported by INTAS under grant 99-00125, the Russian State Program "Physics of Solid State Nanosrtuctures", under grant 97-1091, and Russian Fond for Basic Research under grant 00-02-16775.

References

1. A.V.Subashiev *et al*, *Phys. Low-Dim. Struct.* **1/2**, 1 (1999).
2. K. Togava *et al*, *Nucl. Instrum. Methods* A **414**, 431 (1998).
3. A.V. Subashiev, J.E. Clendenin, *J. Modern Physics* A **15**, 2519 (2000).
4. A.V. Subashiev *et al*, *Semiconductors* **33**, 1182 (1999).
5. Yu.A. Mamaev *et al*, *Phys. Low-Dim. Struct.* **10/11**, 61 (1995).
6. A.D. Andreev, D.V. Donetsky, *Appl. Phys. Let.* **74**, 2743 (1999).

A PULSED POLARIZED ELECTRON SOURCE FOR ATOMIC PHYSICS EXPERIMENTS

C. D. SCHRÖTER, A. DORN, R. MOSHAMMER, C. HÖHR AND J. ULLRICH

Max-Planck-Institut für Kernphysik, Saupfercheckweg 1, 69117 Heidelberg, Germany

Recently, a pulsed photoelectron source has been set up to satisfy the beam specifications required for atomic collision studies using so-called "reaction microscopes". The gun is in operation and delivers short electron pulses ($\sim 500\,$ps) using GaAs/AlGaAs heterostructure crystals. Illuminating the photocathode with light from a pulsed laser diode an average electron current of $60\,\mu$A has been obtained. A quantum efficiency of $3\,\%$ and a cathode lifetime of one week have been achieved.

Planned future atomic collision measurements require a cold beam of transversely polarized electrons. First design changes have been made to produce short pulses of monochromatic and spin-polarized electrons in the future.

1 Introduction

Kinematically complete experiments for multiple ionization of atoms and molecules by electron impact give detailed information about the role of the electron-electron interaction. In our group great efforts are undertaken to study these collisions systematically at incident electron energies ($E_0 \sim 100\,$eV) near ionization threshold. Furthermore first ionization experiments in the presence of a strong laser field ($\sim 10^{13}\,$W/cm^2)[1,2,3] as well as quantum mechanically complete experiments, like ionization and/or excitation in collisions of spin-polarized electrons with polarized targets, are in preparation.

All these electron-impact ionization experiments are performed using the cold-target recoil-ion momentum spectroscopy[4,5,6,7]. This successful experimental method demands, however, short-pulsed electron beams of high quality. Optimum conditions require pulse widths of less than $500\,$ps (for "laser-assisted" experiments even as short as $\sim 100\,$ps), a beam diameter of $100\,\mu$m at the target and a pulse-repetition rate of up to $3\,$MHz.

In the recent past, a pulsed photoelectron source has been developed and set up[8,9] satisfying the beam specifications required by our "reaction microscope" experiments[6,7].

2 Experimental Set-up

The experimental set-up of the electron source and details of the preparation technique of the photocathodes are described elsewhere[8,9]. Therefore only a short description will be given here. A schematic diagram of the set-up is shown in figure 1. The source consists of a three-chamber UHV system, i.e. a

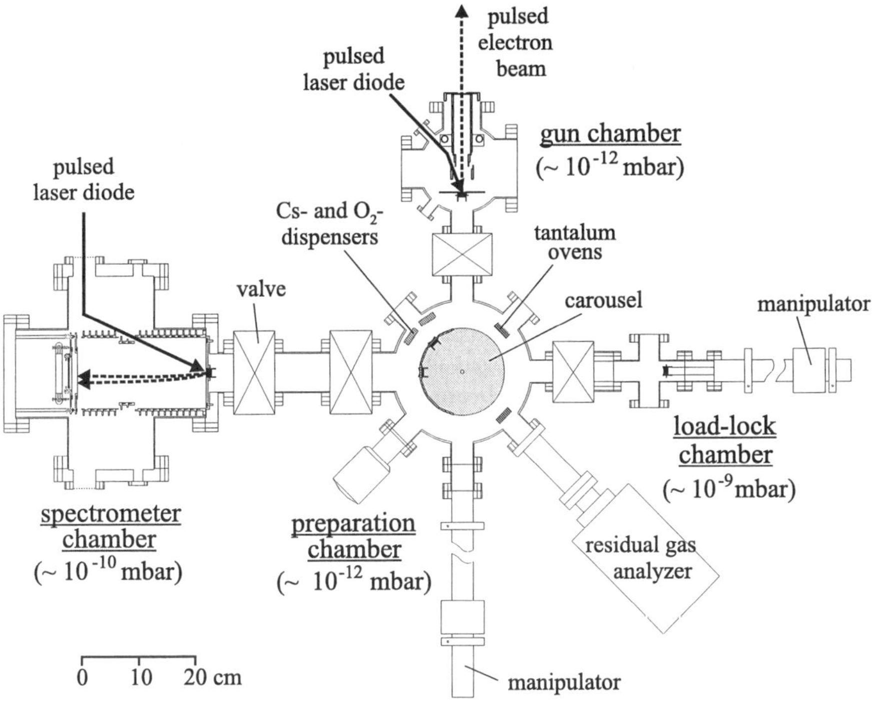

Figure 1. Experimental set-up of the pulsed electron source.

load-lock chamber, a photocathode-preparation chamber and a gun chamber. Both, the preparation and gun chamber are operating at a base pressure in the 10^{-12} mbar regime, the load-lock chamber at $\sim 10^{-9}$ mbar. This three-chamber UHV system allows the fast exchange of photocathodes within a few minutes without breaking the ultra high vacuum of the gun chamber.

The photocathodes are prepared by an oxide-free technique[10] and are transported without exposure to air to the preparation chamber. After a heat-cleaning procedure the GaAs crystals are activated with cesium and oxygen to obtain a negative electron affinity. The activated photocathodes are finally transferred to the gun chamber.

In the gun chamber the cathode is illuminated in reflection mode with a pulsed laser diode (up to $3.3\,\text{mW}$) operating at a wavelength of $650\,\text{nm}$. An average emission current of $60\,\mu\text{A}$, i.e. a quantum efficiency of $3\,\%$, and a photocathode lifetime of one week have been achieved. Short pulses of electrons are produced using GaAs/AlGaAs heterostructure crystals with a thin active GaAs layer. At the layer thickness of $0.9\,\mu\text{m}$ the long tail of the electron pulse, generated by a δ-pulse light excitation, extends out to less than $200\,\text{ps}$[11]. Adding the laser diode pulse width of 100 to $300\,\text{ps}$, which depends on the laser power, a total electron-pulse width of less than $500\,\text{ps}$ is expected.

3 Perspectives

Atomic collision experiments using cold electrons as projectiles, as well as transversely polarized electrons, are planned in the future. Design changes of the electron gun are already in progress to produce a monochromatic and even shorter pulsed ($\sim 100\,\text{ps}$) electron beam. The gun chamber will be replaced by a new UHV chamber, which recently has been designed. A $90°$-electrostatic deflector for the production of transversely polarized electrons has been built and will be mounted inside the new chamber together with an electron optics for the beam transport. The new construction should allow in a further step even cooling of the photocathodes.

The production of an intense cold electron beam with a high degree of polarization demands a powerful infrared laser system. We have built a tuneable Ti:Sapphire laser ($670\text{-}1070\,\text{nm}$) that fits these requirements, delivering $600\,\text{mW}$ at a pulse width of 100 to $200\,\text{fs}$. Using strained GaAs photocathodes together with circularly polarized light from this laser system the source should provide in the future a beam of spin-polarized electrons with a high degree of polarization (up to $80\,\%$).

Acknowledgments

This work was partially supported by the Deutsche Forschungsgemeinschaft within the Leibniz-program and the SFB 276. The heterostructure material was kindly put at our disposal by A. S. Terekhov. We are grateful to A. S. Terekhov as well as to A. Wolf and D. Orlov for their continuous support.

References

1. D. Khalil, A. Maquet, and R. Taïeb, C. J. Joachain and A. Makhoute, *Phys. Rev.* A **56**, 4918 (1997).
2. F. Ehlotzky, A. Jaroń, J. Z. Kamiński, *Phys. Rep.* **297**, 63 (1998).
3. C. J. Joachain, M. Dörr, N. Klystra, *Adv. At. Mol. Opt. Phys.* **42**, 225 (2000).
4. R. Moshammer, M. Unverzagt, W. Schmitt, J. Ullrich, H. Schmidt-Böcking, *Nucl. Instr. and Meth.* B **108**, 425 (1996).
5. J. Ullrich, R. Moshammer, R. Dörner, O. Jagutzki, V. Mergel, H. Schmidt-Böcking and L. Spielberger, *J. Phys.* B **30**, 2917 (1997).
6. A. Dorn, R. Moshammer, C. D. Schröter, T. J. M. Zouros, W. Schmitt, H. Kollmus, R. Mann, and J. Ullrich, *Phys. Rev. Lett.* **82**, 2496 (1999).
7. A. Dorn, A. Kheifets, C. D. Schröter, B. Najjari, C. Höhr, R. Moshammer, and J. Ullrich, *Phys. Rev. Lett.* **86**, 3755 (2001).
8. C. D. Schröter, A. Dorn, R. Moshammer and J. Ullrich, International Workshop on Polarized Sources and Targets, Erlangen 1999, *Conference Proceedings*, 292 (1999).
9. C. D. Schröter, A. Dorn, J. Deipenwisch, C. Höhr, R. Moshammer and J. Ullrich, International Workshop on Polarized Electron Source and Polarimeters, Nagoya 2000, in *SPIN 2000, AIP Conference Proceedings* **570**, 996 (2001).
10. A. S. Terekhov, private communication.
11. P. Hartmann, J. Bermuth , D. v. Harrach, J. Hoffmann, S. Köbis, E. Reichert, K. Aulenbacher, J. Schuler, and M. Steigerwald, *J. Appl. Phys.* **86**, 2245 (1999).

EMISSION FROM SEMICONDUCTOR PHOTOCATHODES AT HIGH INTENSITY INHOMOGENEOUS EXCITATION

BORIS I.REZNIKOV,

A.F. Ioffe Physico-technical Institute RAS, Polytekhnicheskaya 26, 194021, St.-Petersburg, Russia
E-mail: Boris.Reznikov@pop.ioffe.rssi.ru

ARSEN V.SUBASHIEV

St. Petersburg State Technical University, Polytekhnicheskaya 29, 195251, St. Petersburg, Russia
E-mail: arsen@subashiev.hop.stu.neva.ru

Kinetics of the photoemission from a semiconductor thin layer with negative electron affinity surface at high power excitation is studied theoretically for the case of local inhomogeneous excitation of the photocathode. The excitation intensity distribution over the excitation region is shown to be essential both for the evaluation of the photocathode parameters from the charge limitation effects and in the studies of the optimum local excitation regime for the photoemission.

High-energy physics experiments require high-intensity electron beams that can be obtained in the photoexcitation of GaAs-based photocathodes with Negative Electron Affinity (NEA) [1]. At high excitation the electron capture to the surface centers flattens the band bending which results in a decrease of the photocathode quantum efficiency and a saturation of the charge emitted in vacuum in an emission pulse. This phenomena was described as surface charge limitation effects [2]. When the excitation spot is less than the activated photocathode area the distribution of the light intensity in the excitation spot modifies the observed effects. Besides, the inhomogeneous photovoltage distribution can also result from inhomogeneities of the activation layer. In this report photoemission from a semiconductor with NEA surface at high power excitation is investigated theoretically for the case of local excitation regime and Gaussian distribution of the light intensity over the sample surface for the long pulse excitation and two-pulse excitation regimes.

We consider the photocathode with a thin film working layer in which normal drift of the electrons to the surface dominates. The local density of the electron current to the surface is $q_{ns}(r,t) = \alpha d(1 - R)J(r,t)/h\nu$, where $J(r,t)$ is the light intensity, α is the optical absorption coefficient, d is an effective thickness of the working layer (which is equal to the layer thickness for the thin layer limit), R is the optical reflection coefficient, $J(r,t)$ is the optical excitation intensity at the distance r from the light beam center as a function of time t, $h\nu$ is the optical excitation energy. The normal component of the electronic flow to the surface calculated straightforwardly using diffusion equation with the appropriate boundary conditions [3]. According to the experimental studies [4,5], the surface escape probability B_n for the electrons from the band bending region is proportional to the NEA value Δ and linearly decreases with the increase of the surface photovoltage U, so that $B_n = 0$ for $U/\Delta > 1$ and $B_n = B_{n0}(1 - U/\Delta)$, when $U/\Delta \leq 1$.

The value of PV is found from the kinetics of the electron capture on the surface centers and the hole restoring current. We assume that the dominating mechanism for the surface discharge is the thermo-activated light-hole tunneling to the surface centers. Then the hole restoring current can be written in form [4]

$$q_p(U) = q_{p0}\left[\exp\left(U/E_0\right) - 1\right], \tag{1}$$

where E_0 is the characteristic electron tunneling energy. Pre-exponential factor q_{p0} is highly sensitive to the cross section for the hole capture by the surface centers which is not well known and is considered as a parameter of the surface model.

The photovoltage kinetics can be written in terms of charging and discharging the effective capacitor formed by the band bending region by the electron current:

$$\frac{C}{q^2\sqrt{1 - U/V}}\frac{dU}{dt} = q_{ns} - q_p(U), \tag{2}$$

Here U is the initial depth of the well in the BBR, C is the surface capacitance per unit square. In stationary excitation regime Eq. (2) yields for the local emission current density

$$q_{emi} = B_{n0}q_{ns}\left[1 - E_0/\Delta \ln\left(1 + q_{ns}/q_{p0}\right)\right]. \tag{3}$$

The emission current density (3) has a maximum as a function of the electron current density q_{ns} at the excitation intensity J_{opt}. For high values of excitation intensity, when $U \gg E_0$ and $q_p(U) \gg q_{p0}$, J_{opt} and the maximum value of the emission current density q_{opt}, are given by

$$J_{\mathrm{opt}} = \frac{h\nu q_{p0}}{\alpha d(1 - R)}\exp\left(\frac{\Delta}{E_0} - 1\right), \quad q_{\mathrm{opt}} = \frac{q_{p0}B_{n0}E_0}{\Delta}\exp\left(\frac{\Delta}{E_0} - 1\right) \tag{4}$$

while the corresponding photovoltage is $U_{\mathrm{opt}} = \Delta - E_0$.

With the intensity increase to J_{cr} the local photoemission is switched off at $V_{\mathrm{cr}} = \Delta$. Both J_{opt} and J_{cr} are exponential functions of Δ/E_0 ratio. The variation of light intensity over the excitation area modifies the emission

172

current dependence on the excitation power. We use a Gaussian distribution for the radial dependence of the light intensity characterized by the ratio $\beta = I(0)/I(r_{\mathrm{max}})$ where r_{max} is the sample radius.

The stationary emission current distribution as a function of the distance r from the light beam center is depicted for the sample parameters of Ref. [6] in Fig. 1 (a). The quantum efficiency grows with the distance from the light beam center whereas the emission current is decreasing with r when $I_0 < I_{\mathrm{opt}}$. At $I_0 > I_{\mathrm{opt}}$ the SCL is much stronger near the beam center which results in non-monotonic dependence of $q_{\mathrm{emi}}(r)$ and a gradual shift of the maximum of the emission current to the periphery of the electronic beam. As a result the central region of the beam corresponds to an area with high photovoltage and low emission, while the main input to the emission comes from the ring with the excitation intensity close to I_{opt} and the emission current density close to q_{opt} values. Thus, at high excitation the emission is not switched off but the emitting area is strongly reduced. In Fig. 1 (b) the total stationary emission current q_{tot} as a function of the total light excitation power is shown for several values of β. The shift of the maximum q_{tot} to lower intensities and the decrease of the emission current with the increase of β is clearly seen. Besides, the intensity dependence of the emission current is considerably smoothed. The calculated full emission current variation as a function of the excitation

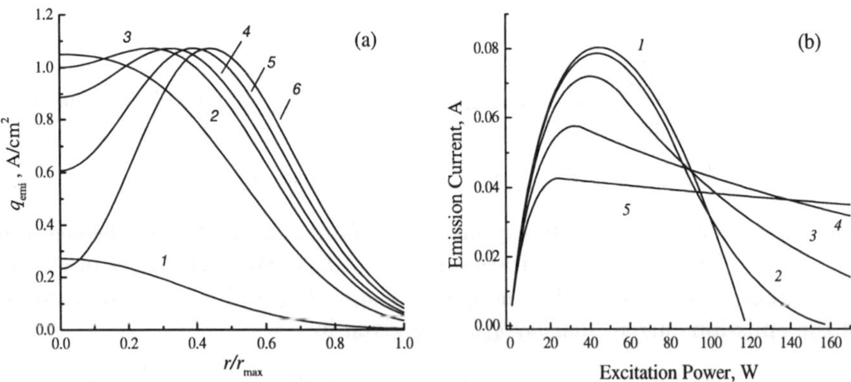

Figure 1. Emission current density $q_{emi}(R)$ as a function of the distance from the beam center R (measured in FWHM-r units) (a) for a Gaussian intensity distribution along the surface $(I_{max}/I_{min} = 100)$; I_{max}, 10^{20} cm^{-2} s^{-1} $1 - 0.1$, $2 - 1$, $3 - 1.7$, $4 - 2$, $5 - 2.5$, $6 - 3$. and emission current in the bunch $q = q_{tot}/q_{opt}$ as a function of the light power $\tilde{I} = I_{max}/I_{opt}$; I_{max}/I_{min}: $1 - 1$, $2 - 2$, $3 - 5$, $4 - 20$, $5 - 100$.

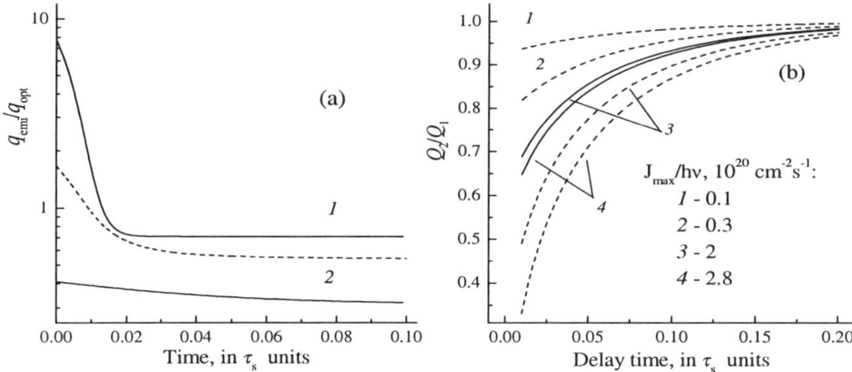

Figure 2. (a) Emission current density at the beam center (1) and at a distance $r = 0.8r_{max}$ (2); dotted line shows the total current variation, and (b) the emitted charge ratio Q_2/Q_1 as a function of the delay time; the dotted lines cotrrespond to homogeneous excitation, solid – to inhomogeneous excitation with $\beta = 100$.

intensity is close to almost saturated dependence of the total emission current observed experimentally at high excitation intensity in [6].

Non-stationary excitation regimes were studied by numerical calculations for the case of long-pulse excitation and a sharp rising of the light intensity and also for the two-pulse excitation regime with equal intensity excitation pulses separated by the delay time t_p. In the long pulse excitation case the photovoltage relaxation time to the stationary state grows with the distance from the excitation center. From Eq. (2) the photovoltage relaxation time τ_s at a small photovoltage is equal to $\tau_s = C/q_{p0}$ and is reduced by the q_{ns}/q_{p0} factor at high photovoltage values. The calculation results for the emitted current density and the total emitted current are shown in Fig 2 (a). Note that the current becomes close to the stationary values at times which are much smaller than the BBR discharge time. The relaxation rate after switching off excitation is found to be much larger in the center of the excitation spot which results in the flattening of the photovoltage distribution over the excitation area. In the pump-probe regime the non-homogeneous excitation manifest itself at short delay times, while the resulting restoration time of the photocathode quantum efficiency is a weak function of the light intensity and its inhomogeneous distribution. The ratio of the charge emitted in the second pulse to that in the first pulse is shown in Fig. 2 (b) for several values of the light intensity in the beam center. At high excitation the restoration rate is noticeably increased for non-homogeneous excitation due to higher discharge rate in the region of high photovoltage. The resulting restoration time of the cathode quantum efficiency is found to be close to the linear-regime restoration time.

We have used the experimental data of Ref. [6] to estimate the parameters of the BBR region and the hole restoration current. The obtained values are $B_{n0} = 0.38$, $E_0 = 44$ meV, $q_{p0} = 2.3 \times 10^{18}$ cm^{-2} s^{-1} , and $\Delta = 200$ meV. These parameters are close to the values determined from the charge limitation experiments on the thin GaAs layer cathodes at homogeneous excitation [4], which shows the possibility to use local high-intensity excitation regime for the characterization of the photocathode activated surface. Note that the variation of band bending along the photocathode surface caused by the surface defects and the spatial fluctuations of the doping centers can cause the spread of the photovoltage and account for the effects similar to inhomogeneous excitation effects, namely, the saturation behavior of the emitted charge observed experimentally.

To summarize, our analysis shows high (exponential) sensitivity of the maximum emission current to the negative electron affinity value obtained in the surface activation. The maximum emitted current is reduced for the inhomogeneous excitation regime, and the dependence of the emitted current on the excitation intensity is smoothed. Rapid relaxation of the photovoltage in the region of high excitation intensity makes the quantum efficiency restoration less sensitive to the excitation intensity and inhomogeneity of the excitation. The excitation intensity distribution is important both for the evaluation of the photocathode parameters from the charge limitation effects and for the studies of the optimum regime for the photoemission.

This work was supported by the INTAS under grant 99-00125, by Russian State Program "Surface Atomic Structures" under grant 107-24(00)-P, and also by the Russian State Program "Physics of Solid State Nanosrtuctures" under grant 97-1091.

References

1. R. Alley *et al,Nucl. Instrum. Methods* A **365**, 1 (1995).
2. M. Woods et al., *Appl. Phys.* **73**, 8531 (1993).
3. B.I. Reznikov, A.V. Subashiev, Semicond. **32**, 1006, 1309 (1998).
4. G.A. Mulhollan *et al, Phys. Lett.* A **282**, 309 (2001).
5. A.S. Terekhov *et al, Solid State Physics* ,, 38 (171) 1996.
6. M.A. Kirillov *et al, Low-energy polarized electron workshop proceedings*, St-Petersburg, 1998, Ed. Yu.A.Mamev et al., p.30-33.

CHARGE LIMITATION EFFECTS IN PHOTOEMISSION FROM GaAsP STRAINED LAYER CATHODE

A. SUBASHIEV, YU. MAMAEV

Department of Experimental Physics,
State Technical University, 195251 St Petersburg, Russia
E-mail: arsen@spes.stu.neva.ru

B. REZNIKOV

Ioffe Physical Technical Institute RAS, 194021 St Petersburg, Russia,
E-mail: Boris.Reznikov@pop.ioffe.rssi.ru

T. NAKANISHI, K. TOGAWA, T. NISHITANI,
M. KUWAHARA, T. HIROSE AND R. MIZUNO

Department of Physics, Nagoya University, Nagoya 464-8602, Japan

A thin layer strained GaAsP photocathode have been used to investigate charge limitations effects in two-pulse excitation mode with Gaussian distribution of the light intensity over the sample surface. The estimated parameters of the band bending region show that the strain does not reduce the hole restoring current, while surface barrier structure and transparency is sensitive to the layer strain.

1 Introduction

For now the most effective sources of highly polarized electron beams are based on photoemission from the strained p-GaAs layers with the surface activated to Negative Electron Affinity (NEA). The structure of the Cs(O) activation layer and the resulting surface escape probability are sensitive to the vacuum conditions, activation technique, doping and the crystalline structure of the emitting layer [1]. These properties can be probed in high-intensity excitation regime when the photoelectrons are rapidly captured in the Band Bending Region (BBR) near the surface so that the NEA is reduced due to the Surface Photovoltage (SPV) build up. Quantum Efficiency (QE) of photoemission is also reduced with SPV growth which is known as Surface Charge Limitation (SCL) effects which are important for the operation of the GaAs photocathodes at linear colliders [2].

In [3] the SCL effects were studied in the unstrained GaAs layers for the homogeneous excitation over the sample surface. It was shown that SCL effects can be well interpreted in a model assuming a linear dependence of the electron surface escape probability B_n on the NEA.

The strained GaAs and GaAsP overlayers are known to have much worse crystalline structure, considerably lower QE and inhomogeneous surface emission. While charge limitation effects were experimentally studied in [4] the hole restoration current mechanism and the role of inhomogeneities were not clarified. Besides, in most of the experiments the QE distribution the light intensity over the sample surface is not homogeneous (see e.g. [5]), which is important in interpretation of the SCL effects [6].

In the present paper, we report on the experimental studies and theoretical analysis of the SPV effects observed in GaAsP strained layer photocathode under Gaussian distribution of the light intensity over the sample surface.

2 Experimental

The set of experiments has been performed with the $GaAs_{0.95}P_{0.05}/GaAs_{0.7}P_{0.3}$ strained MOCVD grown heterostructure, with the parameters of the sample improved to have high coherent strain and, thus, sufficient energy splitting ($\Delta_{def} \approx 58$ meV) of the Heavy Hole (HH) and the Light Hole (LH) bands. The modification consisted of more sharp interfaces fabrication and gradual p-doping of 140 nm thick $GaAs_{0.95}P_{0.05}$ strained overlayer ($p=1\cdot10^{18}$ cm^{-3} in the main part of it, but several times enlarged at the top 20 nm).

All measurements were performed at the Department of Physics, Nagoya University. Experimental set-up has been earlier described in [7]. Prior the deposition into the vacuum chamber GaAsP cathode was chemically cleaned using HCl-isopropanol and then exposed in the UHV camber at the temperature 550 Celsius during 1 hour. During the NEA activation the photocathode was biased at $\approx - 200$ V. To test the quality of activation layer QE was measured using low-intensity HeNe laser to give 4.5 % which is a typical value for this type of cathode.

The low-intensity polarization and QE spectral dependencies were studied at room temperature, using a low-intensity CW Ti:sapphire laser. Maximum quantum efficiency in the high-polarization band at 812 nm excitation was about 0.1 % which is close to previously reported values.

A Nd:YAG-pumped Ti:sapphire laser system was used as the pulsed high-intensity light source. The laser wavelength has been tuned to 812 nm (i.e. the short-wavelength border of the polarization maximum with high QE). The Full Width at Half Maximum (FWHM) of the laser pulse was 6 − 7 ns and the bunch separation was tuned to be 25 ns. The surface of 14 mm diameter photocathode was illuminated with the laser light. For the high-excitation regime the laser light distribution across the sample was close to Gaussian with the FWHM-r = 5 mm.

Figure 1 (a) displays the bunch shapes of the photoemission current in

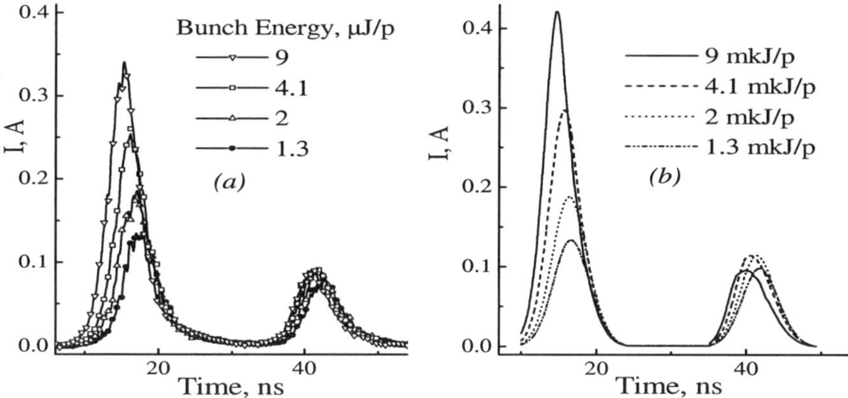

Figure 1. The electron photocurrent bunch shapes for two-pulse excitation (a) ; pulse energy: *1* -1. 2 mkJ *2* - 2.1 mkJ, *3* - .3 mkJ *4* - 9.0 mkJ compared with the results of the calculations (b).

two-pulse excitation for a set of excitation energy in a bunch. The calculated current temporal profiles are shown for comparison in Fig. 1 (b). Charge limitation effects manifested in the decrease of the photoemission current in the second bunch (compared with the current in the first bunch) and shift of the pulse maxima to the head portion of the bunches. Note that the observed values of the current are in the region where the vacuum space charge current limitations [7] are negligible.

The integrated charge in the first and the second pulse as a function of excitation energy in a bunch is presented in Fig. 2 together with the calculation results. Note nonlinear variation of the emitted charge starting from the first experimental points and a surprisingly small variation of the charge emitted in the second pulse on the energy of the excitation pulse.

3 Charge limit model

We have analyzed the observed charge limitation effects in the two-pulse excitation photoemission using a model described in Refs. [3,6,8]. The photoemission current density q_{emi} equals to $q_{\mathrm{emi}} = B_n q_{ns}(r, t)$, where B_n is the electron surface escape probability in vacuum and $q_{ns}(r, t)$ the electronic flow to the surface. The total emission current is obtained by integration of q_{emi} over the sample area.

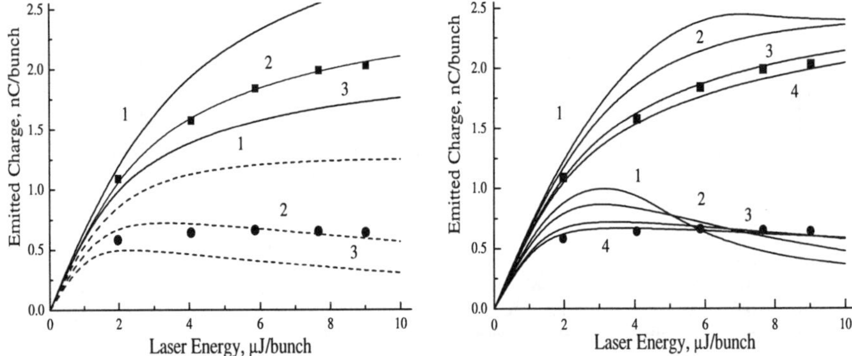

Figure 2. The charge emitted in the first (squares) and the second (circles) bunch as a function of the excitation pulse energy and the curves (a) calculated for $I_{max}/I_{min}=20$ and Δ values, meV: *1* - 90; *2* - 110; *3* - 150; (b) light distribution variation for $\Delta = 100$ meV and I_{max}/I_{min}: *1* — 1, *2* — 5, *3* — 16, *4* — 25.

The working layer strain affects mainly the valence band states, so that the surface escape probability B_n can be assumed to decrease linearly with the photovoltage U, as in the case of the unstrained GaAs layer [3], $B_n = B_{n0}\left(1 - U/\Delta\right)$, for $U \leq \Delta$, while $B_n = 0$ for $U > \Delta$. Here Δ is the initial value of negative electron affinity.

Since the electron spread due to the current along the BBR well is negligible [6], the equation for the kinetics of the surface photovoltage can be written in terms of the local current flow of the electrons and holes to the surface as

$$\frac{C}{q^2\sqrt{1 - U/V}}\frac{dU}{dt} = q_{ns} - q_p(U), \qquad (1)$$

where V is the initial band bending and C is the BBR layer capacitance per unit square area. In (1) we employed that for the strained GaAs layers $B_n \ll 1$, so that the electron surface recombination current is close to q_{ns}. The hole restoring $q_p(U)$ current is controlled by the thermo-activated tunneling through the band bending barrier can be written as [3] $q_p(U) = q_{p0}\left[\exp\left(U/E_0\right) - 1\right]$, where E_0 is a characteristic energy of tunneling. The pre-exponential factor q_{p0} is highly sensitive to the model of surface centers and is taken as a fitting parameter of the model.

4   Results and Discussion.

We have integrated the kinetics equation (1) using for $I(r,t)$ a Gaussian distributions for the temporal profiles of the excitation pulses and a Gaussian for the radial dependence of the light intensity for FWHM-r=0.7 cm, $N_a = 2 \cdot 10^{18}$ cm^{-3}, T=300 K.

In evaluation of the cathode parameters we used the values $V = 0.3$ eV, $E_0 = 34 \div 40$ meV $q_{p0} = (1 \div 2) \cdot 10^{18}$ cm^{-2}s^{-1} as known from the former experimental and theoretical investigations to determine the values of Δ, B_{n0}.

The calculated dependencies of the emitted charge in the first and the second pulse $Q_1(E)$ and $Q_2(E)$ for a several values of Δ and B_{n0} values are displayed in Fig. 2. A high sensitivity of the results to the choice of Δ is clearly seen, namely, (i) growth and saturation of the $Q_1(E)$ dependence, (ii) the growth and shift of the maximum for larger Δ and B_{n0}, (iii) weak energy dependence of the charge emitted in the second bunch. For $\beta = 20$ and $\alpha = 4500$ cm^{-1} the obtained values are $\Delta = 100$ meV, $B_{n0} = 0.035$.

Note high sensitivity of the $Q_2(E)$ on the choice of the main parameters (Δ, α and B_{n0}). Close values of E_0 and q_{p0} for the strained and the values for the unstrained GaAs layers, estimated recently in Ref. [3], are in line with the thermally-activated tunneling model (since for this case the tunneling energy exceeds by far the valence band splitting energy). A very smooth $Q_2(E)$ dependence at high excitation energy is a manifestation of inhomogeneous distribution of the excitation over the sample surface. Good agreement with experimental $Q_2(E)$ curve can be obtained for $I_{\max}/I_{\min} \approx 16 \div 25$.

The features of the kinetics of SCL manifest in the temporal profiles of the photocurrent bunches shown in Fig. 1. The shift of the current maxima to the front side of the first and the second bunch reflects the increase of the photovoltage growth rate with the increase of light intensity. The estimated photovoltage relaxation time $\tau_{s0} = CE_0/q^2 q_{p0} \approx 130$ ns exceeds the time spacing between the pulses resulting in a decrease of the photocurrent in the second bunch.

In the bunch center the estimated maximum of the ratio $U/\Delta > 1$ so that the emission current in the beam center should be close to zero. Note that the calculations give some decrease of the photocurrent in the second bunch with the pulse energy, which was observed in similar conditions in [7].

Finally, the value of B_{n0} estimated in [3,6] for the unstrained thin layer GaAs cathodes was considerably larger $B_{n0} \geq 0.15$. In the model of the electron emission by the tunneling through the residual surface barrier during the energy relaxation in BBR [3] $B_0 \propto \nu_{emi}\Delta/V$, where ν_{emi} is the electron emission rate from the BBR in vacuum. The decrease of ν_{emi} and a variation of band bending along the degraded surface will reduce B_{n0} and may cause the spread of the photovoltage, accounting for the effects similar to inhomogeneous excitation effects, namely, the saturation behavior of the emitted charge observed in [4].

5   Conclusions

The study of the surface charge limitation effects in photoemission from strained layer high-quality photocathode showed that the SCL in the strained layers is well described by the model previously developed for the unstrained GaAs cathodes, so that the tunneling of the light holes dominates in the restoring current. The estimated surface escape probability is considerably lower in strained layers. It suggests that B_n is inhomogeneous along the strained surface and is not only NEA-dependent. We show that an adequate interpretation of the experiment is impossible without accounting for the inhomogeneity of the light intensity distribution over the sample. In spite of the averaging of the photo-response over the bunch the bunch profiles and the emitted charge in the bunches remain sensitive to the main parameters of the cathode structure which makes possible a local characterization of the cathode structure by the SCL effects.

Acknowledgments

This work was supported by INTAS under grant 99-00125, RFBR under grant 00-02-16775 and also supported partially by Grant-in-Aid for Science Research from Japan Ministry of Education and Culture (No. 10354003) and a research fund of KEK for cooperative development (No. 2000-2).

References

1. A.V.Subashiev *et al*, Phys. Low-Dim. Struct. **1/2**, 1 (1999).
2. R. Alley *et al* NIMA **365**, 1 (1995).
3. G. Mulhollan *et al*, *Phys. Let. A* **282**, 309 (2001).
4. H. Tang *et al*, in *Proc. Workshop on Photocathodes for Polarized ELectron Emission*, (SLAC, 1993) SLAC-Rep-433 Rev., 1994, p. 344.
5. M.A. Kirillov *et al*, in *Proc. Low-Energy Polarized Electron Workshop* (St. Petersburg, 1998) p. 30.
6. B.I. Reznikov, A.V. Subashiev, *this volume, p. ....* AIP 2001.
7. K. Togawa *et al* NIMA **414**, 431 (1998).
8. B.I. Reznikov, A.V. Subashiev, *Semicond.* **32**, 1006 (1309)1998.

IV. Polarized Ion Sources

POLARIZED ION SOURCE PROGRESS:
PAST ACHIEVEMENTS! FUTURE ASPIRATIONS?

THOMAS B. CLEGG

*Department of Physics, University of North Carolina, Chapel Hill, NC, 27599-3255, USA,
and Triangle Universities Nuclear Laboratory,[†] Durham, NC, 27708-0308 USA*

Abstract: Sources of nuclear spin-polarized H and D ions have been developed for over 40 years. These exist today in ~20 laboratories worldwide and are now highly refined to suit the requirements of individual accelerators and physics programs. Differences include capabilities for: particular ion species, positive and/or negative ions, dc and/or pulsed operation, and the variety of vector and (for D) tensor polarizations available. A review of the basic types of polarized ion sources used today is presented, with summaries of performances for each source type. Reference is made to features which make each type of source interesting and to limitations which its users still face. Possible future developments which hold particular promise will be presented.

1   Introduction

Development and use of nuclear-spin-polarized H and D beams over the past 40 years has provided a wealth of technical knowledge and a stable of high-quality polarized ion sources. Today such sources can be found in some 20 laboratories worldwide. At specialized workshops like this, it seems appropriate to step back, review what we have learned in developing these sources, examine the challenges we still face, and then suggest possible productive directions for future research.

Thus, in this review I will attempt to summarize and compare the performances of existing polarized ion source systems. In doing so, I will highlight both the key physics ideas learned and the important technical advances accomplished during their development. Finally, I will discuss known opportunities for, and remaining challenges to, achieving further polarized ion source improvement. Those interested can find much more information about past developments in the proceedings of prior specialized conferences and workshops [1-4].

With lots of help this past week from 'polarized sourcerers' worldwide, many of whom are attending this workshop, I was able quickly to gather the latest information about polarized source use and performance. Figure 1 was prepared from replies to my recent survey and provides our first glimpse at the current situation.

[†]Work supported in part by the US DoE Office of High Energy and Nuclear Physics under Grants #DE-FG02-01ER41041 and DE-FG02-01ER41033.

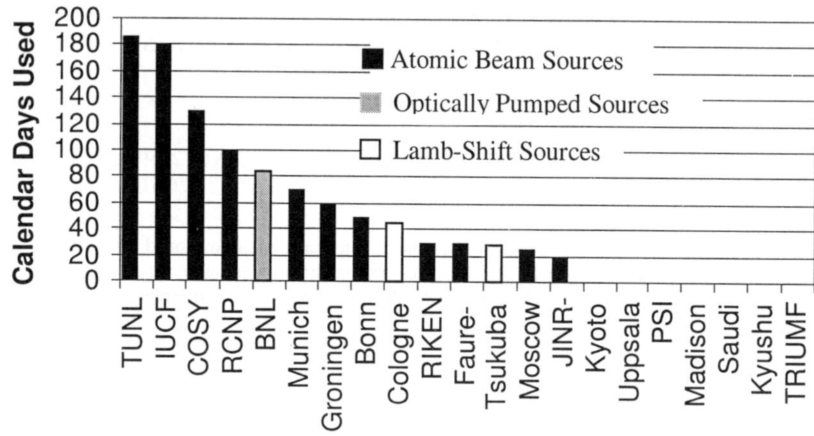

Figure 1 - Number of calendar days in past year when existing polarized ion sources were used.

Operational polarized sources today are of three basic types: Lamb-shift sources, optically pumped sources (OPPIS), and atomic beam sources (ABPIS). While the oldest of these systems were built 25 years ago, three new systems were just completed and placed in operation in the last 18 months. Individual source designs, even among those of the same type, are often quite different. Features provided are always dictated by the explicit requirements of a laboratory's accelerator and by needs of the local experimental program. Thus, these factors usually dictate the particular ion beam species provided (e.g. H^+, H^-, D^+, or D^-), the beam currents and polarizations used, and the explicit beam time structure (DC or pulsed) and the beam emittance required. With so many technical differences, it is not meaningful in many cases to compare performances of polarized sources head-to-head.

While my recent survey revealed a high general level of local user satisfaction with polarized beams in most laboratories, some reports mentioned remaining challenges if improvements needed by future experiments are to be attained. For example, scientists at COSY/Jülich and at JINR/Dubna want more beam for new accelerators. Our laboratory at TUNL needs more polarized beam for experiments below 600 keV. And, labs in Münich and Groningen want improved deuteron beam polarization and on-line polarimetry.

Collectively, operating polarized sources were scheduled for use in experiments during ~920 days in the past 12 months. However, the distribution of this use varies widely. Surprisingly to me, some very good polarized ion sources which were used extensively in the past have recently fallen into disuse. Among them, at the end of 2001 the source which has long supported the polarization studies program at PSI-Villigen is to be decommissioned with their low-energy cyclotron. The polarized source at Kyushu soon will be moved to a different campus location. Polarized sources at TRIUMF and Wisconsin lie unused awaiting new experimental demand for their beams. The polarized source and tandem accelerator at Kyoto are now instructional tools and are no longer used for basic research. I have no recent information about the source at Uppsala.

In the material which follows, I will consider separately the three ion source types presently in use, the specific design variations employed for each, and their individual performances.

2   Overview of present systems

2.1   Lamb-shift Polarized Ion Sources [5]

The oldest sources in use today are Lamb-shift sources at the Universities of Tsukuba [6], and Köln [7]. These are based on a polarization method [8] which requires the production of $H_0(2S)$ [or $D_0(2S)$] metastable atoms by charge exchange of 550eV H^+ [or 1100eV D^+] beams in cesium vapor. The metastable atoms then enter a region of carefully designed electric and magnetic fields which cause selective destruction of unwanted hyperfine states by quenching them to the ground state. Metastable atoms in the remaining hyperfine state(s) are then selectively ionized by charge-exchange in argon to provide the polarized H^- [or D^-} beams needed. Ionization of unpolarized ground-state background atoms in the beam is suppressed.

Two variations of this source are used to polarize the beam. The Tsukuba and Kyushu [9] sources employ a Los Alamos-style [10] 'spin-filter' to obtain the desired hyperfine states. The Köln source is more versatile. Though it was constructed originally to utilize the Sona field-reversal method [11,12], it is now designed to interchange this system with a 'spin filter' when needed. The latter system has the advantage of producing beams in a single hyperfine state with a wider variety of beam polarizations; the former system provides somewhat higher beam intensity because multiple hyperfine states are used. Detailed performance specifications for these sources are included below in Table 1.

Laboratory	Max. Intensity (mA)				Polarization % of Max	Ionizer Type	Ref. #
	H-	H+	D-	D+			
Cologne	0.0005	-	0.0005	-	0.7 to 0.8	Sona, Spin Filter	7
Kyushu	0.0003		0.0003	-	70(p), 65(d)	Spin Filter	9
Tsukuba	0.0003	-	0.0003	-	80(p), 75(d)	Spin Filter	6

Table 1 - Performance of Lamb-Shift Polarized Ion Sources

Lamb-shift polarized sources enjoy the advantages of being relatively simple, reliable, and inexpensive. They also provide DC beams having excellent emittance, a feature which makes them well suited for use with tandem accelerators which have tight phase space acceptance requirements. Users of Lamb-shift sources also enjoy the convenience of intrinsic capabilities for on-line beam polarimetry [13]. However, they often suffer from its one major disadvantage. Lamb-shift sources' output beam intensity is very small, rarely exceeding 0.5 μA.

An important physics lesson should be learned from this limitation. In Lamb-shift sources, the process used to produce metastable atoms, $H^+(550\ eV) + Cs \rightarrow H_0(2S) + Cs^+$, also creates intense internal space-charge electric fields from the slow Cs^+ ions introduced into the beam. This E-field quickly destroys the desired (2S) metastable atoms by Stark quenching them to the ground state. These 'lost' atoms then cannot contribute to the Lamb-shift source's output polarized beam intensity.

The 'spin-filter' mentioned above is an important technical legacy of these sources that has been adopted elsewhere for polarimetry [14], as shown in Figure 2. When it is used in the Lamb-shift source, atoms in individual hyperfine states are selected to provide an output beam with the desired polarization. When it is used as a polarimeter, the relative hyperfine state populations of the entering polarized beam are measured easily [15]. At TUNL, we routinely obtain 1% polarimetry accuracy for H^+ and D^+ beams, but accuracy with H^- or D^- beams is diminished by dilution of the beam polarization with unpolarized molecular contributions, for which the spin-filter lacks sensitivity. Accuracy of a spin-filter polarimeter at COSY/Anke is reported [16] to be better than the 1% accuracy reported at HERMES for Briet-Rabi polarimetry [17]. Another spin-filter polarimeter is under construction for the polarized source at Groningen [18].

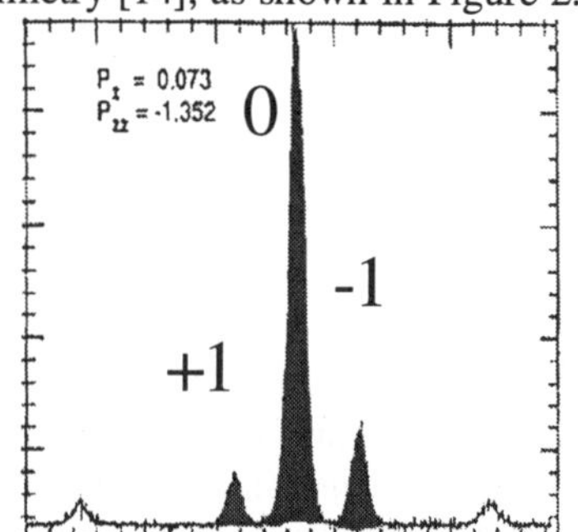

Figure 2 – Spin-filter polarimeter scan of the hyperfine state population distribution for a polarized deuterium beam from the TUNL atomic beam source. Areas beneath the $m_I = +1, 0$, and -1 peaks can be determined in tens of seconds, allowing calculation of the beam's vector and tensor polarizations.

2.2   *Optically Pumped Polarized Ion Sources (OPPIS) [19]*

Two optically pumped sources of polarized H^- ions exist today, an excellent DC source [20] at TRIUMF which has not operated this year, and an extensively used pulsed source [21, 22] which was first placed into operation at BNL/RHIC in 2000. The OPPIS polarization method is based on pickup by 3 keV H^+ ions of a polarized electron in optically pumped, polarized rubidium vapor. To preserve the nuclear polarization, this charge-exchange must occur inside a strong magnetic field. The emerging electron-polarized H_0 atomic beam then experiences a rapid magnetic field flip (a Sona reversal, like that used in the Lamb-shift source) which transfers the electronic to nuclear polarization. Atoms in the beam then pick up a second electron in sodium vapor, producing output nuclear-polarized H^- ions.

The two existing optically pumped sources have been highly successful. Table 2 summarizes their specifications and performance. Remarkably at TRIUMF, DC H^- beams of intensity up to 0.6 mA with $P_z = 85\%$ have been attained within a normalized emittance of 2π mm-mrad. For pulsed beams having 500μs pulse length at repetition rates up to 4 Hz, 1.6 mA was attained with $P_z = 80\%$ with a similar emittance. Most impressive is the very high stability of these beams [23], exemplified by recent successful use in a series of experimental tests for parity violation.

At BNL/RHIC, polarized H^- beam intensities of 0.5 to 1.1 mA with $P_z = 80\%$ are now available in 400 μs pulses at 1 Hz from the source. Up to 0.58 mA has been measured after acceleration to 200 MeV in the injector linac. Earlier TRIUMF tests mentioned above indicate that further improvement in both current and P_z are possible with a longer solenoid to produce a more uniform B-field over the ECR H^+ source and Rb vapor canal.

Laboratory	Maximum Intensity (mA)				Normalized Emittance (80%) π mm-mrad	Polarizaton % of Max	Duty Factor Fraction	Ref. #
	H-	H+	D-	D+				
TRIUMF	0.6	-	-	-	2	<85	DC	25
TRIUMF	1.2	-	-	-	2	<75	DC	25
TRIUMF	1.6	-	-	-	2	<60, 80	DC, 0.002	25
TRIUMF/RHIC	0.5 to 1	-	-	-	2	75	0.002	25
TRIUMF/BINR	8	50	-	-	2	42+/- 5%	0.100	25
KEK*	-	-	0.38	-		70	0.002	45

*** This KEK source has been decommisioned.**

Table 2 – Performance of Optically Pumped Polarized Ion Sources

The principal technical advance that enabled major improvement in OPPIS performance is the development of an improved optical pumping laser. A Cr doped LiSrAl F crystal laser (Cr:LiSAF) developed at Livermore [24] is now commercially available and provides up to 1 kW in 500 μs pulses. Highest simultaneous values of both current and beam polarization P_z become available only when this pulsed laser is used for pumping the Rb vapor.

A smaller, but nevertheless very significant OPPIS performance improvement was obtained when a recirculating sodium jet system was developed for a new ionizer [25]. This has a larger aperture than the previous ionizer cell, providing higher beam current. Biasing the cell at -32 kV improves overall reliability by reducing the loss of Na^+ ions into other parts of the source. It also accelerates quickly the desired polarized H^- ions, thereby reducing the effects of space charge and improving the output beam's emittance. The larger aperture also provides for better illumination of the Rb vapor, resulting in higher beam polarization.

What future improvements are still possible? Tests at TRIUMF in 1999 showed that still higher intensity can be achieved by replacing the ECR source with a new intense pulsed H^+ plasma source whose design is based on developments at the Budker Institute. In this arrangement [22] shown below in Figure 3, up to 10 A of pulsed H^+ beam from the plasma source was neutralized in H_2, entered the strong axial magnetic field, and then was re-ionized in He before entering the Rb cell. Output currents I (for H⁻) of 8 mA and I (for H^+) of 50 mA with $P_z = 42\pm5\%$ were obtained. Both beam polarization and beam intensity would have been higher with a larger B-field over the He ionizer and Rb neutralizer cells.

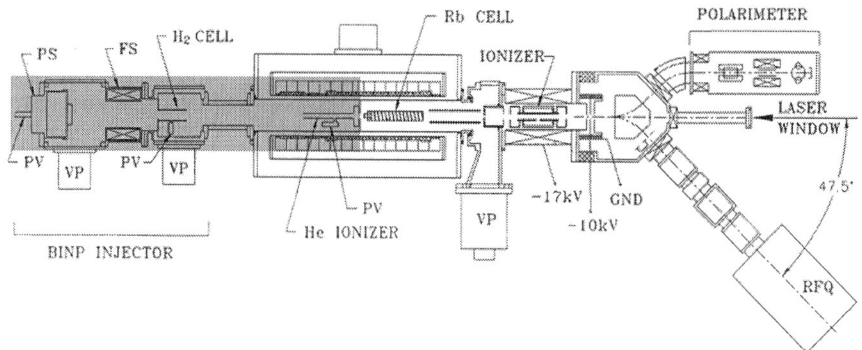

Figure 3 – The modified TRIUMF OPPIS is shown. The region in gray indicates components of the high-intensity pulsed positive source (PS). Emerging H^+ ions are focused by a solenoid (FS) first into a H_2 neutralizer canal. Fast neutral H_0 atoms then enter the magnetic field without emittance growth, are ionized again to make H^+ in a He ionizer cell, and then enter the Rb cell.

2.3 Atomic Beam Polarized Ion Sources (ABPIS)[26]

Atomic-beam sources are the most common type of polarized ion source today, with ~15 operational sources in existence. A summary of their performance specifications for these sources is provided in Table 3. The oldest such sources, at JINR/ Dubna and Bonn, were built in the early 1980's. The newest, at IUCF and at Munich, were first used for experiments only in the past 18 months. The ABPIS polarization method is based on Stern-Gerlach magnetic focusing to electron-polarize H or D atoms in a slow beam. These atoms then undergo adiabatic-fast-passage RF transitions [27] to convert their electronic polarization to nuclear polarization, and subsequently are ionized and accelerated.

The atomic beam stage for these sources shares most design features with many atomic beam systems developed to inject polarized atoms into polarized H or D targets in storage rings, for example at IUCF [28], or at HERMES [29]. Though older atomic beam sources employ electromagnetic sextupoles for atomic beam focusing, all recently developed atomic beam systems have employed NeFeB permanent magnet sextupoles. The latest have improved pole-tip fields up to 1.6 Tesla [30] and are capable in DC operation of focusing up to ~8 x 10^{16} nuclear polarized H atoms/s in two hyperfine states into the ionizer downstream. Further gains of a factor of 2 or 3 can be achieved when the power and gas to the dissociator which produces the atomic beam are pulsed. This increase is enabled by improved vacuum and reduced gas scattering of the slow atoms near the dissociator nozzle.

For ABPIS systems, the most important determining factors for extracted ion current and beam polarization are the type of ionizer used and the time structure

Laboratory	Maximum Intensity (mA)				Normalized Emittance (80%) p mm-mrad	Polarizaton % of Max	Duty Factor Fraction	Dissociator Noz. Temp deg. K	Ionizer Type*	Ref. #
	H-	H+	D-	D+						
Bonn	-	0.05	-	0.05	0.44 (est.)	80(p), 60-80(d)	DC	70	ECR	47
Faure-NAC	-	0.06	-	-	0.21 (25%)	82	DC	30	EB	53
Groningen	-	0.5	-	0.4		60-70(p), 60(d)	DC	35	ECR	18
INR-Moscow	1	-	-	-	1.7 (90%)	85	0.002	77	PI	39
INR-Moscow	-	10	-	-	1 (90%, est.)	80	0.002	77	PI	38
IUCF	1.5	-	1.5	-	1.2 (90%)	80	0.0006	80	PI	40
JINR-Dubna	-	-	-	0.3-0.4		75-85	2 to 3 x 10^{-4}	30	PEI	48
Juelich	0.02	-	in 2002	-	<0.5 (90%)	86	0.01	35	CCB	36
Wisconsin	0.001	-	0.001	-		90	DC	80	CCB	35
Munich	-	-	0.01	0.009	0.5 (100%)	65 (p), 70(d)	DC	80	ECR	34
PSI-Villigen	-	0.3	-	0.3	1.2	75(p), 80(d)	DC	35	ECR	31
RCNP-Osaka	-	0.1	-	0.1	<0.5 (90%)	75(p), 70(d)	DC	30	ECR	32
RIKEN	-	-	-	0.14		80 to 85	DC	35	ECR	49
Saudia Arabia	-	0.023	-	0.02		69 (d)	DC	35	EB	50
TUNL	0.008	0.05	0.008	0.07	0.6 (est.)	75 (p), 80(d)	DC	35	ECR	33
Uppsala							DC	35	ECR	51
* Ionizer Type: EB, Electron Bombardment; ECR, Electron Cyclotron Resonance; CCB, Colliding Cesium Beam; PEI, Penning Ionizer, PI, Plasma Ionizer										

Table 3 – Performance of Atomic Beam Polarized Ion Sources

demanded by the local accelerator. For roughly two-thirds of existing systems, ionization occurs by collisions with fast electrons by the process H + e → H+ + 2e. The oldest electron impact ionization systems employ an electron beam traveling coaxially with the atomic beam inside a strong solenoidal magnetic field, and the extracted H+ (D+) currents range between 25 and 50 μA. More numerous are ionizers which instead cause the electrons to be accelerated by electron cyclotron resonance. Typical ECR ionizers of the passing atomic beam [31,32] provide 50-500μA of polarized H+ (or D+) current, with the extracted ion intensity dependent largely on the injected polarized atomic beam flux and on the aperture allowed for the extracted beam. When needed, as at TUNL [33] or in the recently completed atomic beam source for the Munich tandem accelerator [34], subsequent charge exchange in cesium vapor can provide 5 to 10 μA of H- (or D-) beam with polarizations up to 70% (or 80%) of the theoretical maxima. These polarization values are limited by ionization of unpolarized background atoms by the ECR discharge, whose active ionization volume extends radially beyond the traversing polarized atomic beam.

All of these electron-beam or ECR ionizer systems presently operate as DC sources. The only pulsed source which uses today electron bombardment ionization exists at JINR/Dubna. There a Penning ionizer produces ~400μA of D^{+} beam in 300μs long pulses at 0.1 Hz. [48]

Two atomic beam sources today utilize collisions with fast Cs atoms, H + Cs (30 keV) →H^{-} + Cs^{+}, to produce the desired polarized H^{-} or D^{-} beam. This technique was originated at Wisconsin [35] where a DC system provides 1 µA of output polarized ion current. A pulsed version of this source in operation at COSY /Jülich [36] provides ~20µA of H⁻ current in 20ms long pulses at 0.5 Hz.

Ionization with a fast cesium beam has both advantages and disadvantages. Since the charge-exchange process is highly selective, the output polarized ion beam is not diluted by unpolarized background from water or hydrocarbons. Thus, output beam polarizations $P_z \approx 90\%$, the highest of *any* operating sources, are commonly achieved. However, a major drawback of ionization with a fast neutral cesium beam is the unwanted sputtering when the cesium impinges on interior parts of the source. Considerable design care is needed to minimize the associated reduction in overall source reliability. Thus, pulsed cesium beam operation which reduces the overall cesium flux is more satisfactory than DC operation.

A third ionization method employs the resonant process, $H(pol) + D^{\pm} \rightarrow H^{\pm}(pol) + D$ at ~10 eV, for which the charge-exchange cross sections of $5 \times 10^{-15} cm^2$ (for +) and $\sim 10^{-14} cm^2$ (for -)are among the largest known. This process was first used successfully at INR-Moscow in the mid-1980s in a plasma ionizer for polarized H^+ [37]. In this and other Moscow work, pulsed currents up to 10 (or 1) mA in 200µs bursts at 10 Hz were reported for H^+ (or H^-) [38, 39]. Further advances were then made by utilizing the Budker Institute pulsed H^+ source technology (which was already associated above in Figure 3 with the recent successful pulsed OPPIS source tests at TRIUMF). The new ABPIS with such a plasma ionizer at IUCF [40] has been providing 1.5 mA of H^- and D^- in 200µs pulses at 4 Hz, with beam polarizations exceeding 80% of the theoretical maximum. The plasma injector for this ionizer is also used to provide up to 25mA (peak) of pulsed unpolarized beam for injection into the IUCF Cooler Injector Synchrotron.

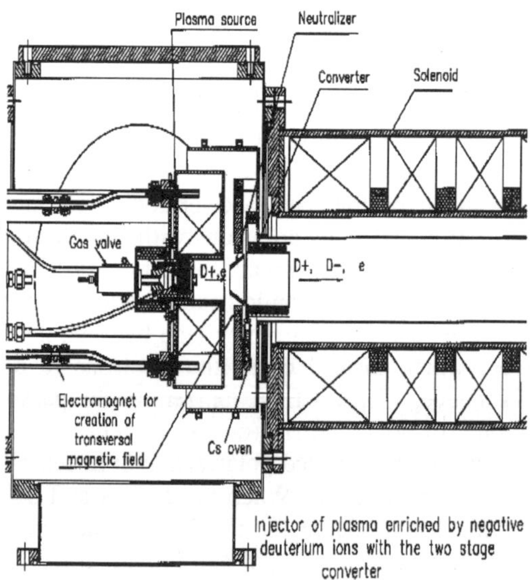

Figure 4 – Schematic of the Moscow D⁻ plasma ionizer with two-stage converter. Polarized H atoms enter the ionization region axially from the right, and polarized H⁻ ions are then extracted at the right and momentum analyzed to form the desired beam.

At this conference we have heard of even further improvement [41], with pulsed polarized H⁻ currents up to 2.5 mA now available. Key to this continuing success is growing detailed understanding of the plasmas used, and growing sophistication of systems developed to control them. Figure 4 illustrates the latest two-stage converter design in which the intense D^+ ion flux emerging from an improved primary source is caused to diverge radially by an axial B-field cusp, so these ions impinge on the interior surface of a conical, cesium-covered electrode. There they are neutralized with approximately 90% efficiency and then are reflected by the shape of the electrode back toward the axis where they collide again with an interior surface, this time that of a heated, cesium-coated molybdenum tube. Ionization efficiencies there of several percent are possible, producing D^- currents up to 90 mA which are then available to ionize polarized H atoms which enter the ionization region. The space charge of the D^- ionizer plasma is compensated by residual D^+ ions which enter this axial region. Also essential is the suppression of fast electrons from the primary D^+ plasma source by using a transverse magnetic field at the entrance of the conical neutralizer. Any such electrons entering the ionization region would otherwise cause substantial loss of the desired negative ions by stripping.

Though these plasma ionizer systems have been extremely successful for providing *pulsed* polarized beams, finding a way to make such systems operate *DC* is still a sizeable challenge because of their very large gas load. At TUNL, we have continued earlier work [42,43] toward developing a DC ionizer system which utilizes a drifting ~1 mA quasi-neutral plasma jet of H^+/e^- from an ECR source, with a measured mean H^+ ion energy of 1.8 eV. This 5 mm diameter jet was recently directed by a 0.1 mT B-field down the axis of a 25 cm long charge-exchange canal filled with D atoms from a dissociator. We found that as the D atom flux increased, the plasma became unstable, which eventually led to complete extinction of the plasma jet [44]. We believe this occurs because the charge exchange replaces an H^+ with a D^+ of lower temperature, leading to an immediate space-charge imbalance in the plasma jet, and to the jet's demise. The earlier key lesson of the Lamb-shift source was apparently not learned. Space-charge resulting from charge-exchange processes in such very low energy ion beams can severely limit their flux density.

3   Future Expectations

Based on past successes and recent developments, it seems likely that H and D polarized beam intensities will continue to improve. Progress is likely to be most rapid for *pulsed* beams. Personally, I believe that pulsed polarized H- currents of 10mA will likely become routinely available, at least at RHIC, within a couple of years. This will only require implementation and refinement of new systems which have already been shown to be successful in early TRIUMF tests.

Those wanting DC $H^{\pm}$ beams should look carefully also at the OPPIS systems which have been developed at TRIUMF. Before ABPIS systems can match their performance, we need better to understand the DC plasma ionizer instabilities.

Indeed, it is not clear yet at all that impressive recent INR-Moscow successes with pulsed plasma ionizer systems for ABPIS sources can be extended to DC operation.

More work will be needed in OPPIS systems to address the need of some laboratories for maximal tensor and vector polarization for D beams. Earlier work at KEK showed [45] that an OPPIS source can produce polarized D beams with useable polarizations if a second polarized electron is attached to the D atom in the Na ionizer. However, development of the much wider range of vector and tensor polarizations available from ABPIS sources is likely possible for an OPPIS system [46], but will require significant further research.

References

1. K. Hatanaka, T. Nakano, K. Imai, and H. Ejiri, Proceedings of SPIN 2000, the 14th International Spin Physics Symposium, Osaka, Oct. 2000, AIP Conf. Proc. **570** (2001).
2. A. Gute, S. Lorenz, and E. Steffens, eds., Proceedings of the International Workshop on Polarized Sources and Targets, Erlangen-Nürnberg, Sept.-Oct. 1999 (Druckerei Langenfelder, Erlangen, 1999, ISBN 3-00-005510-X).
3. R.J. Holt, and M.A. Miller, eds., Proceedings of the Seventh International Workshop on Polarized Gas Targets and Polarized Beams, Urbana, IL, Aug. 1997, AIP Conf. Proc. **421** (1998).
4. H. Paetz gen. Schieck, and L. Sydow,, eds., Proceedings of International Workshop on Polarized Beams and Polarized Gas Targets, Cologne, June 1995 (World Scientific, Singapore, 1996).
5. T.B. Clegg, Proc. of Conf. on Polarized Proton Ion Sources, Vancouver, 1983, eds. G. Roy and P. Schmor, AIP Conf. Proc., **117**, 63 (1984).
6. Y. Tagishi and J. Sanada, Nucl. Instr. Meth. **114**, 411 (1979).
7. V. Bechtold, *et al.*, Nucl. Instr. Meth., 150, 407 (1978).
8. B.L. Donnally and W. Sawyer, Phys. Rev. Lett. **15**, 439 (1965).
9. A. Isoya, *et al.*, Proc. of 6th Intern. Symp. on Polarization Phenomena in Nuclear Physics, Osaka, 1985,eds. M. Kondo, *et al.*, J. of Physical Soc. of Japan, Supplement **55**, 1054 (1986).
10. J.L. McKibben, G.P. Lawrence, and G.G. Ohlsen, Phys. Rev. Lett., **20**, 1180 (1968).
11. P.G. Sona, Energie Nucleaire, **14**, 295 (1967).
12. T. B. Clegg, G. R. Plattner, and W. Haeberli, Nucl. Instr. Meth. **62**, 343 (1968).
13. G.G. Ohlsen, *et al.*, Phys. Rev. Lett. **27**, 599 (1971).
14. A.J. Mendez, *et al.*,, Rev. Scien. Instrum., **67** (9), 3073 (1996).
15. C.D. Roper, Ph. D. Thesis, Duke University, 1998.
16. R. Engels, *et al.,* contribution to this workshop.
17. H. Kolster, in ref. 2, p. 172.
18. H.R. Kremers, *et al.*, in ref.2, p. 82; ref. 3, p. 507, and private communication.
19. A.N. Zelenski, in ref. 4, p. 111.

20. C.D.P. Levy, in ref. 4, p. 120.
21. Y. Mori, *et al.*, Rev. Scien. Instrum. **71**, 1237 (2000);
22. A.N. Zelenski, *et al.*, in ref. 2, p. 94.
23. A.N. Zelenski, *et al.*, in ref. 3, p. 372.
24. S.A. Payne, *et al.*, J. Appl. Phys. 66 (3), 1, (August 1989); M.D. Perry, *et al.*, Laser Focus World , 29 (9) (1993) p. 85.
25. A.N. Zelenski, in ref. 1, p. 179.
26. T. B. Clegg, in ref. 4, p. 155.
27. A. Abragam and J.M. Winter, Phys. Rev. Lett. **1**, 374 (1958).
28. M.A. Ross, *et al.*, Nucl. Instrum. Meth. **A344**, 307 (1994).
29. H. Kolster for the HERMES Collaboration, in ref. 3, p.162.
30. R. Vasilliev *et al.*, Rev. Sci. Instrum. **71** (2000) 3331.
31. L. Friedrich, E. Huttel, and P.A. Schmelzbach, Nucl. Instrum. Meth. **A272**, 906 (1988); P.A. Schmelzbach, Am. Inst. Physics Conf. Proc. **293**, 65 (1994).
32. K. Hatanaka, K Takahisa, and H. Tamura, in ref. 3, p. 352.
33. T.B. Clegg, *et al.*, Nucl. Instrum. Meth. **A357,** 200(1995).
34. R. Hertenberger, *et al.*, in ref. 1, p. 825.
35. T. Wise, A. Roberts, and W. Haeberli, Nucl. Instrum. Meth. **336**, 906 (1993).
36. R. Gebel, *et al.*, contribution to this workshop.
37. A.S. Belov, *et al.,* Nucl. Instrum. Meth. **A255**, 442 (1987).
38. A.S. Belov, *et al.*, Rev. Sci. Instrum. **67** 1293, (1996); also in ref. 4, p. 218.
39. A.S. Belov, in ref. 1, p. 835.
40. V.P. Derenchuk and A.S. Belov, contribution to this workshop.
41. A.S. Belov, contribution to this workshop.
42. S. Lemaitre, T.B. Clegg, and W.B. Hooke, in ref. 2, p. 60.
43. S. Lemaitre and T.B. Clegg, Proc. European Particle Accel. Conf, EPAC2000, Vienna, Austria, p. 1613.
<http://accelconf.web.cern.ch/accelconf/e00/PAPERS/THP3A02.pdf>
44. S. Lemaitre and T.B. Clegg, to be published.
45. M. Kinsho, *et al.*, in ref. 4, p. 126.
46. T.B. Clegg, in ref. 3, p. 336.
47. D. Eversheim, et al., Phys Lett B256, 11 (1991).
48. V. Angelov, *et al.*, in ref. 2, p. 168.
49. H. Okamura, *et al.*, Am. Inst. Physics Conf. Proc. **293**, 84 (1994); also http://iris.riken.go.jp/pis/
50. A.A. Naqvi, in ref. 4, p. 175.
51. 52. G. Clausnitzer, *et al.*, Am. Inst. Physics Conf. Proc. **293**, 80 (1994).
52. D.M. van Rooyen, *et al.*, in ref. 2. p. 78.

AN OPTICALLY-PUMPED POLARIZED H⁻ ION SOURCE FOR RHIC SPIN PHYSICS

A.ZELENSKI[1,2], J.ALESSI[1], B.BRISCOE[1], G.DUTTO[3], H.HUANG[1], A.KPONOU[1], S.KOKHANOVSKI[2], V.KLENOV[2], A.LEHRACH[1], P.LEVY[3], V.LODESTRO[1], Y.MORI[4], M.OKAMURA[5], D.RAPARIA[1], J.RITTER[1], T.TAKEUCHI[5] , G.WIGHT[3], V.ZOUBETS[2].

[1]Brookhaven National Laboratory, Upton, New York 11973-5000, USA

[2]INR, Moscow, Russia

[3]TRIUMF, Vancouver, British Columbia, Canada

[4]KEK, Japan

[5]RIKEN, Japan

A new Optically-Pumped Polarized H⁻ Ion Source (OPPIS) was developed for the RHIC polarization program and successfully used for the first polarized beam commissioning at RHIC. The OPPIS produces in excess of 1.0 mA H⁻ ion current at about 80 % polarization. An ECR primary proton source development and a new 29 GHz microwave power supply are described. A new type of sodium-jet ionizer cell is biased to - 32 kV to produce a 35 keV polarized beam for injection to the RFQ. Higher current and higher polarization were also obtained with the biased jet-cell in comparison with an older oven-type ionizer cell.

1 Introduction

Collider experiments at RHIC, using polarized protons at up to $\sqrt{S} = 500$ GeV energy and $2 \cdot 10^{32}$ cm^{-2} sec^{-1} luminosity, will provide unique complimentary information on proton structure, QCD and electroweak interaction of quarks [1, 2]. To meet the required luminosity the intensity must be $2 \cdot 10^{11}$ protons per bunch in RHIC. The intensity of the polarized H⁻ ion source has to be about $9 \cdot 10^{11}$ H⁻ ions/pulse, which corresponds to a 150 mA-μs integral intensity (for example, a 300 μs pulsed current of a 0.5 mA). Such intensity can be easily produced in the OPPIS, where up to 2.0 mA polarized ion current was obtained in a dc operation and up to 10.0 mA in a pulsed operation [3, 4].

The OPPIS technique is based on spin-transfer collisions between a primary proton or atomic hydrogen beam of a few keV energy and optically-pumped alkali-metal vapors. The RHIC OPPIS was developed in a collaboration between BNL, KEK, INR Moscow and TRIUMF. It is based on the KEK OPPIS, which was successfully used for production of a vector polarized D⁻ beam at KEK, and upgraded at TRIUMF for high current H⁻ operation. The new OPPIS routinely produces 0.5 mA current in a 300 μs pulse, which can be expanded to 500 μs - about the maximum linac pulse duration. This intensity is 30 times higher than the typical intensity of the old BNL atomic beam source, and it is sufficient to produce the required RHIC bunch intensity with a single source pulse. Polarized H⁻ ions are produced in the OPPIS at 35 keV beam energy. The beam is accelerated to 200 MeV with an RFQ and linac for strip-injection to the Booster. About 50% of the OPPIS beam intensity can be accelerated to 200 MeV. The 300 μs H⁻ ion pulse, which contains about $4 \cdot 10^{11}$ polarized protons, is captured in a single Booster bunch, which is accelerated to a kinetic energy of 1.5 GeV in the Booster, then transferred to the AGS, where it accelerated to 25 GeV for injection into RHIC.

The initial longitudinal polarization of the beam in OPPIS is converted to transverse polarization as it passes through two bending magnets (see Fig.1). The second 47.4° bending magnet, which is pulsed, switches linac injection between polarized and unpolarized high intensity (up to 100 mA) H⁻ ion beams, and either beam can be accelerated in the same RFQ on a pulse-to-pulse basis. A pulsed focusing solenoid in front of the RFQ is tuned for the optimum transmission of either beam. This solenoid rotates the polarization direction about 420°, but still keeps it in the transverse plane. A final polarization alignment to the vertical direction, before injection into the linac, can be made using the spin-rotator solenoid in the 750 keV beam transport line.

The duration of the AGS cycle for polarized beam operation is 3 seconds, while OPPIS operates at 1 Hz. Hence, using a pulsed bending magnet in the high energy beam transport line, two pulses can be directed to the 200 MeV p-Carbon polarimeter to monitor beam polarization. Polarization alignment to the vertical plane is done using this polarimeter and the spin-rotator solenoid in the 750 keV line.

2   ECR Primary Proton Source

The OPPIS current is mainly determined by the ECR primary proton source, since the optimal neutralization efficiency in the optically-pumped Rb vapor is 50-70%, the sodium ionizer efficiency is constant (about 9% at 3.0-4.0 keV beam energy) and the ionizer diameter is limited by emittance specification. Obtaining a high polarization requires a magnetic field, in excess of 25 kG, in the optically pumped cell and ion extraction region. In the KEK OPPIS the protons were produced in a 6.4 kG resonance field and extracted in a 27 kG field. With the 28 GHz microwave power supply at TRIUMF, the resonance magnetic field is 10 kG. This gives a factor of 2-3 current gain, other conditions being similar. In the BNL OPPIS, the ECR ion source operates at a 29 GHz frequency. It is situated in the superconducting solenoid having three separately adjusted coils, which produce a 10 kG resonant field for ECR source operation and up to 27 kG flat top field for the optically pumped Rb vapor cell. The 29 GHz microwave power of

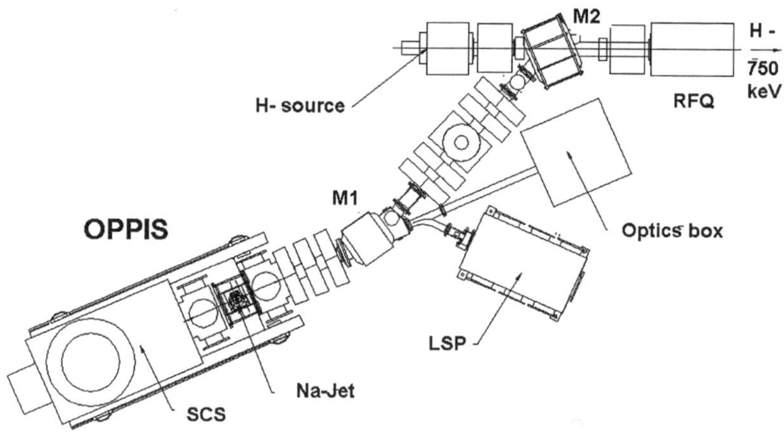

Fig. 1: A schematic layout of the polarized RHIC injector. ECR-electron-cyclotron resonance proton source; SCS- superconducting solenoid; Na-jet - sodium-jet ionizer cell; LSP- Lamb-shift polarimeter; M1, M2-dipole bending magnets.

about 0.8 kW, required for saturating the proton current yield, is produced by a CW extended interaction oscillator (CPI VKQ2453L1 maximum power up to 1.2 kW). Similar to the TRIUMF OPPIS prototype, the BNL OPPIS is practically a dc source, except that it has 35 keV pulsed beam acceleration, and optical pumping is by a pulsed laser. Therefore, the current pulses are very stable and flat, and pulse duration can be extended if necessary. The microwave power can be pulsed by anode voltage modulation (a Behlke HTS-300 switch is used for this purpose). Pulsed ECR source operation was tested at 1 Hz repetition rate and a pulse duration of 10 -100 ms. It was observed that about 10-20% higher current can be obtained in a pulsed operation at the start of the pulse with about 50% less hydrogen supply to the discharge tube. Perhaps hydrogen desorption from the quartz liner produces sufficient gas density at the start of the pulse, which allows to reduce the hydrogen load to the vacuum pumps and improve the residual vacuum in the Rb cell. This also reduces the depolarization due to neutralization in the residual gas.

The proton beam produces positive Rb^+ ions when it captures electrons from Rb atoms. These ions are radially confined by the high solenoid field, and can travel only longitudinally out of the cell. Rb ion flow to the extraction grid of the ECR source is suppressed by a positive voltage of about a 100 V which is applied to the third grid, and Rb ions travel downstream to negatively biased deflecting plates. In the high current source, the primary proton current is about 80 mA, and when 50% of the beam is neutralized a 40 mA Rb ion beam is produced in dc operation. This ionization causes more Rb losses than thermal vapor diffusion out of the source. In pulsed operation the average current is greatly reduced, and Rb consumption is reduced.

3 Sodium-Jet Ionizer Cell

The neutral 3 keV atomic hydrogen beam enters the ionizer cell, and H$^-$ ions produced in the cell are then accelerated to 35.0 keV (which is required for injection to RFQ) by biasing the ionizer to -32.0 kV. A large Na cell aperture is essential for this purpose because the neutral beam, collimated to 2.0 cm in diameter before the cell, should not touch the biased cell parts, since secondary emission will cause sparking. A new jet-type ionizer cell with transverse sodium flow was developed to allow large apertures (see Fig. 2) [4]. The reservoir is loaded with 150 g of sodium metal through the loading port at the top of the cell.

The operating temperature of the reservoir is about 480° C. At this temperature the sodium vapor density is about 10^{17} atoms/cm^3. The vapor is delivered through a hot transport line to the nozzle which produces a horizontal vapor jet having an effective thickness of about $5 \cdot 10^{14}$ atoms/cm^2, sufficient for H$^-$ yield saturation. The nozzle slit is 3 mm wide and 20 mm tall. The transport tube and nozzle are heated to 480° C. The sodium vapor condenses on the collector walls, which are cooled to 120° C (above the melting point of sodium) by hot water circulation. Liquid sodium flows down the return tube and back to reservoir. The return tube temperature is kept at 150° C by an attached cooling line with controlled hot water flow. Backstreaming vapor flow through the return line is negligible due to the low conductance at 150° C. Sodium in the jet-cell circulates along the path reservoir—nozzle—collector—return-line-reservoir and the system provides continuous, stable operation for hundreds of hours with 150 g sodium. While a large amount of sodium is circulated inside the cell, the sodium flow outside is much less than with the oven-type cell.

The entire ionizer assembly and the solenoid magnet are electrically isolated from the rest of the source by 50 mm thick Delrin flanges. The beam is accelerated to 35 keV in a two-gap extraction system of 25 mm and 75 mm lengths. The extraction voltage applied across the first gap is 4.0 kV and across the second gap is 28.0 kV. The immediate acceleration reduces polarized current losses and improves polarization, since the energy for the spin-transfer collisions can be set optimal for efficient polarization transfer and ionization.

4   Experimental Results

The final tests at TRIUMF were performed with the TRIUMF Oxford Instruments superconducting solenoid. The proton polarization was optimized at 3 keV using a Lamb-shift polarimeter, and measured with a nuclear scattering polarimeter based on the $Li^6(\,p,\,He^3)He^4$ reaction, after acceleration to 300 keV. At low Rb thickness, a polarization of 90% was achieved in dc mode with a cw Ti:sapphire laser. At higher Rb density, polarization drops, but the use of a pulsed Cr:LiSAF laser restores the polarization to 85% at 1.6 mA H^- ion beam intensity (see Fig. 2). A long pulse, flashlamp pumped Cr:LiSAF laser had been developed for the BNL OPPIS. This laser produces in excess of 1.0 kW power in a 400 μs pulse.

The OPPIS was shipped to BNL in September of 1999, installed at the 200 MeV linac injector, and used for the RHIC spin commissioning in August-September of 2000. The OPPIS worked continuously and reliably during the three-week final run. So far, a maximum current of 1.1 mA has been obtained, as measured at 35 keV in a Faraday cup two meters from the source exit. About 50% of this current was accelerated to 200 MeV (see Fig. 3). The source current is about 70% of the current obtained in the tests with the TRIUMF solenoid. The Oxford Instruments solenoid has a cold yoke, in contrast with the KEK solenoid, which has a large hole in the room temperature yoke. This disturbs the magnetic field symmetry and introduces a transverse field, which mis-steers the low energy proton beam. The displacement was observed by direct measurements of the atomic H beam profile at the entrance of the ionizer cell.

The Lamb-shift polarimeter was used for initial polarization measurements and polarization optimization. For these measurements, the H^- beam was converted to a proton beam in a pulsed helium ionizer cell. In another mode of operation, the polarization of the protons produced together with the H^- ions in the sodium ionizer cell was measured [5]. Finally, the polarization was measured in the inclusive p-Carbon polarimeter at 200 MeV, which was recently calibrated by comparison with elastic proton-deutron scattering. The proton beam out of the ECR source was determined to have two components: one originating from the dissociation of H^2, and the other from H^{2+}. The former comprises 80% of the output current and has an energy of 3 KeV. The smaller component has a kinetic energy of 1.5 keV. The polarization of the weaker component is about 20%, hence there is a dilution of the polarization of the mixed beam. Molecular ion production in the ECR source can be reduced to less than 5% by pulsing the ECR, by magnetic field shape tuning, and by controlling gas composition. So far about 82 +/- 2% polarization has been achieved at 200 MeV. Work is in progress to identify polarization losses in the source, in LEBT, and in the spin rotating magnets.

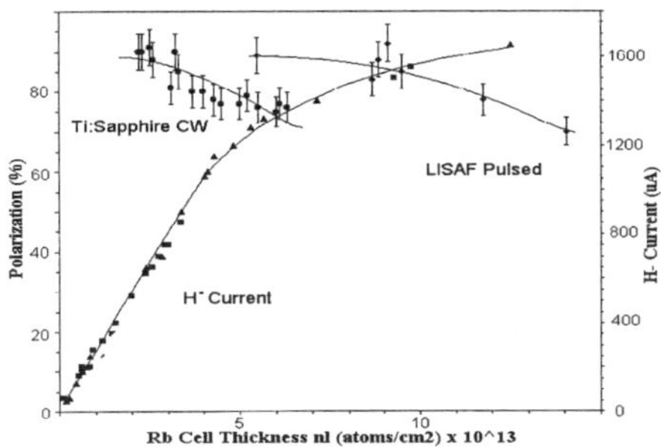

Fig. 2: The H⁻ beam current and polarization vs the optically-pumped Rb vapor thickness.

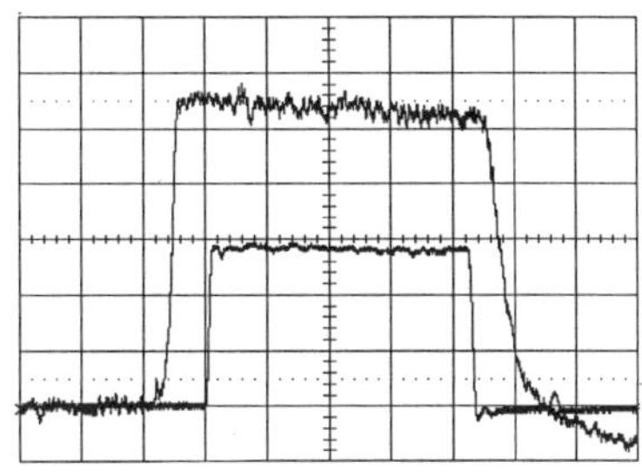

Fig. 3: The polarized H⁻ ion beam current pulse: top trace – the OPPIS beam current at 35 keV energy; bottom trace – the 200 MeV beam current. A vertical scale is 0.2 mA/div, a horizontal scale is 100 μs/div.

Acknowledgments

We acknowledge the important contributions of E. Stephenson from IUCF, and K.Kurita from the RIKEN-BNL Research Center, to the 200 MeV Polarimeter calibration experiment.

References

1. G.Bunce et al., "Polarized Protons at RHIC", Particle World, 3, p.1, (1992).
2. "Prospects of the Spin Physics at HERA", DESY Report 200-95, (1995).
3. A.Zelenski et al., "The TRIUMF High Current dc OPPIS", Proc. 1995 IEEE PAC,Dallas, p.864, (1995).
4. A.Zelenski, "Polarized Ion Sources for High Energy Accelerators and Colliders", SPIN-2000, Osaka, AIP Conf. Proc. 570, p.179, (2001).
5. A.Zelenski et al., NIM A, **334,** p.285, (1993).

NEW DEVELOPMENTS AT THE POLARIZED ION SOURCE OF COSY-JÜLICH

O. FELDEN, R. GEBEL, M. GLENDE, R. MAIER, P. VON ROSSEN

IKP, Leo Brandt Strasse, Forschungszentrum Jülich, 52428 Jülich, Germany

E-mail: r.gebel@fz-juelich.de

P.D. EVERSHEIM

ISKP, Nussallee 14-16, Universität Bonn, 53115 Bonn, Germany

The polarized ion source, originally built by the universities of Bonn, Erlangen and Köln[1,3], is based on the colliding beams principle[2]. The polarized source is in routine operation at the cooler synchrotron COSY[5] after a redesign of major components. Inside COSY a polarisation of 70 % at maximum momentum was reached[4]. In recent experimental runs up to 2.5×10^{10} protons with a polarization of 86 % were stored at the injection momentum of 295 MeV/c in the cooler synchrotron. The intensity of the accelerated beam in the cooler synchrotron on an internal target exceeded 3.5 mA at a momentum of 3200 MeV/c.

Introduction

In the year 2000 polarized beams were used in two blocks of about 3000 hours total beam time for the experiments EDDA and TOF. The total beam time in 2001 sums up to 2600 hours. For the first time the polarized H^- beam intensity exceeded 1 μA after cyclotron extraction. At the same time the number of the polarized protons in COSY[6] reached values up to 1.5×10^{10} at maximum energy with a polarisation of about 70%. A large number of significant modifications resulted in an operation with a reliability hitherto inaccessible.

Pulsed operation

This fundamental improvement was a consequence of a complete redesign of the original cesium ionizer[9] and a new pulsed operation that had been developed with pulse widths of about 20 ms to the point of routine operation over months. The electrically pulsed operation[11] of a cesium ionizer was already demonstrated in 1998[10]. The new cesium ionizer was set in operation for the first time in spring 2000. An additional pulsed electrode in front of the tungsten ionizer, that works similar to a grid of a triode, and an air cooling circuit were integrated in the new set-up. The unit for the cesium ionizer is shown in figure (2). The cesium vapour is fed through a tube to the porous tungsten

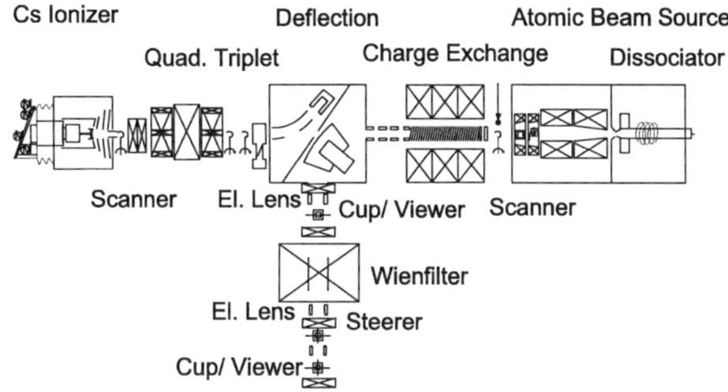

Figure 1. Schematic view of the COSY polarized ion source.

ionizer. The ionizer surface is operated at a temperature of typically 1050°C. The cooling system is able to decrease rapidly the Cs reservoir temperature from 400°C to 40°C in about 6 minutes, an essential prerequisite for efficient operation. The cooling capacity is sufficient to operate the Cs-reservoir down to 40°C with the thermal ionizer having its full operational temperature.

For the pulsed operation the temperature of the cesium reservoir had to be precisely controlled in the range from 35°C to 80°C, depending on the characteristics of the tungsten contact ionizer and the desired intensity. In the example shown in figure (3) the temperature was increased from 55°C to 65°C in steps of 2.5° until the pulse shape reached the desired 20 ms pulse length. Figure (3) (right) shows examples for the variation of the pulse shape and height depending on the pulsed extraction potential for a given setting of the ionizer. The voltage of the pulsed electrode was varied from 4 kV to 8 kV relative to the beam potential of 45 kV. Over 12 mA Cs^+ peak beam intensity was extracted. The full width half maximum of the highest beam

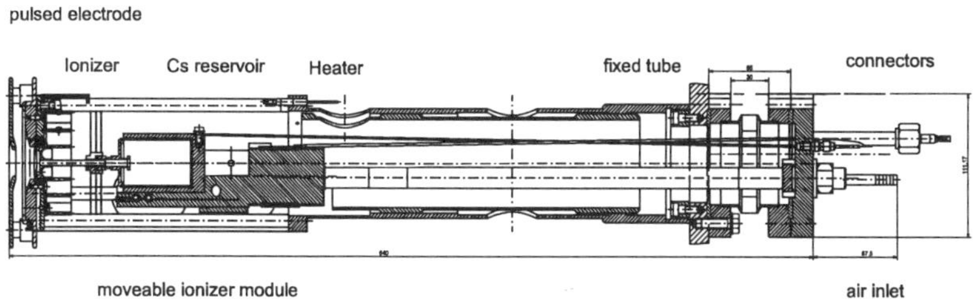

Figure 2. The redesigned cesium ionizer

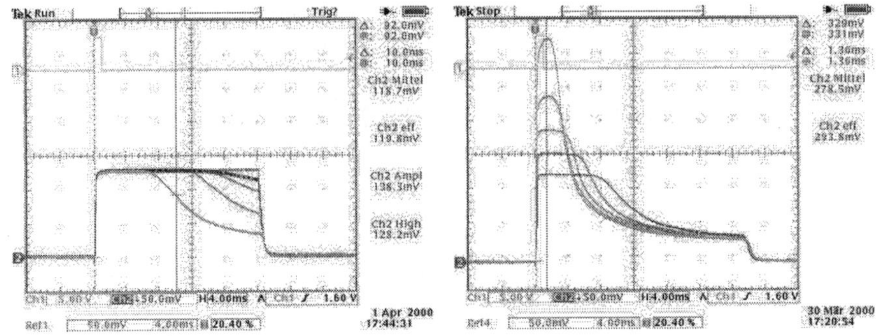

Figure 3. Left: The temperature dependance of the cesium pulse shape. Trace 2 shows the emitted Cs$^+$ current. The pulse width is 20 ms. Right: The voltage dependance of the cesium pulse shape. The pulse width is set to 25 ms. The vertical scale for the current monitor is 2 mA per division. Trace 1 is the trigger signal.

pulse is 4 ms. For an extracted Cs$^+$ current of 5 mA the length and the shape of the Cs$^+$ pulse was closely studied by varying the temperature of the cesium reservoir. Operation takes place at about 0.5 Hz to ensure a steady equilibrium for the delicate processes taking place on the tungsten ionizer surface. The pulsed operation allowed to reduce the damaging effect of the Cs beam by nearly two orders of magnitude. This allowed an operation of 9 weeks without servicing of the cesium ionizer or other parts of the source.

The startup procedure of the cesium beam part is simplified by a set of beam diagnostic elements (see figure (1)). Figure (4) (right) gives an example for online monitoring at the polarized source. The cesium beam is monitored at the power supplies (Trace 3), including the load on the extraction (Trace 4). The beam is neutralized and the remaining charged part is dumped in a faraday cup (Trace 2).

Four beam profile scanner modules,shown in figure (1), allow the online control of the cesium beam envelopes. The scanner modules were described at the Erlangen workshop[8]. For the pulsed Cs beam the scanners are operated by a microcontroller system. This system provides a readout of up to four scanner positions, the integrated charge per pulse and also synchronize the positioning of the wire scanner to the timing of the pulsed ionizer. The scanners for the charged cesium beam are identical. They scan the beam over a width of 60 mm in both directions during one revolution. The fourth scanner for the neutral Cs beam scans over 40 mm. The scan is taken in vertical direction from bottom to top and in the horizontal direction from right to left. Figure (4)

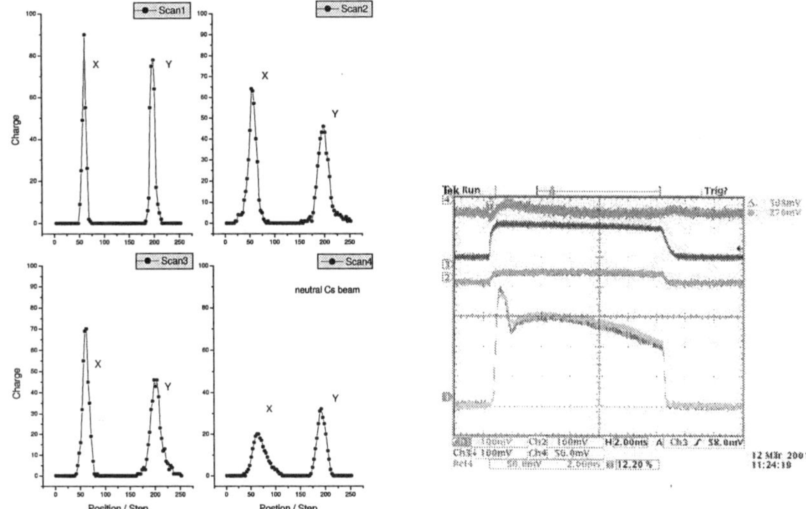

Figure 4. Left: Beam profiles of the pulsed Cs beam taken at four different positions. The Scans 1 to 3 are taken with a resolution of 0.47 mm per step. The neutral Cs beam is scanned with 0.31 mm per step. Right: Trace 1 shows the achievable H^- intensity measured at the first cup in the source beam line. The scale from bottom to top is 100 μA.

(left) shows the result of the measurement of horizontal and vertical profiles at the four positions. The online acquired profiles give the size and the beam shapes. The ca. 5 mm diameter beam, at a distance of 0.25 m after the ionizer, is imaged to a beam size of ca. 9 mm at the entrance to the atomic beam part 3.5 m downstream. Profiles taken at different positions before and behind the quadrupole triplet allow to determine the alignment and the emittance of the cesium beam. This provides reliable information to judge the performance of the thermal ionizer module. Data taken with this system was also used to determine the parameters of the quadrupole system and to optimize the geometry of the cesium beam line for high transmission. A set of correcting dipoles have been added to remove residual beam displacements in the charge exchange region.

Conclusion

All original parts in the cesium beam section, figure (1), have been replaced by improved components. The new parts have been optimized in accordance with the operational experience. The performance of the neutralizer has been advanced, thus reducing the loss of cesium and the contamination of the nearby sections. A new water cooled deflector chamber was put into operation. Its lower deflection angle for Cs^+ and the optimization of the deflector removed the former limitation of the maximum cesium beam momentum allowing to reduce the geometric emittance of the cesium beam.

Due to the improved reliability and quality of the cesium beam the charge exchange and extraction system will be further optimized. Figure (4) shows the peak current measured at the the first cup in beam line to the injector cyclotron to demonstrate the potential of the new source. The beam was extracted at a potential of 4.5 keV and magnetically deflected by 90° to the cup, where an intensity of 30 μA was reached.

Acknowledgments

We are grateful for the support of the ISKP of the University Bonn in machining parts for the improved polarized ion source. The authors like to express their gratitude to W.Haeberli for his interest and helpful discussions.

References

1. P.D. Eversheim *et al*, Proceedings of Polarized Beams and Polarized Gas Targets, Köln; 224,1995.
2. W. Haeberli, *Nucl. Instrum. Methods* **62**, 355 (1968).
3. R. Weidmann *et al*, *Rev. Sci. Instrum.* **67 p.II**, 1357 (1996).
4. A. Lehrach *et al.*, *Nucl. Instrum. Methods* A **439**, 26 (2000).
5. D. Prasuhn *et al.*, *Nucl. Instrum. Methods* A **441**, 167 (2000).
6. D. Prasuhn *et al.*, Proceedings of PAC2001 Chicago, 2001, to be published.
7. R. Gebel *et al*, Proceedings of Polarized Gas Targets and Polarized Beams, Urbana; 503, 1997.
8. R. Gebel *et al*, Proceedings of Polarized Sources and Targets, Erlangen; 86, 1999.
9. S. Lemaitre *et al.*, *Nucl. Instrum. Methods* A **408**, 345 (1998).
10. M. Eggert *et al*, Proceedings of Polarized Sources and Targets, Erlangen; 90,1999.
11. M. Eggert *et al.*, *Nucl. Instrum. Methods* A **453**, 514 (2000).

DEVELOPMENT OF POLARIZED HYDROGEN ION SOURCE WITH RESONANT CHARGE-EXCHANGE PLASMA IONIZER AT INR, MOSCOW

A.S. BELOV, L. P. NETCHAEVA, A.V. TURBABIN, AND G.A.VASIL'EV

Institute for Nuclear Research of Russian Academy of Sciences, Moscow, 117312, Russia

Density of unpolarized negative ions in a charge-exchange region of a plasma ionizer of polarized ion source of INR, Moscow has been increased significantly due to development of a two-stage converter of plasma particles into negative ions. Unpolarized D^- ion current with a peak intensity of 45 mA has been obtained from the plasma ionizer with the two-stage converter. Respectively, polarized H^- ion current of 2.5 mA peak with pulse duration of 150 μs at 5 Hz rep. rate has been obtained. Further increase of unpolarized negative ion density in the charge-exchange region has been achieved due to improvements in the plasma source design. 90 mA of unpolarized D^- ion current and 150 mA of unpolarized H^- ion current have been extracted recently from the ionizer. Problems that are necessary to overcome for corresponding increase of polarized ion beam intensity are discussed. It is expected that 4 mA of polarized H^- ion current will be obtained from the source after implementation of the improvements described.

1   Introduction

A distinctive feature of the atomic-beam type polarized hydrogen ion source developed at Institute of Nuclear Research, Moscow is the use of the resonant charge-exchange reaction for direct conversion of polarized thermal hydrogen atoms into polarized positive or negative ions:

$$H^\circ\uparrow + D^+ \rightarrow H^+\uparrow + D^\circ \qquad (1)$$
$$H^\circ\uparrow + D^- \rightarrow H^-\uparrow + D^\circ \qquad (2)$$

For a collision energy of 10 eV the cross-section for reaction (1) is $5 \cdot 10^{-15}$ cm^2 and for the reaction (2) is 10^{-14} cm^2 [1]. Low colliding energy is achieved by using a deuterium plasma for production and transport of positive and negative deuterium ions to the colliding area.

Ionizer scheme with 1 - 2 keV unpolarized negative deuterium ions colliding with polarized atomic hydrogen beam (reaction (2)) has been proposed by W. Haeberli [2]. The concept of using the resonant charge-exchange reaction (2) in a deuterium plasma has been mentioned in a paper of Yu. Bel'chenko et al. [3]. Later a ring-magnetron ionizer based on the same reaction has been proposed by J.G. Alessi et al., [4].

A pulsed atomic beam-type polarized proton source with the resonant charge - exchange plasma ionizer has been designed and tested for the first time at INR, Moscow [5-6]. A peak polarized proton current of 6 mA with polarization $p_z = 0.76$ has been obtained from the source [6]. The respective D^+ ion current was about 500 mA.

Later the polarized proton beam parameters were improved by using, for the first time, of storage of polarized hydrogen atoms in the charge-exchange region of the polarized proton source [7-8]. With 45 mA of unpolarized D^+ ion current passing through the storage cell the polarized proton beam intensity was increased to 11 mA [8]. Polarization of the beam was measured to be $p_z = 0.8$.

It was proposed to use the resonant charge-exchange ionizer developed at INR for polarized H^- ion production [6]. The first test of the ionizer has been made at INR with a low density of D^- ions in the plasma [9]. Then an external surface-plasma converter was developed to enrich the plasma in the ionizer with negative ions. As a result the unpolarized D^- ion current of 1.2 mA and polarized H^- ion current of 0.15 mA have been obtained from the source [10].

A significant increase of density of unpolarized D^- ions in plasma was achieved due to the placement of a cylindrical converter at the ionizer solenoid entrance, use of transversal magnetic field for decrease of electron density in plasma, heating of the converter to 400°C temperature and optimization of geometry of the converter. Polarized ion beam with current of 1 mA and $p_z = 0.86$ has been obtained for the first time due to these improvements [11].

As a result IUCF decided to build a similar source for their new Cooler Injector Synchrotron [12]. In this paper we describe the recent development of the new two-stage converter and the plasma injector producing dense plasma with negative ions (2.1) and discuss possible modifications of the ionizer (2.2).

2 Polarized negative hydrogen ion source development

The source and atomic hydrogen beam parameters produced have been described in detail in refs. [6, 10,11].

2.1 Plasma injector

Unpolarized D^- ion current has been increased due to improvements made with the plasma injector. The new two-stage converter consists of a neutralizer cone and a converter cylinder and works as follows: deuterium plasma jet goes from the plasma source to the internal surface of the neutralizer guided by magnetic field of the plasma source coil and electromagnet of the converter. In collisions with the neutralizer surface the plasma ions are neutralized with high efficiency (about 90%). The neutral atoms reflected from the neutralizer have a temperature of several eV. The shape of the neutralizer cone was chosen to direct most of the reflected particles to the internal surface of the converter cylinder. These hot atoms are converted into negative ions in collisions with the molybdenum converter cylinder. Efficiency of the conversion is increased due to the decrease of the converter surface work function by Cs atoms that are driven up into the system by a Cs oven. The converter cylinder is heated up to 400°C to prevent absorption of molecules of residual gas by the converter surface. Efficiency of conversion of hot neutral atoms into negative ions according to direct measurements of such a process [13] should be on the order of 10% for neutral atoms energy of several eV.

With the two-stage converter described, 45 mA of D^- ion current has been extracted from the source. It was found that polarized H^- current is still proportional to the D^- ion current showing no significant destruction of polarized H^- ions in plasma. 2.5 mA peak polarized H^- current has been obtained as a result of the development [14].

Further increase of the D^- ion density has been obtained due to improvements of the plasma source construction. The new source has a longer discharge channel and a larger volume near the anode area in comparison with the plasma source of previous design. This increases gas capacity of the plasma source working with pulsed gas supply and suppresses instability in the plasma flux developed during the discharge pulse. This instability restricted the pulse length and forced increase of D_2 gas flux per pulse. This restriction has been eliminated by the new design of the plasma source.

Design of a plasma source cathode was also changed. Magnetic field in the cathode region has been attenuated by an iron cylinder around the cathode acting as a magnetic shield. The shape of the internal surface of the cathode has been chosen by such way to get magnetic field lines to be perpendicular to the cathode internal surface. This leads to more homogeneous movements of the arc-discharge cathode spots on the cathode surface and increases the cold cathode lifetime. Continuous operation of the plasma source was increased two times to 300 hrs at discharge pulse length of 250 μs and repetition rate of 10 Hz.

Intensity of the D^- ion current has been increased to 90 mA due to improvements made with the plasma source and due to the two-stage converter and 30 keV extraction potential. When molecular hydrogen was injected into the plasma source instead of deuterium unpolarized H^- ion current of 150 mA has been obtained. The duration of ion pulses is 200 μs at half height with repetition rate up to 10 Hz. Also, gaseous neutralization of space charge of the intense D^- (H^-) ion beams was used to reduce divergence of the beams. The gaseous neutralization needs residual vacuum pressure of order of 10^{-4} Torr. Unfortunately, it cannot be used for operation of the source with polarized beam because of attenuation of atomic hydrogen beam by scattering with residual gas molecules.

2.2 *Discussion of future modifications of the ionizer*

In order to obtain a respective increase in polarized H^- ion beam intensity in accordance with an increase of unpolarized negative ion current it is necessary at first to solve the problem of the polarized beam divergence due to space charge effect during simultaneous transport of the polarized beam and of the intense unpolarized beam in high vacuum after extraction from the ionizer. We plan to use neutralization of space charge by D^+ plasma ions from the ionizer to overcome the problem. In a pulsed mode of operation this seems to be possible if the extraction voltage pulse will be applied to the extraction electrode system after the plasma pulse. A time shift about 10 µs would be sufficient to inject enough positive ions for space-charge neutralization into a region where the extracted beams propagate. It will be necessary to use all voltages pulsed for the extraction electrode system to ensure drift space free of electric field in the extraction electrode system area during the rise time of the plasma pulse.

An increase of the ion beam energy to 30 keV is also important to reduce the space-charge effect in the bending magnet of the ionizer where the polarized and the unpolarized beams are separated.

For the plasma density achieved with the present ionizer, destruction of polarized H^- ions by the plasma inside the ionizer should be noticed. Attenuation of a polarized H^- ion flux in plasma about of 20% for processes of stripping of negative ions by plasma electrons and mutual neutralization of negative and positive ions is estimated. Attenuation due to negative ion neutralization in collisions with neutral deuterium atoms can vary depending on gas efficiency of the plasma source. This process is determined also by the total number of plasma ions neutralized at the converter. Influence of this process can be decreased by increasing the internal volume of the charge-exchange region. This should also reduce scattering of polarized hydrogen atoms in collisions with deuterium atoms and molecules coming out of the plasma source. We did not take into account these processes up to now but they can be important for production of polarized ion pulses with large pulse duration.

Taking into account that 2.5 mA of polarized H^- ion beam current was obtained with 45 mA D^- ion beam, we expect to increase intensity of polarized H^- ion beam to level of 4 mA due to implementation of the improvements described.

Acknowledgements

We are thankful to S.K.Esin, V.P. Yakushev, V.G. Dudnikov and V.P Derenchuk for many helpful discussions. This work was supported in part by Indiana University Cyclotron Facility.

References

1. Smirnov B.M., Asymptotic Methods in Theory of Atomic Collisions, Atomizdat, Moscow, 1973, (in Russian).
2. Haebeli W., Nucl. Instr. and Meth. **62** (1968) 335.
3. Bel'chenko Yu.I, Budker G.I.,.Derevyankin G.E, Dimov G.I, Dudnikov V.G., Roslyakov G.V., Chupriyanov V. E. and Shamovskii V.G., "Raboti po sil'notochnim puchkam v Novosibirske", in Proc. of 10th Int. Conf. on Particle Accelerators, 1977, Protvino, Russia, p. 287.
4. Alessi J.G., Slyuters Th., and Hershcovitch A., AIP Conf. Proc. **117** (1984) 32.
5. Belov A.S., Esin S.K., Kubalov S.A., Kuzik V.E., Stepanov A.A., and Yakushev V.P, Pis'ma v JETP, **42** (1985) 319.
6. Belov A.S., Esin S.K., Kubalov S.A., Kuzik V.E., Stepanov A.A., and Yakushev V.P., Nucl. Instr. and Meth. In Phys. Research **A255** (1987) 442.
7. Belov A. S., Esin S. K., Netchaeva L. P., Klenov V. P., Turbabin G. A., and Vasil'ev G.A., "Polarized ions from a storage cell", in Proc. of 7th Int. Workshop on Polarized Gas Targets and Polarized Beams, Urbana, IL, 1997, eds. Roy J. Holt and M.A. Milner, AIP Conf. Proc. **421** (1998) 362.
8. Belov A.S., Esin S.K., Netchaeva L.P., Klenov V.P., Turbabin G.A., and Vasil'ev G.A., "Polarized Ion Source with Resonant Charge-Exchange Plasma Ionizer", in Proc. of 13th Int. Symp. on High Energy Spin Physics, Protvino, Russia, 1998, eds. N.E.Tyurin et. al., World Scientific, 1999, p.622.
9. Belov A.S., Kuzik V.E. and Yakushev V.P., JTP, **60** (1990) 179.
10. Belov A.S., Dudnikov V.G., Kuzik V.E., Plokhinsky Yu.V., Yakushev V.P., Nucl. Instr. and Methods in Phys. Res., **A333** (1993) 256.
11. Belov A.S., Esin S.K., Netchaeva L.P., Plokhinsky Yu.V., Vasil'ev G.A., Klenov V.S., Turbabin A.V., Yakushev V.P., and Dudnikov V.G., Rev. of Sci. Instr. **67** (1996) 1293.
12. Derenchuk V.P., Belov A.S., and Lozowski W. B., "First operation of the IUCF pulsed H$^-$/D$^-$ Ion Source", in Proc. of Int. Workshop on Polarized Sources and Targets, Erlangen, Germany, 1999.
13. Seidl M., Cui H. L., Isenberg J. D., Kwon H. J., Lee B.S., "Surface production of negative hydrogen ions", Proc. of 6th Int. Symp. on Production and Neutralization of Negative Ions, eds. Alessi J.G., and Hershcovitch A., 1992, AIP Conf. Proc. **287** (1994) 25.
14. Belov A. S., Esin S. K., Netchaeva L. P., Turbabin A.V. and Vasil'ev G. A., "Development of Polarized Negative Hydrogen Ion Source with Resonant Charge-Exchange Plasma Ionizer", in Proc. of 14th Int. Symp. on High Energy Spin Physics, Osaka, Japan, 2000, AIP Conf. Proc., **570** (2001) 835.

RECENT IMPROVEMENTS IN CIPIOS INTENSITY AND OPERATION

V.P.DERENCHUK[1] AND A.S.BELOV[2]

[1]*IUCF, Milo B. Samson Lane, Bloomington, IN 47405, USA*
[2]*Institute for Nuclear Research of Russian Academy of Sciences, 60th
October Anniversary Prospect, 7A, Moscow, 117312, Russia*

The intensity of the Cooler Injector Polarized Ion Source (CIPIOS) operating in the polarized D⁻ mode is being developed to increase the amount of beam stored in the IUCF Cooler ring. To this end, we are installing improvements and parts to the ionizer that have been developed and constructed at INR, Troitsk. The goal is to accelerate 1 mA or more polarized beam to 7 MeV for H⁻ and 4 MeV for D⁻ through the RFQ to the Cooler Injector Synchrotron (CIS). A planned increase in the ionizer plasma density should improve the ionization efficiency and increase the polarized beam brightness. The average polarized beam current during a 300 μs to 400 μs wide pulse from the source is expected to be improved to 2 mA at 1 Hz to 4 Hz operation from the 1.5 mA peak current per 200 μs wide pulse that was reported previously. Recent polarized beam operating experience with H⁻ and D⁻ will be described.

1 Introduction

Since September of 1999, the Cooler Injector Polarized IOn Source (CIPIOS) [1] has been delivering polarized and unpolarized beams to experimenters at the Indiana University Cyclotron Facility (IUCF) Cooler synchrotron for over 4000 hours per year. The source consists of an atomic beam section that uses permanent magnet sextupoles for focusing and radio frequency transition units to polarize the atoms before they are focused into the ionizer. The resonant charge-exchange ionizer produces pulses of negative ion plasma inside a solenoid. The atomic beam pulse is focused through the extraction system into the solenoid where the atoms are ionized by a highly efficient charge exchange reaction [2]. Tens of milliamperes of unpolarized negative ions may be extracted directly from the ionizer without using the ABS. The ion beam is formed with a 25 kV extraction potential in a 100 μs to 400 μs wide pulse at a rate of 0.8 Hz to 4 Hz. The beam is transported to the RFQ-DTL entrance and accelerated to 4 MeV for D⁻ ions or 7 MeV for H⁻ ions. After the RFQ-DTL, the H⁻ beam polarization is measured at 7 MeV using elastic scattering on a ^{4}He gas target, the D⁻ polarization is measured at 4 MeV using the ^{3}He(d,p)^{4}He reaction [3,9]. The beam is then transported and strip injected into the Cooler Injector Synchrotron (CIS).

The goal of the most recent effort is to develop the intensity and polarization of D⁻ beam for the next major IUCF experiment CE-82, a search for the isospin-forbidden $d+d \Rightarrow \alpha + \pi^0$ [4]. This experiment is mounted in the Cooler synchrotron and will require a luminosity of 3×10^{31} cm^{-2}s^{-1}. To achieve this luminosity, the

Cooler must be filled to nearly 2 mA of tensor or vector polarized deuterons during each cycle. The amount of deuteron beam injected into the Cooler is limited compared to protons for two main reasons. The RFQ used before strip injection into CIS can only accelerate D⁻ up to 4 MeV compared to 7 MeV for H⁻ by using an RFQ-DTL combination [5]. Strip injection into CIS is less efficient for the D⁻ beam and stripping by residual gas in the beam transport lines and during the first orbits in CIS make filling the ring more difficult. Secondly, high tensor polarization requires filtering one of the three atomic deuterium states in the ABS of CIPIOS and as a result reduces the intensity by 33%. To reach the required luminosity, at least 1 mA of polarized beam must be available for strip injection into CIS. Peak beam intensity greater than 2 mA from CIPIOS is necessary to meet this requirement.

2 Operation

During the past two years of operation, improvements have been made to increase the reliability and decrease the maintenance requirements for the ionizer. To this end, the INR team has designed and tested modifications to the plasma source cathode [6] that has resulted in cathode lifetimes that exceed 1400 hours at typical IUCF operating parameters. Typically, the only maintenance required during a scheduled run is to recondition the cold nozzle on the dissociator every two weeks.

The ion source has operated for more than 50% of the time since it was put into operation (Table I). This time includes about 15% source development and the remainder scheduled accelerator running time. Less than 2% of the total time can be attributed to failures.

Table 1. CIPIOS Hours of Operation

Year	Total Hours of Operation	Unpolarized Operation
1999	3,100	672
2000	4,500	840
2001 (to Oct)	3,900	1200

3 ABS Operation and Development

The atomic flux of hydrogen and deuterium atoms was measured at the entrance of the ionizer using a time-of-flight (TOF) mass spectrometer constructed at INR. When CIPIOS was first put into operation in 1999, the permanent magnet sextupoles were significantly weaker than expected. Using the TOF, the flux was measured to be $\Phi_{H1} = 1.35 \times 10^{17}$ at/s/cm² into a 1 cm² aperture. Since then, two new sextupoles with the expected field strength were received from Vacuumschmelze in Germany [7], were installed and an improvement in the atomic beam flux was verified. The flux was improved to $\Phi_{HI}= 1.84 \times 10^{17}$ at/s/cm^2 for hydrogen atoms and $\Phi_{DI}= 1.37 \times 10^{17}$ at/s/cm^2 for deuterium. This corresponds to a density at the ionizer of $\rho_{HI}= 0.97 \times 10^{12}$ at/cm^3 and $\rho_{DI}= 0.96 \times 10^{12}$ at/cm^3 respectively. Figure 1 shows the expected and measured magnetic field gradient for the second permanent magnet sextupole. The first sextupole was also replaced.

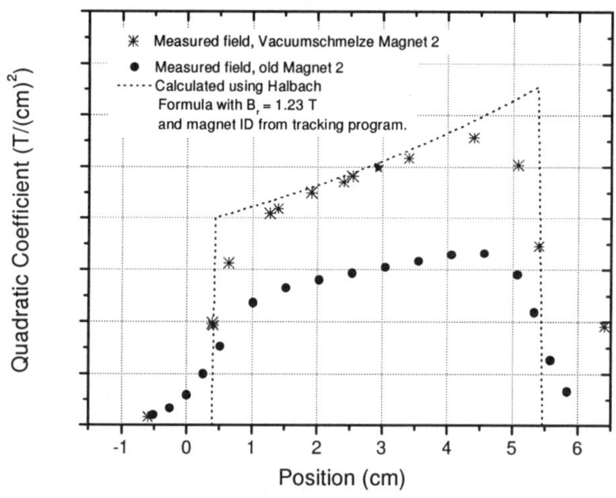

Figure 1. A plot of the field gradient versus axial position for the second ABS sextupole. The dots show the old magnet field strength, the stars the Vacuumschmelze gradient and the straight line the calculated field.

4 Ionizer Operation and Development

Several goals for the development of the resonant charge-exchange ionizer at INR were: to improve the lifetime of the cathode and plasma injector, increase the efficiency of negative ion production the plasma, and to design a new extraction geometry suitable for tens of milliamperes of extracted beam current. This development work which has resulted in a new lengthened plasma source and neutralizer/converter design is described elsewhere [6,8]. At IUCF, ionizer development occurred only between scheduled runs. It was found that a new ionizer solenoid tune decreased the extracted electron current thus decreasing destruction of the negative ions. Figure 2 shows the longitudinal field along the ionizer axis. Best operation occurs when the field passes through zero in the

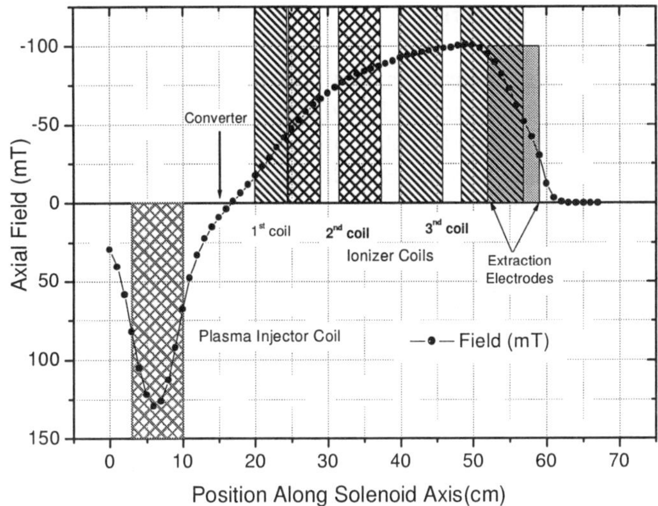

Figure 2. This shows the ionizer solenoid field plotted along the axis of the plasma injector coil and solenoid coils. The solenoid is split into three separately adjustable sets of coils. Atoms enter the ionizer from the right. Most are ionized within the first 20 cm of the extraction electrodes.

converter. In additional tests, higher operating temperatures of the converter and Cesium oven allowed us to operate the ionizer at 0.8 Hz with no loss in beam intensity. With the addition of the new INR cathode, it is expected that maintenance intervals will exceed thousands of hours for the plasma source.

The addition of a new lengthened plasma source developed at INR in combination with a redesigned extraction system has resulted in an improvement in plasma density and therefore the intensity of unpolarized beam extracted from the ionizer. The unpolarized D^- current now exceeds 30 mA and is expected to exceed 40 mA with further improvements in the converter. A similar gain in the unpolarized H^- current is expected from 40 mA now to 50 mA in the future. Polarized beam current scales linearly with the extracted unpolarized current.

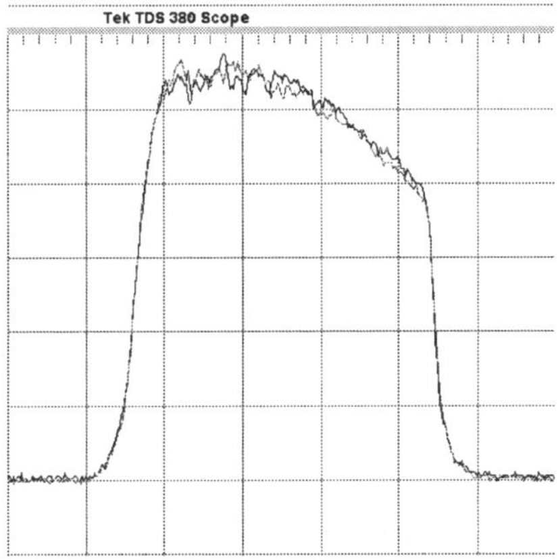

Figure 3. More than 1.6 mA peak of polarized D^- current is shown after mass analysis. Scale is 0.3 mA/div and 100 μs/div.

5   Polarized Beam Operation

The H⁻ beam current intensity is now 1.6 mA peak, more than 0.80 mA is accelerated to 7 MeV through the RFQ/DTL. The polarized D⁻ beam intensity is comparable (Figure 3). The D⁻ polarization was measured and is better than 85% in most states [9]. The average H⁻ polarization is measured to be $p_z = 0.83$ for 0.8 mA of accelerated beam current.

6   Conclusions and Future

Regular operation of CIPIOS is now very reliable and relatively low maintenance. Peak currents of 1.6 mA with polarization exceeding 80% of nominal are reproducible. Further improvements in the ionizer should result in polarized D⁻ beams with more than 2.0 mA peak current and nearly 1.0 mA accelerated to 4 MeV for 2 states.

The results reported would have been impossible without the help of Ron Kupper, Bill Lozowski and many other technical and professional staff at IUCF. This work is funded by NSF grants PHY-97-24216 and PHY-93-14783 and Indiana University.

References

1. V.P.Derenchuk et al, "A Multi-Milliampere Polarized and Unpolarized Negative Ion Source for IUCF", presented at PAC 2001, Chicago, IL. http://pacwebserver.fnal.gov/papers/Wednesday/AM_Poster/WPAH008.pdf
2. A.S.Belov, et al, *Nucl. Instr. Methods* **A333**, (1993)256.
3. E.J.Stephenson, et al, "A Polarimeter for Proton Beams with a Very Low Duty Factor", these proceedings.
4. Browse to http://www.iucf.indiana.edu/Experiments/COOLCSB/
5. D.L.Friesel, et al, 1997 Particle Acc. Conf., *IEEE* **97CH36167**, 2811(1997).
6. A.S.Belov, et al, "Development of Polarized Hydrogen Ion Source with Resonant Charge-Exchange Plasma Ionizer at INR, Moscow", this proceedings.
7. Vacuumschmelze Gmbh D-63412 Hanau, Germany, www.vacuumschmelze.de VAC Corp. USA. Oklahoma City, OK 73108, www.vaccorp.com.
8. A.S.Belov, et al, Proc. of 14th International Spin Physics Symposium, eds. K. Hatanaka et al, Osaka, Japan, 2000, *AIP Conf. Proc.*, **570** (2001) 835.
9. E.J.Stephenson, et al, "Development of a Polarized Deuteron Beam for the IUCF Cooler", these proceedings.

A HERMES TYPE STERN-GERLACH SOURCE WITH ECR IONIZER AND CHARGE EXCHANGE IN CESIUM VAPOUR FOR INTENSE ION BEAMS OF NEGATIVE DEUTERIUM

R. HERTENBERGER, Y. EISERMANN, A. METZ, P. SCHIEMENZ[†],
H.-F. WIRTH AND G. GRAW

Ludwig-Maximilians-Universität, Am Coulombwall 1, D-85748 Garching, Germany
E-mail: ralf.hertenberger@physik.uni-muenchen.de

Supported by BMBF and DFG

Motivated by the excellent performance of the Hermes atomic beam source, we try to combine for the production of negative polarized hydrogen ion beams an 80 K atomic beam source with an electron cyclotron resonance ionizer and with subsequent double electron pickup in cesium vapour. Our atomic beam source provides an intensity of $6.4\times10^{16}\frac{H}{sec}$ for hydrogen and of about $5\times10^{16}\frac{D}{sec}$ for deuterium. Beam intensities of 9 μA were observed for negative vectorpolarized $\vec{D^-}$ ion beams in two hyperfine substates with an emittance of about 5 π mm rad $\sqrt{eV}$. Vectorpolarizations of 65 % and 70 % have been observed for accelerated deuterium ions and protons, respectively.

1 Introduction

Internal polarized hydrogen targets have been pioneered by the use of high intensity Stern-Gerlach type atomic beam sources (ABS) in the HERMES internal polarized hydrogen target at the HERA/DESY electron storage ring and the PINTEX target at the IUCF cooler ring[1]. Both sources are optimized in regard to particle flux, they are operated at nozzle temperatures of about 80 K with atomic hydrogen jets of high velocity and use systems of 5 or 6 permanent sextupole magnets of FeNdB[2] as a spin filter for the production of electron polarized atomic hydrogen beams. Nuclear polarized beams are then created by induced adiabatic radiofrequency transitions between Zeeman splitted hyperfine levels. A flux of 6.8×10^{16} $\frac{H}{sec}$ into a compression tube of 10 mm diameter and 100 mm length has been reported.

Sources for beams of charged $H^{\pm}$ or $D^{\pm}$ ions require density optimized atomic beams and use generally ABS sources with nozzles at 30 K and respective slow atomic beam velocities. For the production of polarized negative ion beams we tried to combine the 80 K FeNdB ABS-technology with ionization in an electron cyclotron resonance plasma (ECR) and subsequent double electron pickup in a dense cesiums vapour jet (Cs). This type of ion source is successfully operating at TUNL using 30 K technology and conventional

216

electric sixtupoles[3]. The higher atomic flux and the better emittance of the 80 K atomic beam source is expected to compensate the density reduction due to higher beam-velocity.

In comparison with polarized hydrogen sources of potentially higher beam intensities the combination of ABS, ECR and Cs is attractive because of its versatility. The use of adiabatic rf-transitions allows to produce $\vec{H}$ and $\vec{D}$ ion beams with all possible vector- and tensor-polarizations. About 80 % of the theoretical maximum polarization can be expected in ion beams with low emittance.

Stern-Gerlach Atomic Beam Source

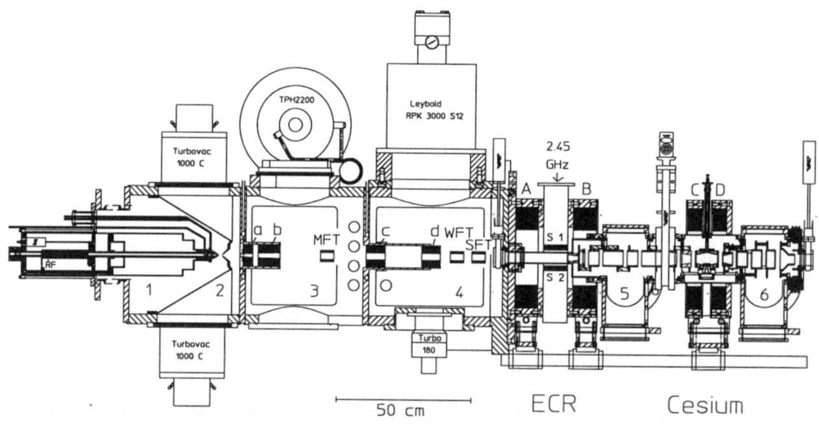

Figure 1. Vertical section of the Stern-Gerlach atomic beam source with four sextupole magnets (a...d), the adiabatic radiofrequency transitions (MFT,WFT,SFT) and the dissociator for the production of a supersonic atomic hydrogen jet (in vacuum chamber 1). The ECR ionizer consists of two pancake coils (A,B) producing an axial magnetic mirror field, the waveguide for the 2.45 GHz rf-wave and the radially focussing sextupole magnet of six permanent FeNdB dipoles S1...S6 (shown are only S1 and S2, which are placed in the waveguide). The positive ion beam after the ECR ionizer is transported and focussed into the cesium vapour target by electrostatic electrodes. In the cesium jet target the negative ion beam is created by successive pickup of two electrons at an ion beam energy of 500 $\frac{eV}{nuc}$.

A vertical section of the ion source is given in figure 1. The first of the four differentially pumped vacuum chambers of the ABS houses the dissociator[4], the third and fourth the Stern-Gerlach sixtupoles of FeNdB permanent magnets. The additional pumping in chamber 2 reduces the scattering of beam atoms with molecules of the residual gas by about 20 %. The hydrogen flux through the dissociator is 2.2 $\frac{mb\,l}{s}$ at 340 Watts of 13.5 MHz rf power and at a degree of dissoziation of 77 $\pm$ 5 %. For deuterium the working points are 1.4 $\frac{mb\,l}{s}$ at 290 Watts with similar degree of dissoziation. Through the entrance of the ECR ionizer, an aperture of 10 mm in diameter, the ABS provides atomic beam intensities of $6.4 \times 10^{16} \frac{H}{sec}$ of polarized hydrogen and of about $5 \times 10^{16} \frac{D}{sec}$ of polarized deuterium[5]. Figure 2 shows an intensity measurement of atomic hydrogen beam into a compression tube of 10 mm diameter and 100 mm length which was installed at the location of the ECR ionizer. The beam intensity is varied as a function of the temperature of the dissociator nozzle. It does not yet reach a maximum at 80 K, the working point given by the liquid nitrogen cooling of the nozzle. This indicates a slight underfocussing of the 80 K atomic beam and would allow to run the source at lower nozzle temperatures without loosing intensity. A Monte-Carlo simulation is based on the parameters for beam-velocity, -temperature and peaking of Lorentz[6]. It does not take into account the beam loss due to scattering of beam atoms with molecules from the residual gas. Applying the absorption data from molecular hydrogen eliminates the overestimate of the measurement, no further renormalization is necessary.

Charged Ion Beams

The atomic beam is ionized with an efficiency of a few percent by electron impact in the electron cyclotron resonance plasma. The plasma of the ECR ionizer is located inside the wave guide and thus directly coupled to the 2.45 GHz rf power of about 20 W. At the location of an E-field maximum of the H_{10} wave, two 40 mm holes in the small sides of the wave guide hold the cylindrical plasma vessel of pyrex. The electric rf field is perpendicular to the axial magnetic mirror field, which is produced by a pair of pancake coils with magnetic flux guidance. It has a minimum of about 875 Gauss located in between maxima of 1600 and 1400 Gauss. The sixtupole field for radial confinement of the plasma is provided by six dipole magnets of FeNdB. Two of them are located inside the waveguide. No absorption of the rf wave or overheating of the magnets has been observed. For stable plasma operation, nitrogen is introduced into the ECR volume by a separate inlet.

Out of the ECR ionizer about 100 μA of positively charged polarized H^+ and D^+ ions were observed in the acceptance of the cesium target of $4\,\pi$ cm rad $\sqrt{eV}$ and at plasma currents of about 400 μA. A compact Wien-filter has been used to determine the composition of the ion beam.

Figure 3 shows an intensity measurement of negative vector polarized deuterium ion beam. The beam intensity varies as a function of the temperature of the dissociator nozzle which defines the velocity of the atomic deuterium beam. The almost linear increase in beam intensity is due to a convolution of the temperature dependent transmission function through the sixtupoles and the density dependence of the ionization. The double electron pickup occurs at a kinetic energy of 500 $\frac{eV}{nucleon}$ and is temperature independent. The maximum D^- beam intensity of 13.5 μA in three hyperfine substates does not yet saturate at a temperature of 80 K, but lower nozzle temperatures are technically not yet feasible. The D^- beam intensity corresponds to 9 μA of vector polarized ion beam in two hyperfine substates with an emittance of about $\epsilon = 5\ \pi$ mm rad $\sqrt{eV}$ defined by two apertures of 5 mm diameter and 13 cm distance. Beam intensities of 2.5 μA are available for experiments after acceleration by the Munich tandem accelerator.

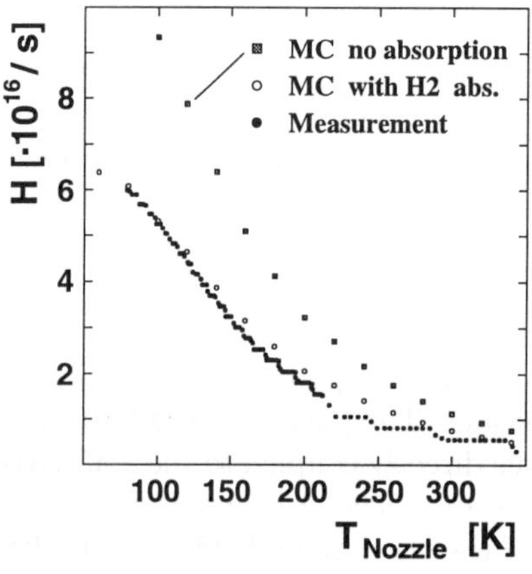

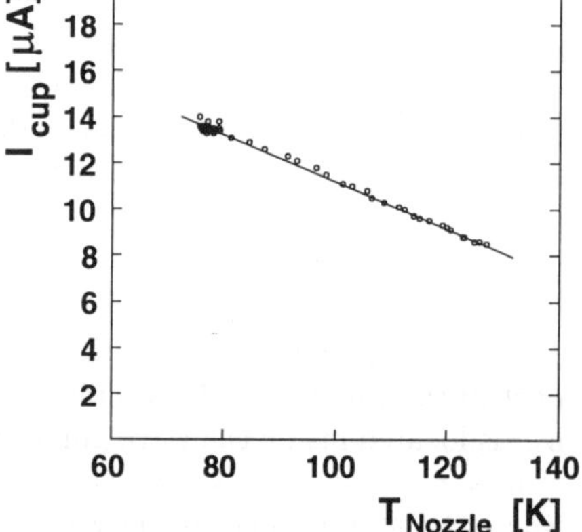

Figure 2. Intensity of atomic hydrogen beam as a function of the temperature of the dissociator nozzle measured in a calibrated compression tube of 10mm diameter and 100 mm length.

Figure 3. Intensity of negative deuterium ion beam in three hyperfine substates as a function of the temperature of the dissociator nozzle.

The double charge exchange in the cesium jet target saturates at a cesium area density of 10^{15} $\frac{atoms}{cm^2}$ [7]. For our jet target this corresponds to a temperature of 310 °C for the evaporation of liquid cesium at 120 Watts of heater power. The efficiency for double charge transfer $D^+ \rightarrow D^-$ was measured to be 22 % for the emittance mentioned above.

The depolarization due to hyperfine coupling in the magnetic holding fields of 1 kGauss during charge exchange in the ECR and Cs regions is negligible for deuterium. For hydrogen it causes a reduction in polarization of about 15 %.

Polarimetry

For tuning of the vectorpolarisation of deuterons the elastic scattering reaction 22 MeV ^{65}Cu($\vec{\mathrm{d}}$,d$_0$) was used at the scattering angle of 75 degrees. It has a vector analysing power of about A_y=−0.705 at low tensor analysing power A_{yy}. For protons we used the elastic scattering reaction 21.1 MeV ^{12}C($\vec{\mathrm{p}}$,p$_0$) with an analysing power of A_y=0.563 at 75 degrees. Both reactions have large cross sections well above 10 $\frac{mb}{sr}$. Countrates in the Q3D-detector above 5 kHz, achievable at target area densities of a few $\frac{mg}{cm^2}$, allow the online tuning of the adiabatic rf transitions. Vectorpolarizations of 65 % and 70 % have been observed for accelerated deuterons and protons, respectively.

Summary

The new polarized hydrogen source is running reliably since July 2000 with intensities of vectorpolarized deuterium beams of 2.5 μA after acceleration and with polarizations of 65 %. Angular distributions of $\frac{d\sigma}{d\Omega}$ and A_y have been accumulated for 7 nuclear reactions with excellent energy resolution of up to 4 keV for 20 MeV bombarding energy. Multinucleon transferreactions for the study of heavy supersymmetric odd-odd nuclei as ^{198}Hg($\vec{\mathrm{d}}$,α)^{196}Au [9] profit considerably from the increase in beam intensity and beam quality.

References

1. F. Stock, et al., *Proc. Int. Workshop on Polarized Beams and Polarized Gas Targets*, Cologne 1995, p. 260
   T. Wise, A.D. Roberts, W. Haeberli, *Nucl. Inst. Meth* **A336**, 410 (1993).
2. P. Schiemenz, A. Ross, G. Graw, *Nucl. Inst. Meth* **A305**, 15 (1991).
3. T.B. Clegg, et al., *Nucl. Inst. Meth* **A357**, 200 (1995).
4. K. El Abiary, *Diploma thesis*, LMU München, 1996, p.66.
5. R. Hertenberger et.al., *Rev. Sci. Inst.* **69,2**, 750 (1998).
6. B. Lorentz, *PHD Thesis*, MPI Heidelberg, 1994.
7. A. S. Schlachter et al., *Phys. Rev.* **A22,6**, 2494 (1980).
8. C. Pertl, *Thesis*, LMU München, 2000.
9. A. Metz, et al., *Phys. Rev. Lett.* **83**, 1542 (1999).
   A. Metz, et al., *Phys. Rev.* **C 61**, 064313-1 (2000).

DESIGN CRITERIA FOR POLARIZED SOURCES

D. EVERSHEIM[1], R. GEBEL[2], O. FELDEN[2], M. GLENDE[2]

[1]*Institut fuer Strahlen-undKernphysik, University Bonn, 53115 Bonn, Germany*

[2]*IKP, Forschungszentrum Juelich, 52425 Juelich, Germany*

A polarized H⁻ source has been installed at the COoler SYnchrotron COSY at Juelich, which provides in the COSY ring about the same number of polarized protons as the unpolarized source. The general design criteria of this colliding-beams-source with respect to the present cyclotron injector machine are discussed, as well as a possible future injection scheme, based on a LINAC. The matching of timing and phase-space turns out to be very crucial.

1 Introduction

Triggered by the demand for physics with polarized particles - namely protons and deuterons - three groups from the universities of Bonn, Cologne and Erlangen joined to plan, build and set in operation a "polarized" source for the COoler SYnchrotron COSY at Juelich [1,2]. The design of this source had to take into account the limited current ($<$ 150 µA at 4.5 KeV) and acceptance ($<$ 0.5 π mm mrad, $\beta \cdot \gamma$ normalized) capability of the injector cyclotron JULIC, the limited transmission (~10%) through this cyclotron and the stripping (non Liouville) injection into the COSY ring. The latter condition implied the construction of an H⁻ (D⁻) source.

The design idea for the injection into the COSY ring was to have accelerated about as many polarized particles as unpolarized particles in COSY. Furthermore, to develop efficient strategies to overcome the unavoidable depolarizing resonances during the acceleration in COSY, the polarization of the beam had to be sizable from the beginning. As a consequence, a proton polarization of $P > 0.7$ should be possible without intensive tuning of the source. The maximum polarization should be limited only by the physics of the source well beyond $P = 0.9$. In addition, in order to maximize the buncher efficiency in the injection of JULIC, the energy spread of the beam had to be small ($<$ 100 eV).

Required parameters for the "polarized" ion Source of COSY:

Polarization:	$>$ 0,75
I($\vec{H}^-$, $\vec{D}^-$):	30 µA
Inj. Energy:	4,5 KeV ($\vec{H}^-$), 9 KeV ($\vec{D}^-$)
Energy spread:	$<$ 100 eV

Emittance: < 0.5 π mm mrad ($\beta \cdot \gamma$ normalized)
Pulse length: < 20 ms
Repetition period: > 2 s

2   The Choice of the Type of Source

At the time of the design (1990) and with the focus on high polarization only sources were considered, which ionized selectively atoms, so that the inevitable hydrogen molecules cannot contribute to an unpolarized background. Since the output of Lamb-Shift sources are limited to about 3 μA only optically pumped sources and colliding-beams-sources (CBS) were taken into consideration. With view to the required polarized deuterons the final choice was a colliding beams source with a neutral, intense (several particle mA), and fast Cs beam meeting a polarized dense atomic hydrogen beam inside a magnetic holding field.

These sources should be capable to deliver $\vec{H}^-$ currents in excess of 30 μA with small emittances and high polarization (> 0,9). Moreover, this type of source allows for a spacious modular design at ground potential, so that maintenance is easily facilitated and measuring devices can be implemented without restricting the performance of the source.

From the design point of view a high modularity is welcome, since combined function elements can be avoided. Single dedicated function elements and numerous measuring devices are the basis for an easy error tracking in case of a malfunctioning. The same arguments hold, in case one considers an automatic tuning of the source. The one to one relation between function and assigned device efficiently decouples parameters, which in turn allows for faster optimization algorithms.

3   The COSY-CBS

The operation scheme of the COSY-CBS is governed by the charge-exchange reaction: $\vec{H}^{\circ} + Cs^{\circ} \rightarrow \vec{H}^- + Cs^+$. A neutral nuclear polarized atomic beam ($\vec{H}^{\circ}$) meets in the charge exchange region a fast (50 KeV) neutral cesium beam (Cs°). An electron is swapped and the resulting $\vec{H}$ beam is guided by a small electric gradient field ($\Delta U \sim 20V$) to the extraction orifice, where the acceleration to the beam potential U_{ex} takes place. Thus the energy spread E of the beam is given by:

$$\Delta E \leq e \cdot \Delta U \qquad (1)$$

A more detailed analysis of the charge exchange region reveals implications for the atomic- and cesium- beam part. In a charge-exchange region of length L = 400 mm the neutral polarized hydrogen beam with the velocity v and intensity I_{H^0} meets the fast neutral Cs^0 beam with its intensity I_{Cs^0} in a mean cross section area F. With the cross section σ given for the charge-exchange reaction $\vec{H}^\circ + Cs^\circ \rightarrow \vec{H}^- + Cs^+$ the intensity of I_{H^-} of the extracted $\vec{H}^-$ beam can be calculated:

$$I_{H^-} = I_{H^0} \cdot \frac{L \cdot \sigma}{F \cdot v} \cdot I_{Cs^0} = n_{H^0} \cdot L \cdot \sigma \cdot I_{Cs^0} \tag{2}$$

Eq.(2) explicitly shows that the atomic-beam velocity v and the overlap area F of the $\vec{H}^\circ$ and Cs° beam should be as small as possible. The velocity v is reduced by a cooled nozzle and F is reduced by focusing elements in the atomic- and cesium-part. The cesium beam potential is defined by the peak in the charge exchange cross-section σ at about 50 keV with a $\sigma = 8 \cdot 10^{-15}\ cm^2$.

In the charge-exchange solenoid the polarization is preserved by the longitudinal field $\vec{B}$ with its vector potential $\vec{A}$. Table 1 shows how much polarization can be expected for a given solenoidal field.

Table 1. Calculated maximum polarization for two states

Solenoid field [Gauss]	Max. polarization [%]
0	50
500	85
1000	94.5

On the other hand the solenoid field defines the emittance of the beam. As a consequence of the conservation of total angular momentum $\vec{M}$ the "spatial part" $\vec{L}$ of the total angular momentum will grow to the extent that the H^- ions leave the solenoid field and the vector potential $\vec{A}$ becomes zero.

$$\vec{M} = e\vec{A}r + \vec{L} \tag{3}$$

At this point the "magnetic part" $e\vec{A}r$ of the total angular momentum has been converted to a macroscopic spatial angular momentum of the beam. The resulting transversal emittance for protons is then given by:

$$\varepsilon = 1.1 \cdot 10^{-2} \cdot \pi \cdot Br^2 \cdot \frac{1}{\sqrt{U_{ex}}} \tag{4}$$

B [Gauss]; r [mm]; U_{ex} [kV]; ε_{H^-} [mm mrad]

For the nominal acceptance of the injector cyclotron $\varepsilon_H^- = 150\ \pi$ mm mrad, $U_{ex} = 4.5$ kV, and $r = 5.5$ mm, B must not exceed 1000 Gauss. By varying the solenoid field, the transversal emittance ε_H^- can be traded for polarization.

Equation (2) imposes also conditions for the cesium part. The fast, neutral Cs^o-beam for the charge-exchange reaction in the solenoid is produced in two steps. First, Cs vapour is thermally ionized on a hot (1200 °C) porous tungsten surface (button) at an appropriate beam potential of about 40-60 kV. Since it turned out to be difficult to transport this Cs^+-beam further than about 450 mm, this beam has to be focused into the charge-exchange solenoid by magnetic quadrupoles. Having passed the quadrupoles the charged Cs beam is neutralized. Behind the solenoid, the Cs beam intensity can be measured by a calorimeter. The Cs-beam source has a well defined perveance $1/\alpha$, which can in principle be calculated from the geometry of the tungsten button and the extraction electrodes. Therefore, the Cs^+ intensity I_{Cs}^+ is related to the ionizer potential U_{But} by:

$$U_{But} = \alpha \cdot (I_{Cs^+})^{2/3} \tag{5}$$

Equations 1-5 show how the various beam parameters are decoupled and can be tuned.

COSY-CBS Performance:

Efficiency of the dissociator:	>	80 %
Efficiency of the RF transitions:	>	97 %
Most probable velocity (H^o):		1200 m/s
Density of the atomic beam:	>	$5.8 \cdot 10^{16}$ atoms/s (two states)
Cs-beam intensity (pulsed):	>	3-13 mA at 45 keV
Neutralizer efficiency:	>	90 %
Polarization:	>	0,85
H^- intensity:		30 µA (max: > 40 µA)

4 Consequences for the LINAC Injection

An alternative injection scheme is presently discussed. This scheme comprises a RFQ, which accelerates H^- from 25 KeV to 2.5 MeV and a LINAC that boosts this energy up to 52 MeV for the injection into COSY. Given the high brightness of the present COSY-CBS no problems with respect to acceptance or beam load of the LINAC cavities are expected, if RFQ and LINAC can be operated over 10-20 ms.

On the other hand it is obvious that COSY injection conditions can be kept the more optimal the shorter the injection periods become. As can be seen from Eq. (2), the same charge can be injected into COSY within less than 20 ms by increasing the Cs-beam intensity in the COSY-CBS, respectively. As a consequence the H⁻ beam intensity will increase accordingly. A limit evolves, once the Cs beam is not longer emittance dominated rather than perveance dominated, i.e. there is a substantial emittance increase caused by the space charge of the beam.

At the other end of the time scale, considering sources that provide high currents over a very short period of time (1 A over 0,1 µs), the problem of an increased emittance due to space charge can be expected inside the RFQ and/or in the LINAC. Thus, the operating conditions of these sources have to be extended in the time domain substantially. Otherwise an order of magnitude of particles can be lost easily. In addition, intensity fluctuations of the source may change the electric load of the LINAC, resulting in an increased momentum spread, which in turn can gives rise to an longitudinal momentum mismatch between LINAC and COSY.

Which type of source in the end meets the injection condition of COSY best can be estimated not before the final design parameters of the RFQ and LINAC are fixed. It is at present not obvious that there is another type of source different from the COSY-CBS, which provides all possible tensor- and vector-polarization states of a deuteron beam and simultaneously increases the number of accelerated particles in COSY by an order of magnitude.

References

1. R. Gebel et al., Polarized Gas Targets and Polarized Beams, Seventh Intl. Workshop, *AIP Conference Proceedings* **421** (1997) pp. 503.
2. P.D. Eversheim, R. Gebel and P.v. Rossen,The Polarized Ion Source of COSY, *COSY NEWS* **8** (2000) pp. 1

CRYOGENIC SOURCE OF POLARIZED DEUTERONS POLARIS FOR THE JINR ACCELERATOR NUCLOTRON

V.P. ERSHOV, V.V.FIMUSHKIN, L.V.KUTUZOVA, YU.K.PILIPENKO,
A.D.STEPANOV, V.P.VADEEV, A.I.VALEVICH

Joint Institute for Nuclear Research, Dubna, Russia
E-mail: pilipen@sunhe.jinr.ru

A.S. BELOV

Institute for Nuclear Research RAN, Troitsk, Russia
E-mail: belov@al20.troitsk.ru

The spin physics program is an important part of the LHE JINR scientific program. An intensive study of polarization phenomena was carried out at the Dubna 4.5 Gev/nucl. synchrophasotron, using cryogenic source of polarized deuterons ($\uparrow D^+$) POLARIS. There is a proposal to make a polarized deuteron beam in the new accelerator Nuclotron to continue the spin physics experiments. It is planned to modify the source POLARIS and use it as a source of polarized deuterons. The Nuclotron has a short one-turn injection (8 μsec) of positive ions at present. To increase the intensity of the accelerated polarized beam up to $0.7\text{-}1*10^{10}\uparrow d$/pulse a multyturn charge exchange injection of negative ions (20-30 turns) should be applied. It is realized by injection into the accelerator of $\uparrow D^-$ ions and stripping them inside the ring. The polarized $\uparrow D^-$ source is required. A new plasma charge exchange ionizer for production of $\uparrow D^-$ polarized beam is designed and manufactured. The tests of the ionizer are in progress.

1   Introduction

For the last decade, an intensive study of polarization phenomena in high-energy spin physics was carried out at the Dubna 10 GeV synchrophasotron using a polarized deuteron beam. The polarized deuterons were produced by the cryogenic source POLARIS. The commissioning of the new superconducting accelerator Nuclotron supposes to continue the spin physics program. There is a project to use the polarization technology at the Nuclotron and to produce a polarized deuteron beam. Simulation shows that depolarizating resonances are absent under polarized deuteron acceleration almost at all energy range of that accelerator [1].

In this project it is supposed to realize the following:
- to upgrade the atomic stage of the source POLARIS,
- to modify the existing $\uparrow D^+$ charge exchange ionizer into an $\uparrow D^-$ ionizer,
-to realize multyturn charge exchange injection in the Nuclotron ring.

It will provide new good possibilities for spin physics experiments.

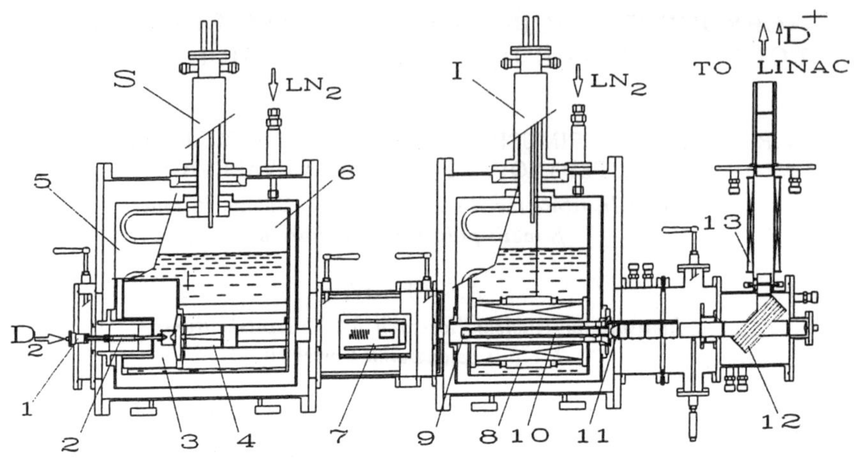

Figure 1. Schematic view of the polarized deuteron source POLARIS. S - polarized atomic source, I - Penning ionizer. 1- electromagnetic gas valve, 2- dissociator, 3- nozzle chamber, 4- SC sixtupole magnets, 5- nitrogen shield, 6- helium cryostat, 7- RF cell, 8- SC solenoid, 9- electron optics, 10- ion optics, 11- vacuum gate, 12- electrostatic mirror, 13- solenoid of spin-precessor, 14- Faraday cup.

2 $\uparrow D^+$ Source POLARIS

Many years the source runs at the JINR 10 GeV synchrophasotron [2,3,4]. It is the cryogenic atomic beam source with two tapered sixtupole superconducting magnets, SC solenoid, internal 4.2 K cryopanels for gas pumping. The source POLARIS consists of two LHe cryostats: a pulsed atomic beam stage and a Penning plasma ionizer (Fig.1). Additional permanent sixtupole magnet (i.d.- 30 mm, l- 60mm) is planned to installed after the SC sixtupoles for better focussing of the atomic beam. The energy of the deuteron beam at the output of the source is about 3 keV, the current 0.3-0.4 mA. The vector and tensor polarizations are: $P_z=\pm0.54$; $P_{zz}=\pm0.76$.

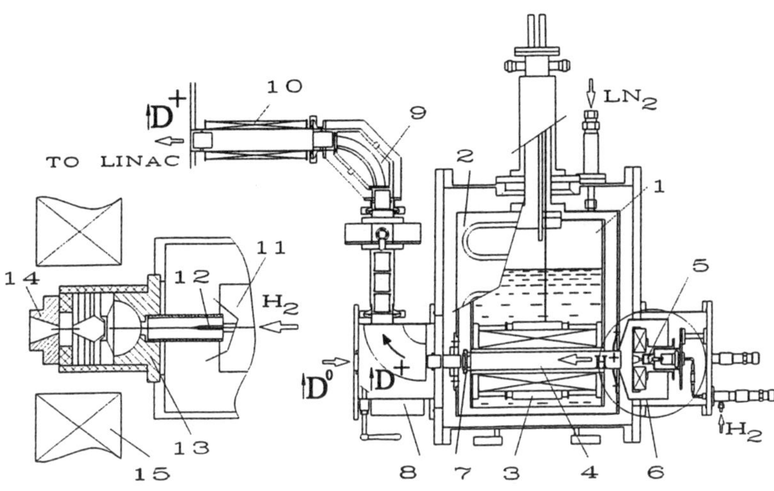

Figure 2. $\uparrow D^+$ charge exchange ionizer. 1- LHe cryostat, 2- LN_2 shield, 3- SC solenoid, 4- ionization space, 5- H^+ plasma source, 6- HV shield, 7- extraction grids, 8- bending magnet, 9- electrostatic mirror, 10- solenoid of spin rorator, 11- H_2 valve, 12- ignition electrode, 13- cathode, 14- anode, 15- magnetic coil.

3   Charge Exchange Ionizer

The new machine Nuclotron has a short one turn injection (8 μsec) of positive ions (factor 50 less compared to the old one). So to get a large intensity of the accelerator a new plasma charge exchange ionizer has been developed [5,6] (Fig.2). It has similar the Penning ionizer LHe cryostat with the 60 mm cold bore SC solenoid. A short pulse (300 μsec) H^+ arc plasma source is installed at the solenoid entrance. The potentials of the plasma source and the HV shield are +12 kV. A nuclear polarized deuterium atomic beam is injected into the solenoid space towards the plasma beam. The charge exchange reaction between polarized deuterium atoms and hydrogen plasma ions $\uparrow D^0 + H^+ = \uparrow D^+ + H^0$ takes place inside the HV shield.

Using POLARIS atomic beam stage, a 0.8 mA polarized D^+ beam, accompanying a 9 mA background H_2^+ plasma current at the deflecting magnet output of the charge exchange ionizer, have been measured . The efficiently of the ionizer was 3-5% instead of 1-2% for the our old Penning ionizer (0.2 mA D^+).

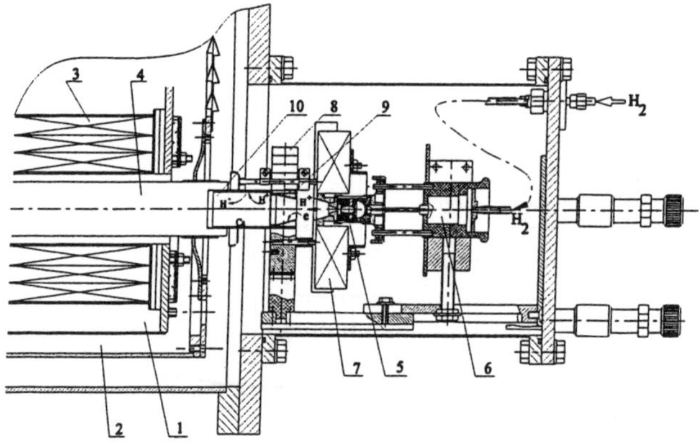

Figure 3. $\uparrow D^-$ charge exchange ionizer version. 1- LHe cryostat, 2- LN_2 shield, 3- SC solenoid, 4- HV shield, 5- H^+ plasma source, 6- H_2 valve, 7- magnetic coil, 8- magnetic filter, 9- Cs pellet capsule, 10- Cs-Mo converter.

Vacuum in the ionizer is provided by cryopumping. Hydrogen vapor pressure at 4.2 K cryostat is 2-4*10⁻6 mbar. An adsorption charcoal panel, attached to the LHe cryostat improves vacuum to 6-7*10⁻7 mbar.

To reach the accelerated polarized beam intensities up to $0.7\text{-}1*10^{10}$ $\uparrow$d/pulse a multyturn charge exchange injection (20-30 turns) should be applied. It is realized by injection into the Nuclotron ring of $\uparrow D^-$ ions and stripping them inside the ring. A polarized $\uparrow D^-$ beam from the source is required.

The existing D^+ plasma charge exchange ionizer has been modified in two versions of $\uparrow D^-$ ionizers. In the first version H^- plasma for charge exchange reaction is produced by the arc multybeam surface-plasma source directly [7]. The second version has H^+ plasma source, but at its output a Cs surface converter is placed to produce H^- ions [8], (Fig3). Titanium-cesium chromate pellets are loaded into the first part-capsule of the converter and heated up to 300-500⁰C. During a plasma source pulse the cesiated molybdenum surfaces of the converter are exposed to an intense flux of positive ions and effectively generate H^- ions.

Fast plasma electrons, accompanying H^+ plasma, are removed at the converter entrance by a transverse 250 Gs magnetic field of a permanent magnet. The H^- ions, neutralized by positive ions fill up a space of the HV shield (-20 kV), where charge exchange reaction $\uparrow D^0 + H^- = \uparrow D^- + H^0$ takes place. Polarized negative deuterium ions, confined in the radial direction by magnetic field of the solenoid, drift to the extracting grids. The 90^0 bending magnet separates accelerated H^- plasma and polarized $\uparrow D^-$ ions. Tests of the $\uparrow D^-$ ionizers are in a progress.

Acknowledgments

This work is supported by the Russian Fund of Fundamental Research, the Grant 01-02-16406.

References

1. A.M. Kondratenko *et al*, - In: The 6th Workshop on High Energy Spin Physics. Protvino 1995, **v.2**, 212 1996.
2. N.G. Anischenko *et al*, -In: The 5th Int. Symp. on High Energy Spin Physics, Brookhaven 1982, AIP Conf. Proc. N.Y. **N95**, 445 1983.
3. N.G. Anischenko *et al*, -In: The 6th Int. Symp. on High Energy Spin Physics, Marseille 1984, Jorn.De Phys., Colloquia C2, Supplement an n° 2, **46**, C2-703 1985.
4. V.P. Ershov *et al*, - In: Int. Workshop on Polarized Beams and Polarized Gas Targets, Cologne 1995, (World Scientific, Singapure) 193 1996.
5. A.S. Belov *et al*, Nucl. Instr. and Meth., vol.**A255**, 3, 442 1987.
6. V.P. Ershov *et al*, -In: The 13th Int. Symp. on High Energy Spin Physics Protvino 1998, (World Scientific, Singapure) 615 1999.
7. V.P. Ershov *et al*, -In: Int. Workshop on Polarized Sources and Targets, Erlangen 1999, (Druckerei Lengenfelder, Erlangen) 456 1999.
8. A.S. Belov *et al*, Rev. Sci. Instr. vol.**67**, 3, 1293 1996.

V. Sources of Polarized Neutrons

OVERVIEW OF POLARIZED NEUTRON SOURCES

T.R. GENTILE

National Institute of Standards and Technology, Gaithersburg, MD 20899, USA

Polarized neutrons have applications in materials science and weak interactions studies. Several methods exist for polarizing neutrons, such as supermirrors for cold neutron energies, Heusler alloy for thermal neutron beams, and polarized proton filters for epithermal energies. While each of these devices are quite useful for different applications, there are also several applications that need a new method. After surveying the basic characteristics of these devices, this talk will focus on the present state of the art for neutron polarizers that are based on the large spin dependence of the cross section for neutron absorption by ^{3}He. The status of applications to neutron scattering and fundamental physics will be also be discussed.

1 Introduction

Experiments with neutrons involve neutron energies anywhere between 10^{-7} eV (ultracold) and 10^7 eV, and as one would expect, the experimental techniques vary substantially across this range. Hence this paper will concentrate on the restricted range of energies between 1 meV and 10 eV. Steady-state neutron beams are produced by nuclear reactors, while spallation sources produce repetitive pulses of neutrons. Spallation neutrons are produced by driving a high energy particle beam into a target of heavy atoms. The typical energy and wavelength of a neutron from a room temperature moderator is 25 meV and 0.18 nm, respectively. "Cold" neutrons are obtained from cryogenic moderators such as liquid hydrogen and have a typical energy of few meV, while epithermal neutrons have a typical energy of 1 eV. We will first survey applications of polarized neutrons and the methods for producing them, followed by greater detail on neutron spin filters based on polarized ^{3}He.

2 Applications of polarized neutrons

Both the magnetic scattering of neutrons by unpaired electrons and the scattering of neutrons by nuclei of non-zero spin can be strong functions of the neutron spin. The pioneering work of Moon, Riste, and Koehler[1] showed that polarized neutron scattering is a powerful technique to uniquely separate and identify the relevant cross sections. Applications for polarized neutrons in neutron scattering span a broad range of issues in physics, chemistry, materials science, and biology. Some important examples are: 1) separation of magnetic from nuclear scattering (spin wave/phonon separation, magnetization profile measurements in thin films and multilayers, magnetic cluster/chemical cluster separation in small angle scattering)[2,3,4], 2) measurements of magnetic moments, magnetization density distributions, and magnetic form factors in paramagnets and ferromagnets from the interference between magnetic and nuclear scattering[5], 3) measurement of magnetic domain sizes in spin glasses, amorphous magnets, etc. by energy dependence of neutron depolarization[6], 4) measurements of small energy transfer using the neutron spin-echo technique[7] and 5) for polarizable targets, contrast variation of the nuclear scattering.[8]

Polarized neutrons are also important in studies of the weak interaction. The neutron differential beta decay distribution can be written as:

$$dW \propto (g_v^2 + 3g_a^2)F(E_e)\left[1 + a\frac{\vec{p}_e \cdot \vec{p}_\nu}{E_e E_\nu} + \vec{\sigma}_n \cdot \left(A\frac{\vec{p}_e}{E_e} + B\frac{\vec{p}_\nu}{E_\nu} + D\frac{\vec{p}_e \times \vec{p}_\nu}{E_e E_\nu}\right)\right]$$
(1)

where $\vec{\sigma}_n$ is the initial spin of the decaying neutron; a, A, and B are the electron-neutrino, spin-electron, and spin-neutrino asymmetry coefficients, respectively; D is the T-odd, triple product coefficient; g_a and g_v are the vector and axial-vector coupling coefficients, respectively; $F(E_e)$ is the beta electron energy spectrum; and $\vec{p}_e$, $\vec{p}_\nu$, E_e and E_ν are the momenta and kinetic energies of the decay electron and anti-neutrino.[9] It can be seen in this expression that polarized neutrons are required for measurements of either the asymmetry coefficients A and B, or the time-reversal violating coefficient D. In addition, measurements of A and B require accurate knowledge of the neutron polarization. Past experiments to measure these fundamental neutron beta-decay parameters have used supermirror polarizers,[10,11,12,13] while ^{3}He-based spin filters are proposed for an experiment to measure A and B at LANSCE.[14] In addition to beta-decay measurements, parity violation experiments also require polarized neutrons, such as the planned measurement of the parity violating asymmetry A_γ in the reaction n+p $\rightarrow$ d+γ[15] and the parity-violating neutron spin-rotation observable.[16]

3 Sources of polarized neutrons

Several devices exist for producing polarized neutrons,[17] too numerous to cover in this paper. Instead we will first discuss the basic properties of the most prevalent device for each of three neutron energy ranges: supermirrors for cold neutrons,[18] Heusler alloy for thermal neutrons,[19,20] and polarized proton spin filters for epithermal neutrons.[21]

Supermirrors rely on selective reflection of one spin state from a magnetized surface for very shallow angles of incidence, and can typically produce 95 % polarization with 30% absolute transmission (60 % transmission of the desired spin state).[22] The critical angle beyond which reflection does not occur depends on wavelength. A series of thin films is typically used to extend to shorter wavelengths (or larger angles), but nevertheless the typical short wavelength limit is 0.2 nm. Their fairly constant high polarization and transmission throughout the cold neutron range make supermirrors quite useful for both condensed matter research and weak interaction studies. Nevertheless, supermirrors also have limitations for both applications. For neutron scattering, their small angular acceptance of about 1 degree is not well-matched to the highly divergent beams encountered in many applications, such as reflectometry and small angle neutron scattering. Although most weak interaction studies have employed supermirrors, a polarizer that allows for more broadband operation, large area, low gamma background, convenient polarization reversal, and, in some cases, highly accurate measurement of the neutron polarization, is desirable.

At thermal neutron energies, polarizing monochromators based on Cu_2MnAl (Heusler alloy) are often employed. This device also typically produces 95% polarization, but with lower reflectivity than supermirrors. Although Heusler alloy is useful for the continuous beams produced by a reactor, it is not desirable for the pulsed beams produced by a spallation source because in this case time-of-flight information is employed to analyze all wavelengths simultaneously. It is perhaps in this application that the greatest need exists for a broadband polarizer.

Finally, in the epithermal neutron range (0.1 eV to 10 eV) spin filters based on the spin dependence of the neutron scattering cross section for hydrogen have been employed. Such polarized proton spin filters, which use dynamic nuclear polarization to provide the polarized protons,[23,24] have been employed for parity violation experiments.[25,26] These apparatus have generally required dilution refrigerators, but recently high proton polarization has been demonstrated by microwave-induced optical nuclear polarization at 77 K.[27]

4 ³He Polarizers

The limitations of these devices lead us to the emerging technology of ³He neutron polarizers, which will constitute the remainder of this paper. These devices are based on the large spin dependence of the cross section for neutron absorption by ³He gas. The primary advantages of ³He-based neutron spin filters for neutron scattering are the ability to function over a broad energy range without energy selection and accomodate large divergence beams. This feature is particularly relevant to spallation sources because time of flight analysis makes it possible to measure the polarization as a function of neutron energy.[28] For weak interaction studies, the primary advantages are the ability to polarize large area beams, to switch from polarized to unpolarized neutrons and to reverse the direction of the ^{3}He polarization, and to determine the neutron polarization accurately by measurements of transmission alone[29,30]. The last feature is best accomplished with either monochromatic beams or time-of-flight analysis of a pulsed beam. Motivated by these features, several neutron laboratories are pursuing ^{3}He-based neutron spin filters for both neutron scattering instruments and fundamental physics experiments.

A ^{3}He spin filter with sufficient thickness and 100% ^{3}He polarization would provide 100% neutron polarization with 50% absolute transmission of the beam. For the typical value of 50% ^{3}He polarization for current spin filters an optimal thickness must be chosen based on a trade-off of neutron polarization and neutron transmission. Since the cross section has a simple linear dependence on the neutron wavelength, polarized ^{3}He can polarize a broad range of wavelengths and the required product of cell pressure and cell length varies linearly with the wavelength. For cold neutrons (wavelength of 0.5 nm) a pressure-length product of 5 bar-cm of 50 % polarized ^{3}He yields 72 % neutron polarization with 23 % absolute transmission. The relationships between these neutron and ^{3}He parameters have been discussed by several authors.[31,32,33] The limits on performance due to the currently available ^{3}He polarization make it clear that conventional polarizers are preferable for applications that do not make use of the capabilities of ^{3}He polarizers. Comparison of ^{3}He spin filters with conventional neutron polarizers have been studied both theoretically[34] and experimentally.[35]

4.1 Production of Polarized ^{3}He

As has been the case for ^{3}He targets for electron scattering experiments,[36,37,38,39] two optical pumping methods are being employed to produce polarized ^{3}He gas: spin-exchange (SEOP)[40,41,42,43] and metastability-exchange (MEOP).[44,45,46] However, the requirements of neutron spin filters are quite different. Electron scattering experiments require thick targets for narrow diameter beams (1 cm), and have typically operated near 10 bar to avoid excessive length. Although there are difficult technical issues with the high pressures and extremely thin windows in such glass cells, the high pressure and relatively small volume is well-matched to the SEOP method. Neutron spin filters are often large in diameter

(10 cm) and require lower gas thickness, making operation near one bar desirable. Unfortunately this pressure is not optimally matched to either optical pumping method. Both methods require glass cells that contain no ^{10}B, which strongly absorbs neutrons.

SEOP can polarize gas directly at pressures of several bar, but the time constant for polarizing a cold neutron spin filter is long (10 h-20 h). The process is most efficient at higher pressures (especially when broadband diode laser arrays are used), which introduces a mechanical stability issue for the large diameters needed for neutron beams. In contrast, MEOP is a rapid process, but the pressure range is $\sim$1 mbar. The gas must then be compressed without excessive loss of polarization. The typical overall time constant for polarizing a cold neutron spin filter is short (1 h-2 h).

The needs of neutron polarizer applications can be significantly different for different applications. Long term fundamental physics experiments require a polarizer that must operate continuously for long periods with minimal maintenance. For this application the simplicity and stability of a sealed SEOP cell is a natural choice. In contrast, neutron scattering laboratories may have several instruments that will periodically need polarized neutrons for relatively short periods of time. This requirement, combined with the space constraints on many instruments, make cells that are polarized off line and transported to the beam line a convenient option. In this case the relaxation time of the polarized gas in the absence of optical pumping becomes relevant, no matter what the time scale of production. The MEOP method has the advantage of rapid production of gas, but either method can be employed.

4.2 Spin-exchange optical pumping

In the SEOP method, electronic polarization is produced in rubidium atoms by optical pumping, and the polarization is transferred to ^{3}He nuclei by the hyperfine interaction during collisions. In current applications, the optical pumping is typically performed with high power diode laser arrays.[47] Although these lasers have a spectral width much greater than the pressure-broadened Rb absorption line (18 GHz/bar), they are effective and convenient. The recent development of narrowband, high power diode laser arrays for SEOP[48,49] should improve the efficiency, primarily for optical pumping at relatively low pressures (near one bar). The SEOP method was used for the first demonstration of a ^{3}He neutron spin filter at Los Alamos National Laboratory.[30] It is currently being pursued for neutron spin filters at the University of Michigan,[50] KEK in Japan[51], ISIS in the United Kingdom[52], and a collaboration between NIST, Indiana University, and Hamilton College.[53,54,55,56] The primary issue in the application of the SEOP method is the need for long relaxation time cells, and the conflicting requirements of small Rb volume and high pressure for the best optical pumping efficiency vs. large ^{3}He volume and small pressure differential for large diameter neutron beams. These issues are being addressed by different approaches: NIST is pursuing various sealed cell designs, while Michigan is pursuing decoupling the optical pumping from the neutron spin filter. In the Michigan system, gas is optically pumped in a high pressure cell that is optimized for optical pumping and then transferred through a valve to spin filter cell. This approach has desirable features for a "filling station" approach.

There are several constraints on the materials and design of sealed SEOP cells for long term experiments: impermeability to helium at temperatures of 450 K, compatibility with alkali metals, high neutron transmission, and relaxation times ∼100 h. NIST has developed cells from GE180,[57] a boron-free, aluminosilicate glass. For relatively modest diameters of up to 6 cm, 1.5 cm long cells that are filled to a pressure of 3.5 bar of ^{3}He have been employed. The most recent design is a shape similar to that of a meniscus lens, which allows for mechanical stability and a reasonably uniform gas thickness. The best cell lifetime is 170 h and typical ^{3}He polarization is presently 50%.

More recently, NIST has constructed cells up to 11 cm in diameter that are filled with 0.85 bar of ^{3}He. A positive feature of operation at such low pressure that the reduced dipole-dipole relaxation[58] allows for very long lifetimes, thus reducing the already significant demand on the laser power. Several large diameter cells that have multihundred hour lifetimes have been tested.[54] Some of these cells have lifetimes that approach the dipole-dipole limit of 930 hours, indicating that wall relaxation is minimal. In a test of a 9.5 cm diameter cell, 50% ^{3}He polarization has been obtained using 40 W from two broadband diode laser beams. A major motivation for this work is the npdγ experiment.[15] The results are significant for other fundamental physics experiments, such as the measurement of the beta decay asymmetry coefficients.

4.3 Metastability-exchange optical pumping

In the MEOP method, electronic polarization is produced by optical pumping of metastable atoms, and the polarization is rapidly transferred to the nucleus of the metastable atom via the hyperfine interaction. The electronic excitation in a metastable atom is transferred to a ground state atom during a collision, resulting in nuclear polarization of the new ground state atom. The newly excited metastable atom is then polarized by laser light. At the optimum pressure of 0.5 mbar, 80% polarization with a rapid polarizing rate (10^{18} s^{-1})

has been achieved using this method.[45] Based on the piston compressor system developed at the Univ. of Mainz for electron scattering experiments,[59,60] a similar apparatus is being employed for neutron applications at the Institut Laue-Langevin in Grenoble, France.[61] ^{3}He polarization of 55 % has been obtained, and a new compressor is under development.[62] Apparatus of this general design have been constructed at the Indiana University Cyclotron Facility[55,32] and the Hahn-Meitner Institute.[63,64]

NIST has developed a compact, simple, and inexpensive apparatus using a modified diaphragm pump.[65,66] Because of the relatively high range for the inlet pressure (1-3 mbar), the best results have been obtained using ^{3}He–^{4}He mixtures. Polarizations of 45%-65% polarization are obtained in the optical pumping cell (OPC), and 75% of this polarization is preserved during compression, yielding 30%-50% polarization in the storage cell. This apparatus has typically been used to fill cells for transport to the neutron beam line. Cells have been filled at liquid nitrogen temperature, which extends the range of achievable gas density, and/or allows for higher polarization. Whereas cells for the SEOP method are always coated with Rb, cells for the MEOP method can either be coated or uncoated. Alkali coatings have been shown to yield longer lifetimes.[67] Typical relaxation times for the uncoated cells used with the diaphragm pump apparatus are about 30 h. Some temporary degradation, after the cell is returned to room temperature following filling at liquid nitrogen temperature, has been observed. Storage cells up to 9.5 cm in diameter have been constructed, with optically sealed windows that permit a highly uniform thickness of ^{3}He. These cells are constructed using either GE180 or ^{10}B-depleted Corning 1720 glass[68] windows sealed to a cylindrical body of normal 1720 glass. Recently two 7 cm diameter, 10 cm long, Rb-coated cells made from "reblown" GE180 glass have been constructed for use with this apparatus, and each of these has a 200 h relaxation time. In a recent test, no significant degradation of the relaxation time was observed following filling with polarized gas at 77 K, and 55 % polarization was obtained in a 5 bar-cm neutron spin filter.

4.4 Neutron applications

Applications of ^{3}He based neutron spin filters are in progress. At the ILL, several neutron scattering experiments have been performed using ^{3}He spin filters filled using their compression apparatus.[69,70,71] This apparatus has also been used for a parity violation experiment in lanthanum[72] and a high accuracy comparison of different techniques for measuring neutron polarization.[73]

At NIST, both SEOP and MEOP spin filters have been tested on the

NG-6M monochromatic beam line[56], and ^{3}He-based polarization analysis has been demonstrated for small angle neutron scattering.[74] Studies of separating magnetic from nuclear scattering in a colossal magnetoresistive material and in $ErNi_2B_2C$ are in progress.

A highly precise test of measuring neutron polarization using time of flight analysis was conducted at LANSCE.[28] At the Intense Pulsed Neutron Source at Argonne National Laboratory, a polarized ^{3}He cell was transported from the IUCF compression apparatus for study of Zeeman splitting in polarized neutron reflectometry.[75,55] Increased applications can be expected as the technology matures and becomes available at more laboratories.

References

1. R.M. Moon, T. Riste, and W.C. Koehler, Phys. Rev. A **181**, 920-931 (1969).
2. G.L. Squires, Introduction to the Theory of Thermal Neutron Scattering, (Cambridge University Press, 1978), p. 189.
3. G. Shirane, Physica B **137**, 43 (1986) and references therein.
4. G.P. Felcher, K E. Gray, R.T. Kampwirth and M.B. Brodsky, Physica B **136**, 59 (1986), and references therein.
5. R. M. Moon, Physica B **137**, 19 (1986), and references therein.
6. M. Rekveldt, Neutron News **4**, 15 (1993), and references therein.
7. F. Mezei, Z. Phys. **255**, 146 (1972).
8. H.B. Stuhrmann, Physica B **156-157**, 444-451 (1989).
9. J.D. Jackson, S.B. Treiman, and J.H.W. Wyld, Phys. Rev. **106**, 517 (1957).
10. Neutron decay experiments are reviewed in B.G. Yerozolimsky, Nucl. Instrum. Meth. **440**, 491-498 (2000).
11. J. Reich *et al*, Nucl Instrum. Meth. **440**, 535-538 (2000).
12. A.P. Serebrov *et al*, JETP **86**, 1074-1082 1998).
13. L.J. Lising *et al*, Phys. Rev. C **62**, 055501 (2000).
14. W.S. Wilburn *et al*, in *Fundamental Physics with Pulsed Neutron Beams, Research Triangle Park, North Carolina, June 1-3, 2000*, edited by C.R. Gould, G.L. Greene, F. Plasil, and W.M. Snow (World Scientific, Singapore, 2001), pp. 214-223.
15. W.M. Snow *et al*, ibid., pp. 203-213; W.M. Snow *et al*, Nucl. Instrum. Meth. A **440**, 729-735 (2000).
16. D.M. Markoff and F.E. Wietfeldt, ibid., pp. 192-202.
17. W. Gavin Williams, Polarized Neutrons, (Oxford, New York, 1988).
18. F. Mezei, Commun. Phys. **1**, 81 (1976).

19. P. Courtois, Physics B **267-268**, 363-366 (1999).

20. A. Freund et al, Physica **120B**, 86-90 (1983).

21. G.A. Keyworth *et al*, Phys. Rev. C **8**, 2352-2363 (1973).

22. T. Krist, C. Lartigue and F. Mezei, Physics B **180-181**, 1005-1006 (1992).

23. Carson D. Jeffries, Dynamic Nuclear Orientation, (Wiley, New York, 1963).

24. A. Abragam and M. Goldman, Nuclear Magnetism: Order and Disorder (Oxford, New York, 1982).

25. P.P.J. Delheij, Nucl. Instrum. Meth. A **356**, 120-121 (1995).

26. N.R. Roberson *et al*, Nucl. Instrum. Meth. A **326**, 540-565 (1993).

27. M. Iinuma *et al*, Phys. Rev. Lett. **84**, 171-174 (2000).

28. D.R. Rich, Ph.D. thesis, Indiana University, 1999; D.R. Rich *et al*, accepted by Nucl Instrum. Meth.

29. G.L. Greene, A.K. Thompson, M.S. Dewey, Nucl. Instrum. Meth. A **356**, 177 (1995).

30. K.P. Coulter *et al*, Nucl. Instrum. Meth. A **288**, 463 (1990).

31. D.J. Goossens and L.D. Cussen, Physics B **267-268**, 348-351 (1999).

32. C.D. Keith *et al.*, Nucl. Instrum. Meth. A **402**, 236 (1998).

33. F. Tasset and E. Ressouche, Nucl. Instrum. Meth. A **359**, 537-541 (1995).

34. L.D. Cussen, D.J. Goossens, and T.J. Hicks, Nucl. Instrum. Meth. **440**, 409-420 (2000).

35. J. Kulda *et al.*, Physica B **241-243**, 136-138 (1998).

36. D. DeSchepper *et al.*, Nucl. Instrum. Meth. A **419**, 16-44 (1998).

37. J.R. Johnson et al, Nucl. Instrum. Meth. A **356**, 148-152 (1995).

38. C.E. Jones *et al.*, Phys. Rev. C **47**, 110 (1993).

39. G. Eckert *et al.*, Nucl. Instrum. Meth. A **320**, 53-65 (1994).

40. T.G. Walker and W. Happer, Rev. Mod. Phys. **69**, 629-642 (1997).

41. W. J. Cummings *et al.*, Phys. Rev. A **51**, 4842-4851 (1995).

42. M.E. Wagshul and T.E. Chupp, Phys. Rev. A **49**, 3854-3868 (1994).

43. M.A. Bouchiat, T.R. Carver, and C.M. Varnum, Phys. Rev. Lett. **5**, 373 (1960).

44. E. Stoltz *et al.*, Appl. Phys. B **63**, 629 (1996).

45. T.R. Gentile and R.D. McKeown, Phys. Rev. A **47**, 456-467 (1993).

46. F.D. Colegrove, L.D. Schearer, and G.K. Walters, Phys. Rev. **132**, 2561-2572 (1963).

47. Coherent Semiconductor Group, 5100 Patrick Henry Drive, Santa Clara, CA, USA 95054. Certain trade names and company products are mentioned in the text or identified in an illustration in order to adequately specify the experimental procedure and equipment used. In no case does

such identification imply recommendation or endorsement by the National Instutue of Standards and Technology, nor does it imply that the products are necessarily the best available for the purpose.

48. B. Chann, I. Nelson, and T.G. Walker, Opt. Lett. **25**, 1352 (2000).

49. J. N. Zerger, M. J. Lim, K. P. Coulter, and T. E. Chupp, Appl. Phys. Lett. **76**, 1798 (2000).

50. T.E. Chupp and K.P. Coulter, in *Fundamental Physics with Pulsed Neutron Beams, Research Triangle Park, North Carolina, June 1-3, 2000*, edited by C.R. Gould, G.L. Greene, F. Plasil, and W.M. Snow (World Scientific, Singapore, 2001), pp. 181-191.

51. K. Sakai *et al.*, Nucl. Instrum. Meth. A **402**, 244-246 (1998).

52. M.J. Barlow and S. Parnell, private communications.

53. T.R. Gentile *et al*, in *Fundamental Physics with Pulsed Neutron Beams, Research Triangle Park, North Carolina, June 1-3, 2000*, edited by C.R. Gould, G.L. Greene, F. Plasil, and W.M. Snow (World Scientific, Singapore, 2001), pp. 230-234.

54. D.R. Rich *et al.*, submitted to Appl. Phys. Lett.

55. D.R. Rich *et al.*, Physica B, in press.

56. G.L. Jones *et al.*, Nucl. Instrum. Meth. A **440**, 772-776 (2000).

57. GE Lighting Component Sales, Bldg. 315D, 1975 Noble Rd., Cleveland, OH 44117.

58. N.R. Newbury *et al.*, Phys. Rev. A **48**, 4411 (1993).

59. J. Becker *et al.*, Nucl. Instrum. Meth. A **402**, 327-336 (1998).

60. J. Becker *et al.*, Nucl. Instrum. Meth. A **346**, 45-51 (1994).

61. W. Heil *et al.*, Physica B **267-268**, 328-335 (1999).

62. T.W. Roberts *et al.*, Physica B **297**, 282-287 (2001).

63. A. Danzig, K. Habicht and A. Rupp, Physica B **276-278**, 185-186 (2000).

64. A. Danzig and A. Rupp, Physica B **267-268**, 344-347 (1999).

65. T.R. Gentile *et al*, J. Res. Natl. Inst. Stand. Technol. **106**, 709-729 (2001).

66. T.R. Gentile *et al.*, Magn. Reson. Med. **43**, 290 (2000).

67. W. Heil *et al.*, Phys. Lett. A **201** 337 (1995).

68. Corning Glass, Corning, NY 14831.

69. B. Nickel *et al.*, Rev. Sci. Instrum. **72**, 163 (2001).

70. M. Acet *et al.*, Physics B **276-278**, 728-729 (2000).

71. J. Kulda *et al.*, Physica B **267-268**, 252-254 (1999).

72. W. Heil *et al.*, Physica B **267-268**, 289-293 (1999).

73. O. Zimmer *et al.*, Nucl. Instrum. Meth. A **440**, 764-771 (2000).

74. T.R. Gentile *et al.*, J. Appl. Crystallog. **33**, 771-774 (2000).

75. D. Hussey *et al.*, submitted to Appl. Phys. A.

POLARIZED NEUTRON FACILITY AT RCNP

H. SAKAI[1], K. YAKO[1], Y. MAEDA[1], A. TAMII[1], H. OKAMURA[2], K. SUDA[2],
T. WAKASA[3], J. KAMIYA[3], AND K. HATANAKA[3]

(1) Department of Physics, University of Tokyo, Tokyo 113-0033, Japan
(2) Department of Physics, Saitama University, Saitama 338-8570, Japan
(3) Research Center for Nuclear Physics (RCNP), Osaka 567-0047, Japan

A polarized neutron facility was constructed at Research Center for Nuclear Physics
(RCNP), Osaka University. The facility provides a quasi-mono-energetic neutron
beam of 200–350 MeV with a flux of 2×10^6/s/3 cm^W $\times$ 2 cm^H at target position.
The experiment for the $\vec{n} + d$ scattering at E_n =250 MeV and ^{90}Zr,^{27}Al$(\vec{n}, p)$
reaction at E_n =293 MeV have been carried out successfully.

1 Introduction

The Gamow-Teller (GT) quenching problem is one of the most interesting
topics. The quenching value provides information on the Landau-Migdal parameters such as g'_{NN} or $g'_{N\Delta}$ which represent the short-range correlations
of isospin-spin interactions in nuclei. We have recently deduced the quenching value of 0.90 ± 0.05 by using our precise measurement[1] of the ^{90}Zr(p, n)
reaction at 295 MeV for the $S_{\beta-}$ strength combined with the TRIUMF $S_{\beta+}$
value[2] derived from the ^{90}Zr(n, p) reaction at 200 MeV. This quenching value
gives $(g'_{NN}, g'_{N\Delta})= (0.6, 0.2)$[3]. Suppose $g'_{N\Delta}= 0.2$, a critical density ρ_c of
pion condensation becomes as small as $\rho_c \sim 2$. However the ρ_c value depends strongly on $g'_{N\Delta}$, consequently the quenching value. Therefore it is of
crucial importance to extract a reliable quenching value. One way to reduce
the systematic uncertainties largely is to measure the ^{90}Zr(n, p) reaction at
300 MeV as discussed in Ref. 3 in which the importance of the choice of the
bombarding energy is explained. For this purpose we have constructed the
$(\vec{n}, p)$ facility at RCNP.

2 $(\vec{n}, p)$ facility

The facility consists of the neutron production target, the clearing magnet to
sweep away the primary proton beam, the $(\vec{n}, p)$ target chamber box and the
existing Large Acceptance Spectrometer (LAS) to detect an ejectile charge
particle. A schematic overview of the $(\vec{n}, p)$ facility is shown in Fig. 1.

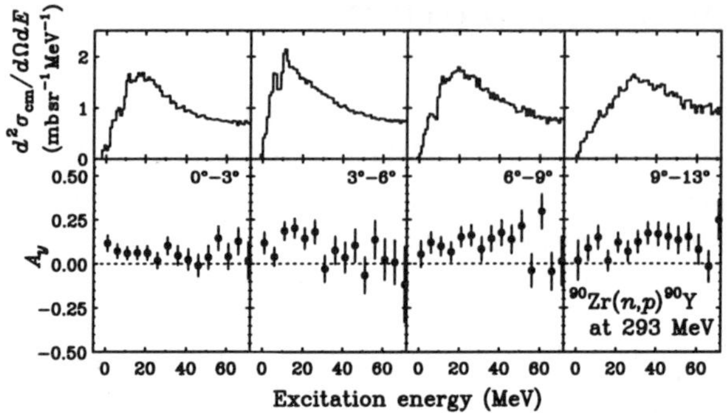

Figure 1. A schematic view of the $(\vec{n}, p)$ facility.

2.1 $\vec{n}$ production, clearing magnet, etc.

The central part of the $(\vec{n}, p)$ facility is shown in Fig.2. The incident proton beam strikes the neutron production target ^{7}Li and it is then bent away by the clearing magnet to a beam dump in the floor of the experimental room about 15 m away.

The clearing magnet has a C-shape and its maximum magnetic field strength is 1.85 T with a 5 cm gap. The deflection angle is 23.2°. A field clamp is attached to reduce the stray field.

The polarized neutron is produced via the ^{7}Li$(\vec{p}, \vec{n})$ reaction. Produced neutrons go through the clearing magnet, the veto scintillator, and then into the $(\vec{n}, p)$ target box which locates 96 cm downstream of the neutron production target.

In order to reduce the background a charged particle blocker as well as collimators are placed inside/outside of the vacuum chamber as shown in the figure.

2.2 (n, p) target box

The (n, p) target box is located over the LAS pivot to enable (n, p) angular distributions to be measured. Figure 3 shows front and side views of the target chamber box which consists of the four (n, p) target layers and multiwire drift chamber planes (MWDC1). Each (n, p) target is sandwiched by MWDC planes to determine the target layer in which the (n, p) reaction occurs. A correction for the energy loss in the downstream target layers can be done to keep a good energy resolution. The position information of MWDC1 and

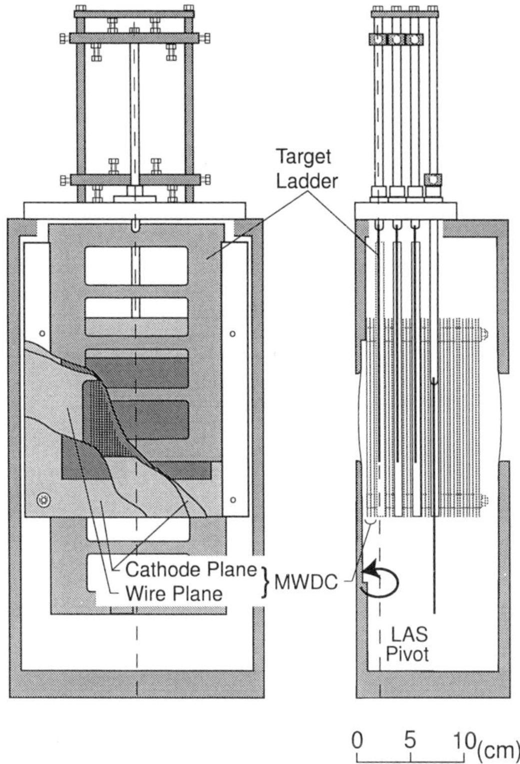

Figure 2. The central part of the $(\vec{n}, p)$ facility.

MWDC2 provides a scattering angle information to an accuracy of $0.1°$.

2.3 Large Acceptance Spectrometer (LAS)

The LAS works as a charged particle recoil spectrometer. It has nice characteristic features such as an intrinsic resolution ΔE of less than 300 keV, a momentum bite $\Delta p/p$ of about 35% and an angular acceptance $\Delta \Omega$ of about 10 msr. Note that the MRS at TRIUMF charge-exchange facility has $\Delta E = 100$ keV, $\Delta p/p = 15\%$ and $\Delta \Omega = 2.5$ msr.

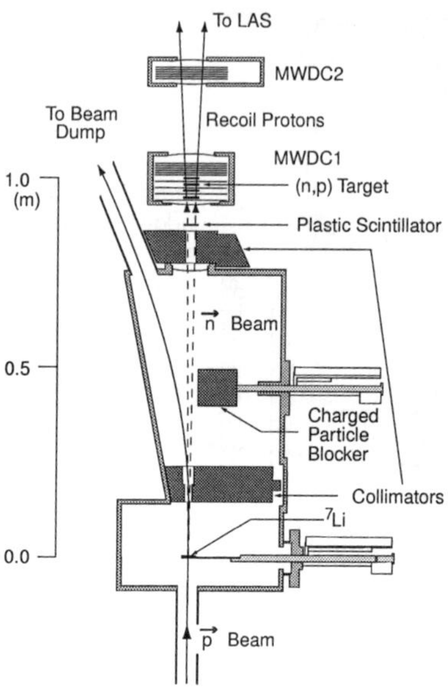

Figure 3. Front and side views of the (n, p) target box.

3 Experiments

The overall (n, p) facility was examined by measuring the $\vec{n}$-p scattering at 245 MeV since its cross section $\frac{d\sigma}{d\Omega}$ and analyzing power A_y are well known. The polyethylene (CH_2) sheet with a thickness of 189 mg/cm^2 was used as a hydrogen target. Obtained angular distributions $\Theta_{cm} = 85° - 180°$ of $\frac{d\sigma}{d\Omega}$ and A_y showed a good agreement with the prediction by the phase-shift analysis[5].

We have completed the measurements for the ^{90}Zr,^{27}Al $(\vec{n}, p)$ reaction at 293 MeV and $\vec{n} + d$ scattering at 245 MeV.

A typical example of energy spectrum $\frac{d^2\sigma}{d\Omega dE}$ and analyzing power A_y is shown in Fig. 4 for the ^{90}Zr$(\vec{n}, p)$ reaction. Experimental conditions are: proton beam intensity $I_p = 450$ nA, proton polarization $P_p = 0.7$, thickness of ^{7}Li target 6 mm, thickness of ^{90}Zr target 230 mg/cm^2 ×3 layers; 690 mg/cm^2 in total. The neutron flux at the (n, p) target position was $\phi = 2 \times 10^6$/s/3 cm^W × 2 cm^H.

An energy resolution of about 1.5 MeV was achieved. The main contributions to the energy resolution are due to the proton beam energy spread of 0.7 MeV, to the proton beam energy loss of 0.9 MeV in the neutron production

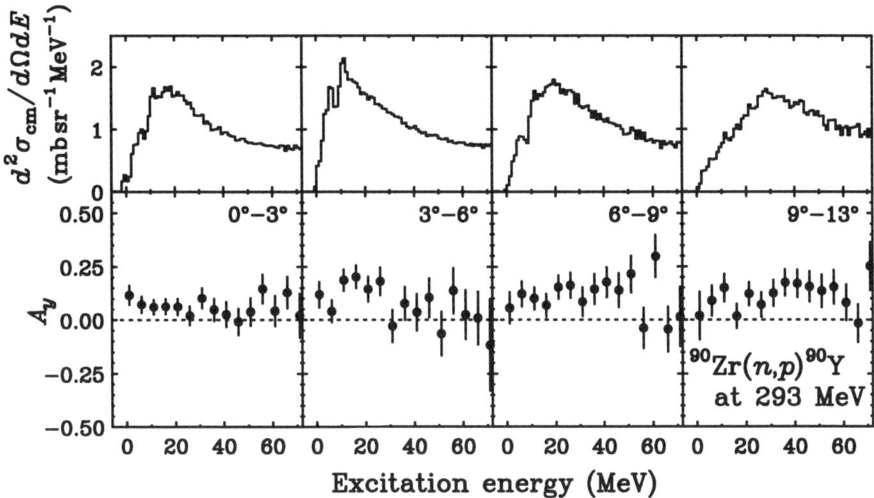

Figure 4. $\frac{d^2\sigma}{d\Omega dE}$ and A_y for the ^{90}Zr$(\vec{n},p)$ reaction at 293 MeV.

target ^{7}Li and to the recoiled-proton energy loss of 0.9 MeV in the ^{90}Zr target.

The final (n,p) energy spectrum is obtained after unfolding the observed raw proton spectrum by the incident neutron energy spectrum which is produced by the ^{7}Li(p,n) reaction and therefore it has a continuum spectrum together with a peak.

The energy spectra up to an excitation of 70 MeV are obtained at once thanks to the large momentum bite of LAS. Neither $\frac{d^2\sigma}{d\Omega dE}$ nor A_y shows a large structure. The statistical accuracy of A_y suffers from the smallness of neutron polarization ~ 0.2. By using these spectra, the precise $S_{\beta+}$ value will be derived[6].

References

1. T. Wakasa *et al.*, Phys. Rev. **C55** (1997) 2909.
2. K.J. Raywood *et al.*, Phys. Rev. **C41** (1990) 2836.
3. T. Suzuki and H. Sakai, Phys. Lett. B**455**(1999) 25.
4. H. Sakai, Nucl. Phys. **A690** (2001) 66c.
5. R.A. Arndt and L.D. Roper, *Scattering Analysis Program* (SAID).
6. K. Yako, PhD thesis. 2001, University of Tokyo.

DEVELOPMENT OF POLARIZED ^{3}He-BASED NEUTRON SPIN FILTERS AT IUCF

W.M. SNOW[1], D. HUSSEY[1], C. BAILEY[1], A. BELOV [1], T.B. SMITH[1], T.R. GENTILE[2], D.R. RICH[2], A.K. THOMPSON[2], AND G.L. JONES[3]

(1) Indiana University, Bloomington, IN
(2) National Institute of Standards and Technology, Gaithersburg, MD
(3) Hamilton College, Clinton, NY

The large spin dependence of the cross section for neutron absorption by ^{3}He meams that a sample of polarized ^{3}He gas is a transmission polarizer. In neutron scattering the main advantage of ^{3}He-based spin filters is the large angular acceptance relative to supermirrors. In addition, ^{3}He-based neutron polarizers have advantages for the measurement of absolute neutron polarization of cold neutron beams. At IUCF we have constructed a system based on metastability-exchange optical pumping to produce polarized ^{3}He gas with subsequent nonmagnetic compression to produce polarized ^{3}He target cells for neutron scattering experiments. We will describe this system, its performance, and its use in a neutron scattering experiment in polarized neutron reflectometry at the Argonne IPNS. We will also discuss a measurement of the absolute neutron polarization produced by a polarized ^{3}He target at LANSCE of 0.3% absolute accuracy.

DEVELOPMENT AND TESTS OF ^{3}He NEUTRON SPIN FILTERS AND CONSTRUCTION OF A FILLING STATION

A. GORZEL, K. HABICHT, V. HUTANU, J. KLENKE, <u>A. RUPP</u> AND A. WIEDENMANN

Hahn-Meitner-Institut Berlin GmbH, Glienicker Str. 100, D-14109 Berlin, Germany

Polarized ^{3}He to be used as a neutron spin filter shows certain advantages, practically resulting in unlimited angular acceptance, useability over a broad range of neutron wavelengths and a total decoupling from other neutron optics (see e.g. [1]). Hence, the development of suitable filter cells and the construction of a filling station to supply them is a focal point in the instrumental development program at HMI.

In our case, ^{3}He becomes polarized by metastability exchange optical pumping [2] useing a cw Nd:LNA laser with 5.8 W output power and 2.7 GHz bandwidth. Optimization of optical pumping has been performed with sealed-off DURAN glass cells. By pumping on the C8 line polarization degrees of 80 $\pm$ 2% have been reached reliably while the time constant for its build-up was $\tau = 17$ - 20 s.

To enhance the ^{3}He pressure in order to enable the use in neutron spin filter cells (p $\leq$ 3 bar), an appropriate piston compressor is under construction, designed to provide a total compression ratio of at least 1 : 3000 via a two stage process while mastering a gas flow of up to 10 mbar l s^{-1}. To integrate it, rebuilding of the filling station for a flow through type of operation is under the way. Building the whole system completely from suitable amagnetic materials and sizing all components carefully promises polarization losses during compression to be at least less than 7% in relative.

After preliminary tests with sealed-off cells, various refillable prototype filter cells made of quartz glass have been prepared, coated with Cs which is known to extend the relaxation time significantly [3]. Transmission and scattering experiments have been performed at the 3-axis spectrometer E1 and at the SANS instrument V4, both at HMI. Decay time constants between 33 h and 50 h have been observed. NMR measurements confirmed these results. The transmission of the empty cell was found to be about 80%. Perturbation due to scattering by the cells themselves turned out to be negligible. The cells have been filled at the University of Mainz and were brought to the experiments immediately. A special transportation unit [4] keeps the cell in a magnetic guide field of high homogeneity during both, transportation and use.

The authors want to thank Prof. W. Heil and J. Schmiedeskamp for the supply with compressed, polarized ^{3}He.

References

1. H. Humblot et al., Neutron News **8** **(4)** (1997) 27.
2. F. D. Colegrove, L. D. Schearer, G. K. Walters, Phys. Rev. **132** (1963) 2561.

3. . Heil et al., Phys. Lett. **A 201** (1995) 337.
4. A. Danzig, A. Rupp, Physica B **267 and 268** (1999) 344.

VI. Polarimetry

MEASURING THE PROTON BEAM POLARIZATION BEYOND 1 GeV

YOUSEF I. MAKDISI

Brookhaven National Laboratory
RHIC Project
Upton, NY 11973

Polarimeters are necessary tools for measuring and maintaining the beam polarization during the acceleration process as well as a yardstick for performing spin physics experiments. In what follows, I will describe the principles of measuring proton beam polarization, and the techniques that are employed at energies above 1 GeV. I will use as a guide, the work done for the Polarized Proton effort at the BNL Relativistic Heavy Ion Collider (RHIC).

1 Introduction

A polarimeter measures the degree of beam polarization. Polarmeters are used at various stages: as diagnostic tools to tune and maintain the polarization of the beam during acceleration, storage, and beam transport. Of course, the goal is the physics measurement and polarmeters are often employed as a first stage of an experiment to measure the polarization of the beam impinging on the target. This then normalizes the final result and the error in the polarization measurement has a direct statistical impact on the data. The desired accuracy of 5%, represents a lofty goal for high energy proton beam polarimeters.

2 Beam Polarization

The degree of beam polarization along a certain direction is defined as:

$$P = \frac{N_\uparrow - N_\downarrow}{N_\uparrow + N_\downarrow}$$

Where $N_\uparrow$ and $N_\downarrow$ are the number of protons with spins parallel and antiparallel to the desired direction. In an accelerator, the stable spin direction is usually transverse to the momentum vector and along the vertical. The task is to find proton induced reactions that are sensitive to the spin alignment of the beam. The yield then depends on whether the beam is polarized up or down and is reflected in the number of scatters or events of particular interest measured in the apparatus.

Typically one measures the number of events with the beam polarized up versus those with the beam polarized down making sure that the experimental conditions remain unchanged. The resulting beam polarization is:

$$P = \frac{1}{A} \frac{n\uparrow + n\downarrow}{(n\uparrow + n\downarrow)}$$

$n\uparrow$ and $n\downarrow$ are the number of scatters, normalized to the incident beam, with the beam polarization up and down respectively and A is the analyzing power. Similarly, an apparatus that can measure the number of scatters to beam-left, n_L, and beam-right, n_R, simultaneously will determine the degree of up and down beam polarization independently and one substitutes n_L and n_R in the above equation. In general both methods are utilized and combined in order to reduce systematic errors and potential dependence on geometrical effects.

The associated statistical error in these measurements is a function of both the analyzing power A and the total number of events N [$(n\uparrow + n\downarrow)$ or $(n_L + n_R)$ or the sum of all four terms if a combined measurement is performed]

$$\Delta P = \frac{1}{A} \frac{1}{\sqrt{N}}$$

It is therefore important to utilize reactions with large analyzing power as well as large cross section. In the design of a polarimeter, the quantity to optimize is the product: $N A^2$. H. Spinka [2] gives a discussion of polarimetry ratios and errors.

3 Desired features of a polarimeter

In general a beam polarimeter should provide the following:

- A polarization monitor capable of several samples over a reasonable period.
- A tool that can be sampled on demand, online and a turnkey operation.
- A tool for machine tuning with feedback provided within a few minutes.
- A large dynamic range is desired specially when dealing with the acceleration cycle. At RHIC the range is from 25 GeV at injection to 250 GeV at top energy.
- A large analyzing power, high cross section and low background contamination.

- Finally, the fiscal constraints require building such apparatus at reasonable cost.

It is not always possible to combine all these features and certain compromises are often necessary.

4   Proton beam polarimeter reactions, the empirical way

Unlike electromagnetic reactions (see the presentation by J. Grames on electron beam polarimeters in these proceedings) the analyzing power and cross section of proton induced nuclear reactions are not precisely calculable especially at high energies. Thus one reverts to experimental results.   These measurements are generally done with polarized proton targets the polarization of which is well measured using NMR techniques and Masers. Until recently, these targets were not pure hydrogen thus the target material presents undesired background especially in inclusive measurements. However, this is not an impediment for exclusive reactions such as p-p elastic scattering when both outgoing particles are measured.

At relatively low energies below 12 GeV, good accuracy 3%-5% and can be attained in beam polarization measurements.   This was done at the Argonne National Laboratory ZGS and Saturne at Saclay. Common wisdom had it that at higher energies spin effects are less pronounced and the analyzing power becomes increasingly small resulting in reduced accuracy.

At Brookhaven National Laboratory, an effort is underway to equip the RHIC, with the capability to accelerate, store, and collide polarized proton beams at high luminosity [1]. Polarimeters will be deployed to cover the energy ranges; the AGS: 3 – 24 GeV; and RHIC: 24 – 250 GeV. I will discuss the choice of the physics processes and associated apparatus for each, but first a few words about RHIC.

5   The Relativistic Heavy Ion Collider

RHIC is in the second year running for physics. For heavy ions, the collider uses the Tandem-Booster-AGS as an injector to accelerate, store and collide beams of various species from light ions to gold. The design luminosity for colliding gold beams is 2×10^{26} cm^{-2} sec^{-1}. The running experiments comprise two large detectors, STAR and PHENIX, and two smaller ones, BRAHMS and PHOBOS. The physics goal is to study the new state of matter under conditions of extreme temperature and pressure, the quark gluon plasma.

The polarized proton capability utilizes a polarized H⁻ source and the 200 MeV Linac-Booster-AGS injector. Funding is provided by RIKEN, Japan, to equip the collider with the necessary hardware. This includes two "Siberian Snakes", in each ring, to preserve the beam polarization. Two sets of spin rotators around the STAR and PHENIX experiments that orient the beam spins in the longitudinal or transverse

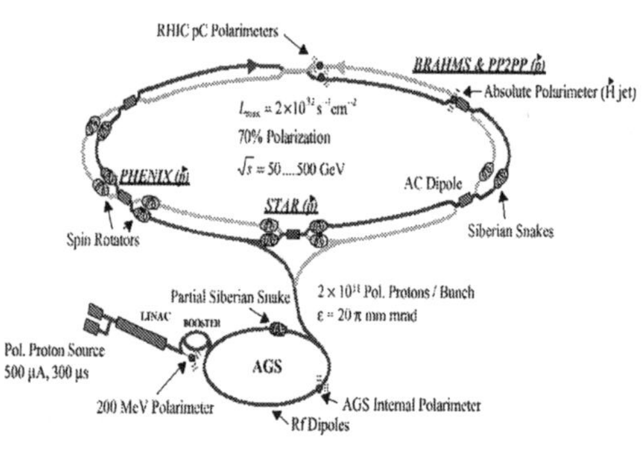

Figure 1. The polarized proton complex at RHIC

directions depending on the physics demands. Beam polarimeters to tune the machine and measure the beam polarization during the store. The goal is the determination of the contribution to the proton spin from the constituent quarks, antiquarks, and gluons. A fifth experiment pp2pp will be installed prior to the start of the polarized proton run this year.

The respective polarized beam intensities are: at the exit of the Linac 300 uA over a 300 µsec spill; the AGS circulating beam intensity of 2×10^{11}; and in RHIC 2×10^{11} polarized protons per bunch with a bunch spacing of approximately 100 nsec. A luminosity at $\sqrt{s}$ of 500 GeV is 2 $\times10^{32}$ cm^{-2} sec^{-1} and 70% beam polarization, figure 1.

6   Candidate processes for beam polarimetery

Several reactions serve as potential candidates for polarimetry. These are:

Transverse beam polarization
- p-p Elastic scattering near t ~ 0.1- 0.3 GeV^2/c^2 . This has been the workhorse at various accelerators at energies below 25 GeV/c.
- Primakoff Production $P + Z \rightarrow \Delta / N^* + Z \rightarrow \pi^o + p + Z$ ($\gamma + p \rightarrow p + \pi^o$) in the Coherent Coulomb region.at t < 0.001. A large analyzing power of -0.57 +/- 0.12 +/- .20 was observed in E704 at Fermilab. A large nuclear background under the coulomb peak can be extracted with data from two targets of high Z material.
- Asymmetry in inclusive Pion Production.
- p-Carbon and p-p elastic scattering in the CNI region.

Longitudinal beam polarization
- p-e in Deep Inelastic Scattering
- p-e elastic scattering. Calculable, large analyzing power A_{LL}, A_{SL}

I will focus on pp elastic scattering, inclusive pion spectrometry, and p-Carbon CNI. Due to lack of space, I will not discuss the longitudinal polarimetery [10].

7 Polarimeters vs energy: pp elastic scattering

Most polarimeters at energies below 25 GeV employ the pp elastic scattering near t ~ 0.1- 0.3 GeV^2/c^2. The elastic cross section is high but falls off with increasing momentum transfer and energy. The analyzing power (A) was initially measured using polarized targets with statistical errors of the order of 10%. The analyzing power falls approximately as 1/p (the incoming beam momentum). The data, selected energies shown in Figure 2, were fitted and parameterized [2]. At the AGS energies the analyzing power ranges between 10 and 3 %.

A typical p-p elastic polarimeter has two arms to simultaneously measure the forward scattered and recoil protons. However, the physical constraints in the AGS [3], the vacuum pipe and the 10ft long straight sections, do not permit the placement of the forward arms that subtend a few degrees from the beam line. The elastic scattered recoil protons, at kinetic energy of approximately 500 MeV and scattering angle of approximatcly 76 degrees with respect to the beam direction, are measured using time of flight, energy range, and energy angle correlation. The detectors are scintillation counters. Thus it is a counting experiment of the events that pass some set constraints.

From figure 2, one can discern larger analyzing powers

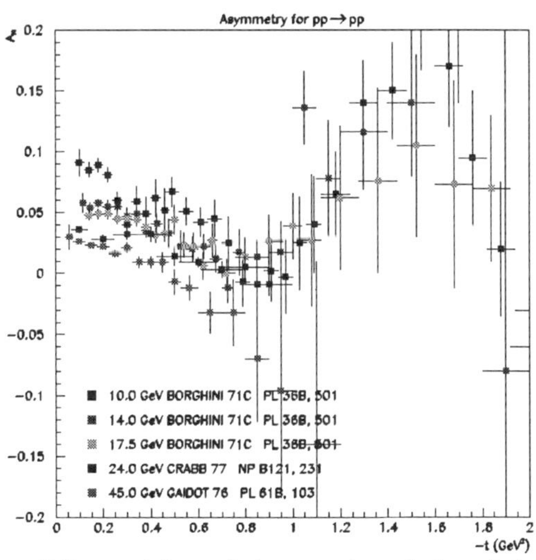

Figure 2. Representative analyzing power in pp elastic scattering.

can be attained at higher momentum transfer. However, the accuracy is progressively worse due to lower cross sections. High beam intensities require special targets such as carbon that can withstand beam heating. The environment in an accelerator such as the AGS is extremely harsh. The high counting rates necessitate debunching the beam during the polarization measurement. Pileup leads to accidental rates of the order of 10-20 % depending on the energy. A computer controlled nylon $C_6H_{11}NO$ fish line target is spooled at a rate of 100 cm/sec and flipped in and out of the beam in order to avoid damage due to heating or localized radiation. The loss of a target results in a few hours downtime. A 1 % statistical measurement takes 5 minutes at low energy to 20 minutes at the higher energies. A second carbon target is periodically substituted for the fish line in order to measure and subtract the carbon contamination. This apparatus serves as a relative polarimeter to tune the 50 imperfection and four intrinsic and depolarizing resonances in the AGS. Lately, it was decided to use the carbon target only and also install forward counters for a better measurement at lower energies. While this relieves the need for spooling and beam debunching, the effective analyzing power is reduced by approximately 50%. For physics purposes this polarimeter is calibrated to an accuracy of 10-12 % against an external polarimeter that measures the fully constrained p-p elastic scattering process. Figure 3 depicts a typical experiment, AGS E925, that combines both a full two-arm elastic scattering polarimeter as well as an inclusive pion spectrometer the results from which are described later.

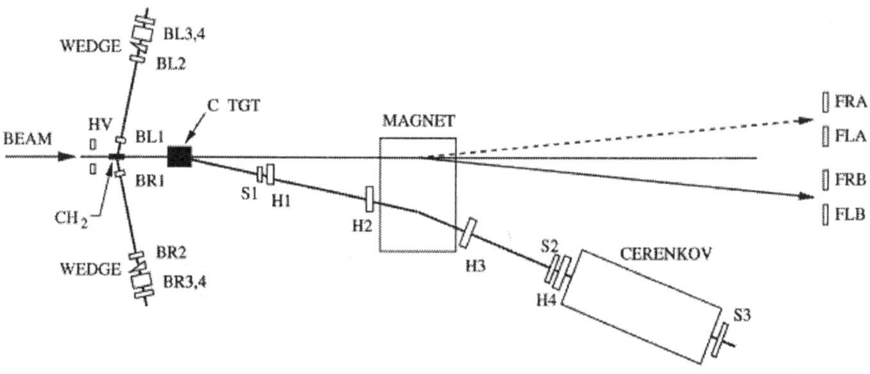

Figure 3. A schematic of the E925 apparatus at the AGS

8   Polarimeters versus energy: inclusive pion production

As the beam energy increases, the reach of the AGS-type pp elastic polarimeter becomes increasingly difficult. Experiments using polarized proton beams scattering from unpolarized hydrogen targets, carried out at 12 GeV/c at the ZGS [4] and at 200 GeV/c at Fermilab [5], observed surprisingly large asymmetries reaching over 30% in the inclusive production of pions, Figure 4. The copious production of pions and the large analyzing power seem to be ideally suited for a polarimeter application. The Fermilab data and associated systematics provide a 7 % absolute measurement which may be good for beam commissioning and the early physics results.

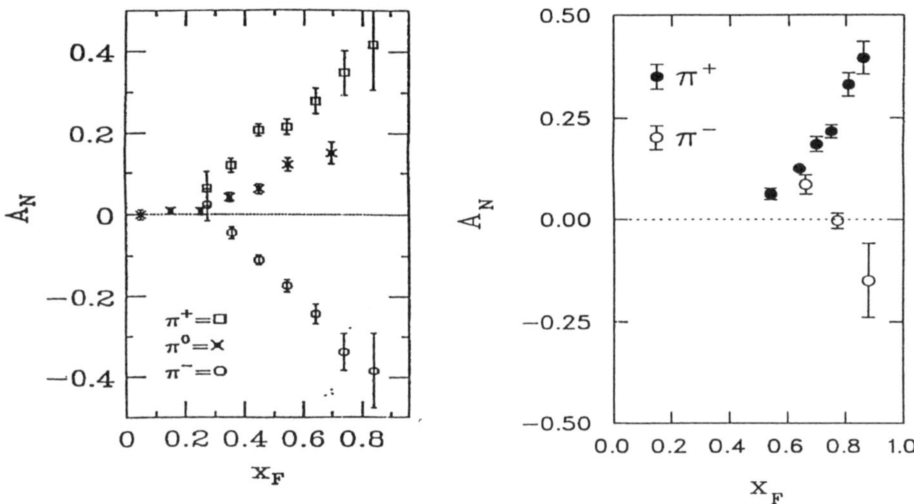

Figure 4. The analyzing power in inclusive pion production at 200 GeV/c and 12 GeV/c

The requirement for carbon targets, presented additional issues. Does the production from a nuclear target dilute the observed asymmetry from hydrogen? Is the asymmetry large enough at lower energies? A measurement of the pion asymmetries [6] using the AGS polarized proton beam on hydrogen, carbon, and CH_2 targets at 22 GeV shows that the observed mirror asymmetries persist at lower energies and there appears to be little or no target dependence. Thus negative pion production is a viable candidate. This simplifies the polarimeter design as no particle identification is needed and backgrounds are negligible.

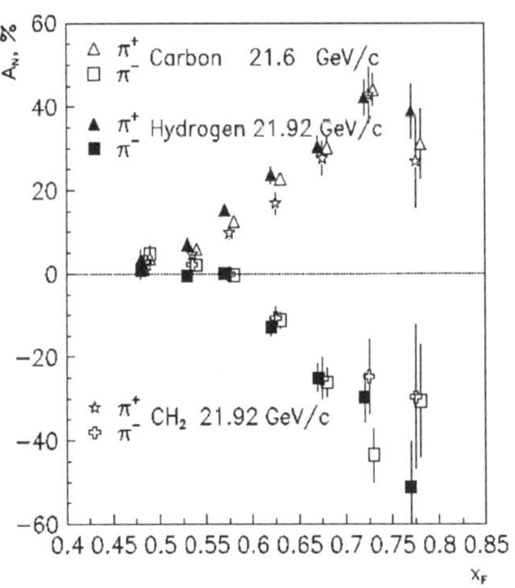

Figure 5. Analyzing power in inclusive pion production at 22 GeV/c on Hydrogen, Carbon , and CH_2.

In the design of an inclusive pion polarimeter for RHIC the choice of parameters utilized the data from Fermilab. One can parameterize both the analyzing power and the cross section data in terms of the kinematic variable x_f (~ the ratio between the scattered pion momentum to the beam momentum) and then optimize the value A^2N in order to determine the ideal conditions. For $p_t > 0.7$ GeV/c this resulted in $x_f \sim 0.5$ and a respective analyzing power of 15%. Choosing an acceptable transverse momentum of the outgoing pion ($p_t = 0.8$ GeV/c) determines the scattering angle, the geometry and the production cross section. The scattering angle ranges from 64 mrad to 6.4 mrad as the energy is raised from 25 to 250 GeV. For a 7% statistical measurement and at 15% analyzing power one requires 10^4 scattered pions in the detector. The measurement takes less than a minute.

9 Polarimeters at high energies: p-p and p-C scattering in the Coulomb Nuclear Interference region

The analyzing power in the Coulomb Nuclear Interference region ($10^{-3} < t < 10^{-2}$) in p-p elastic scattering arises from the interference between the real electromagnetic helicity-flip amplitude and the imaginary hadronic helicity-nonflip amplitude. This is shown in Figure 6. Unlike the previous process, this is

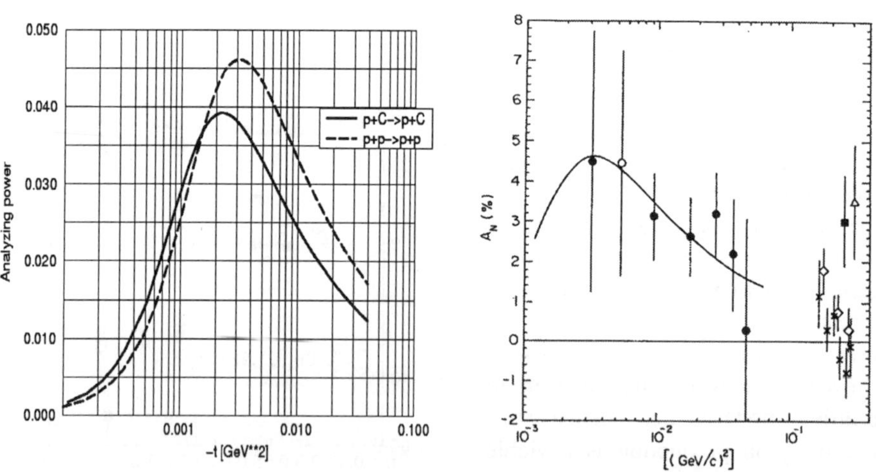

Figure 6. Analyzing power in pp and p-carbon CNI theoretical. The pp experimental data from Fermilab E704.

calculable and the analyzing power peaks at a respectable 4.5% for p-p and it is approximately 10% lower for p-Carbon. There appears to be little energy dependence. However, the accompanying hadronic interaction need not conserve helicity in the small t region and the inclusion of the single-flip hadronic amplitude may be relevant. The associated uncertainties with the latter render this measurement less certain. The fit to the experimental data measured at Fermilab [7] is good to approximately 15%.

AGS experiment E950 (see I. Alekseev in these proceedings) measured the analyzing power of p-Carbon scattering in the CNI region. The slow recoil carbon (100 – 800 KeV) was clearly observed using silicon detectors placed near 90 degrees on either side of the beam. Due to the long AGS bunch length, good timing resolution was achieved by detecting, using micro channel plates, prompt electrons generated by the passage of the recoil carbon through a thin 5 $\mu g/cm^2$ carbon foil. The preliminary data indicate an asymmetry dependence Vs t similar to what is expected [8] but of somewhat lower magnitude. The results were normalized to the external polarization measured in the E925 pp elastic scattering polarimeter. The analysis continues to try to extract the single flip amplitude using the theoretical parameterization of Trueman et al [9].

With this, we developed and installed a polarimeter in the RHIC Blue ring based on the same p-Carbon CNI technique and used it during the commissioning run in September 2000, figure 7. The asymmetry was normalized to the E950 results. A similar unit was installed in the Yellow ring and is ready for the polarized proton run starting November 2001. The RHIC polarimeters utilize the tight bunch crossing for timing. Also the higher rates necessitate the use of Wave-Form Digitizers for the DAQ. A 1 % statistical measurement will take less than a minute.

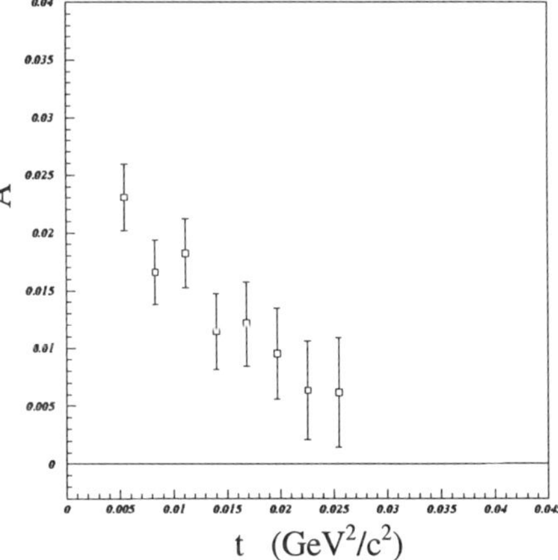

Figure 7. Analyzing power p-carbon CNI at RHIC injection.

10 Absolute beam polarization calibration

The CNI polarimeters discussed so far will serve as a relative polarimeter. We anticipate that within few years the physics output will approach the statistical significance that will require the absolute beam polarization be known to better than 5%. The plan is to develop and install a polarized jet target of intensity approaching 3×10^{11} per cm^2. The jet polarization will be measured using a Breit Rabi Polarimeter to an accuracy of 3% (see T. Wise in these proceedings). This then will be used to calibrate the CNI analyzing power and polarimeters at the desired energy. It is estimated that a 3% statistical measurement is expected to take a good portion of one store.

References

1. I. Alekseev et al., Design Manual for the Polarized Proton Collider at RHIC 1996, unpublished.
2. H. Spinka et al., Nucl. Instr. And Meth. 211 (1983) 239.
3. F. Z. Khiari et al., Phys. Rev. **D39** (1989) 45
4. W. H. Dragoset et al., Phys. Rev. **D18** (1976) 3939.
5. D.L. Adams et al., Phys. Lett. **B264** (1991) 462.
6. K. Krueger et al., Phys. Lett. **B459** (1999) 412, and C.E. Allgower et al., Sumbitted to Phys. Rev. D (2001)
7. N. Akchurin et al., Phys. Rev. **D48** (1993) 3026-3036
8. J. Tojo et al. Proceedings of the Spin 2000 14th International Spin Physics Symposium, Osaka, Japan. pp 790-794
9. B.Z. Kopeliovich and T.L. Trueman, Phys. Rev. **D64** (2001) 34004
10. See for example D.M. Nikolenko et al. Proceedings of the Spin 98 13th International Symposium on High Energy Spin Physics, Protvino, Russia pp 541. Also G. Igo and F. Messer private communications.

AN OVERVIEW OF ELECTRON POLARIMETERS AND AN INTERCOMPARISON

J. M. GRAMES

Thomas Jefferson National Accelerator Facility,
12000 Jefferson Ave, Newport News, VA 23606, USA

Electron beam polarimetry may be described as the technique of separating scattered particles for detection using some physical interaction between the polarization of the electron beam and the total analyzing power of a target. The target is itself polarized in many polarimeter designs. A careful intercomparison of the relative analyzing power of five electron beam polarimeters was performed at the laboratory during a dedicated two day machine development period. A Wien-style spin manipulator at the injector was used to vary the spin orientation of the electron beam at each polarimeter. A series of measurements as a function of spin orientation provides determination of the relative analyzing power of each polarimeter, and more importantly, quantifies the instrumental systematics that help realize high precision absolute electron polarimetry. In addition, a comparison of the value of the injector spin angle that provides precise longitudinal beam polarization at each experimental hall leads to an independent and potentially high precision absolute measurement ($dE/E \sim 10^{-4}$) of the final electron beam energy. Discussion of the experiment will be presented.

1 Overview

1.1 Purpose and Outline

The purpose of this report is to first provide an overview of electron polarimetry as is generally found at electron accelerator facilities and second to report on a comparison of five electron polarimeters made at Jefferson Lab (JLab). The overview is not inclusive to all types of electron polarimeter designs, nor is it historical in nature. The goal of the first section is to indicate how electron polarimetry is often done and where present capabilities may be improved. The purpose of the second section is to report on a measurement comparing the relative analyzing power of the JLab polarimeters and the use of spin precession to make a high precision energy measurement. The report can be briefly summarized:

- Describe the purpose and desirable parameters of electron polarimeters
- Define analyzing power and discuss why it is difficult to measure
- Detail unique *Spin Dance 2000* experiment at JLab to compare five polarimeters
- Present polarimeter comparison and spin-based energy measurement results
- Summarize the goals and role of electron polarimetry for the future

2 Electron Beam Polarimetry

Electron beam polarimetry is the technique of separating scattered particles for detection using some physical interaction between the polarization of the beam (P_b) and the total analyzing power of a polarimeter's target (A_{tot}). The target is itself often polarized in many polarimeters and A_{tot} is then proportional to the product of the target polarization and the analyzing power of the interaction. The experimental asymmetry is defined as $\varepsilon = A_{tot} \cdot P_b$. This asymmetry can be measured by either reversing the polarity of the beam or target polarization while detecting the scattered particles from each state. For example, measuring N^+ events with electron polarization $+P_b$ and then N^- events with electron polarization $-P_b$ yields,

$$\varepsilon = A_{tot} \cdot P_b = \frac{N^+ - N^-}{N^+ + N^-}.$$

As an example consider a single detector from the 5 MeV Mott polarimeter at JLab.

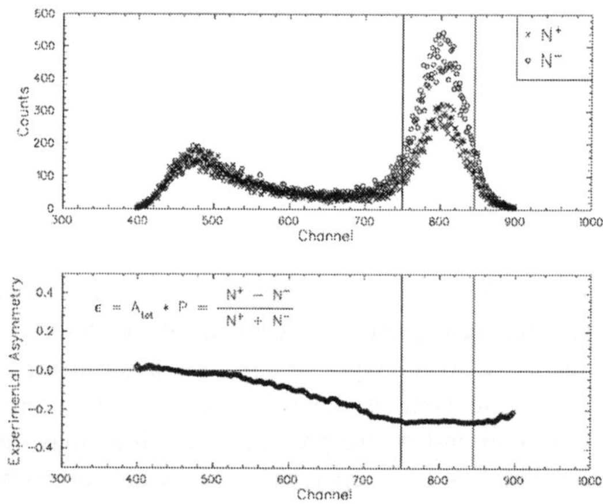

Figure 1. Energy spectra for both polarization states is shown in the upper plot. The experimental asymmetry for a running average 5 channels wide is shown in the lower plot.

The energy spectra are shown in Figure 1 for both polarization states along with the calculated experimental asymmetry. The plots indicate that the elastic region (peaks) of the Mott asymmetry carry a (constant) analyzing power, while the inelastic events do not. In this case, the beam polarization is 70% and A_{tot} is −40%.

Electron polarimeters serve a variety of purposes at accelerator laboratories:

- keV and MeV region for sources and nuclear parity experiments
- GeV and greater for parity and nucleon spin structure functions
- Future experiments are pushing toward 1% absolute polarimetry
- Eager users!

Consequently, the desirable (necessary) features are:

- Large, precise analyzing power
- Polarimeter designs with reduced sensitivity to inherent systematics
- Non-invasive techniques that do not disrupt the experiment
- High luminosity to rapidly achieve small statistical uncertainty

However, knowledge of the total analyzing power is limited. This is because the total analyzing is not a directly measured quantity. For example, unpolarized double-scattering experiments to measure the analyzing power are difficult because of limitations due to cross-section and background discrimination. Consequently, the analyzing power is most often determined by both theory and simulation.

From this point of view the experimental quantities sought are those used to model the analyzing power. These include measuring target polarizations (often inferred) and having a careful understanding of detector acceptance, energy resolution, and multiple scattering or background effects that dilute the experimental asymmetry.

A description of the five electron polarimeters used at JLab and the relevant parameters used to describe their analyzing powers are described in the next section.

3   Electron Polarimeters

3.1   Mott Polarimetry

The Mott scattering asymmetry is derived from the spin-orbit coupling between a beam electron and target nucleus. The targets are unpolarized, high-Z materials such as gold or lead. The analyzing power is described by the Sherman function [1] which can be quite large (30-50%). In reality, Coulomb screening [2] is important for keV energies and finite nuclear size [3] becomes important for MeV energies. These contribute to the main uncertainty in the theoretical value. Experimentally, multiple and plural target scattering reduces the experimental asymmetry by diluting the elastic events detected resulting in what is known as an effective Sherman function, dependent upon target thickness.

Operationally, Mott polarimeters are relatively simple in design and are often used near polarized electron sources (100 keV) or at lower injector energies (MeV). Yet, they employ solid targets that are invasive to experiments.

3.2 Moller Polarimetry

The Moller scattering asymmetry arises from the spin-spin interaction of a polarized beam electron and a polarized target electron. The QED calculated analyzing power is a 3x3 matrix which becomes diagonal and essentially energy independent at energies > 1 GeV with maxima at the 90 degree center of mass scattering angle (few degrees in lab frame).

Operationally, this type of polarimeter is useful at higher beam energies used to probe nucleon spin structure. The luminosity (10-100 kHz/μA/μm) is sufficient for rapid statistics. To discriminate against the Mott background Moller polarimeters are often operated in coincidence, detecting both the incident and target electron, achieving a signal to noise of >1000:1.

The uncertainty in analyzing power is dominated by acceptance effects. These include detector finite energy acceptance, Mott background in single-arm polarimeters, and multiple scattering. The Levchuk effect, which describes the sensitivity of detector acceptance to scattering off of the unpolarized target electrons, has incurred corrections of 10%. These factors are usually considered in simulation codes.

The targets are typically iron or iron-alloy and require a polarizing magnetic field. Two types of polarized target designs are shown in Figure 2. Conventional Moller polarimeters use iron-alloy targets tilted at a small angle with respect to the beam and reside in a low magnetic field (100 G). The targets easily magnetize "in-plane" and have a spin polarization component along the beam direction. Measuring the target polarization accurately at the beam location is difficult and inhomogeneities in target thickness lead to uncertainty in the flux-magnetization relationship.

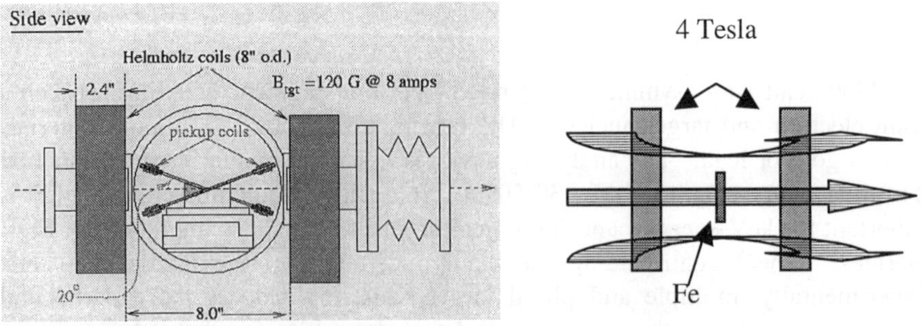

Figure 2. Conventional and novel Moller polarimeter target designs shown side by side.

A novel alternative design uses superconducting coils (3 Tesla) to saturate the foil magnetization normal to the surface of the foil. This design has the advantage of being insensitive to target thickness. Moreover, pure iron provides the best known magnetization-spin polarization relationship of such target materials (0.25%), so that the target polarization is already known prior to the beam environment.

Either polarimeter design is susceptible to beam heating that can reduce the target magnetization if not cared for by maintaining low beam intensities (few μA).

3.3 Compton Polarimetry

The Compton backscattered photon asymmetry results from the interaction between longitudinally polarized electrons against incident circularly polarized photons. The QED calculated Compton asymmetry for this reaction grows significantly for high beam energies and this design is preferable for its non-invasive method.

Calculating the Compton analyzing power requires accurate knowledge of the backscattered photon spectrum, shown in Figure 3. Uncertainty in the analyzing power is often dominated by a) the energy calibration of the Compton edge (maximum photon energy) and b) the low energy threshold of the detector and data acquisition system. Additionally, synchrotron radiation (magnets) and bremsstrahlung (residual gas) are significant sources of background. In the case of the JLab polarimeter, beam scraping of the optical cavity was an important concern.

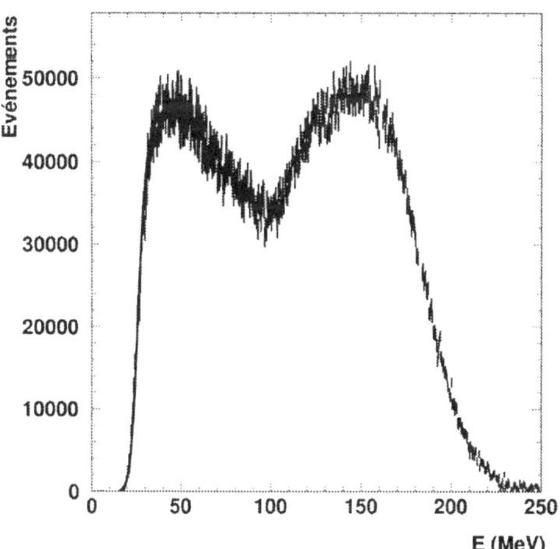

Figure 3. A Compton backscattered photon energy spectrum shows the low energy threshold and the high energy Compton edge.

Operationally, the photons are generated using a laser beam and thus high luminosity is achieved with either high peak intensity pulsed electron/laser beams, or high intensity CW electron/laser beams. At JLab, a low power diode laser locks in frequency to a high gain (5000) Fabry-Perot optical cavity located on a beam line chicane, providing 1-2 kW of optical power. The laser polarization is >99% and can be remotely changed using a wave plate for systematic study.

4 The *Spin Dance 2000* Experiment

The *Spin Dance 2000* experiment made use of the five electron polarimeters at Jefferson Lab [4-8] to compare their analyzing powers. The experiment lasted 56 hours. Of this, 40% was used for polarimeter checkout. The remainder was dedicated to setting the injector (laser, spin angle), the accelerator (beam extraction, energy measurements), and for polarization measurements. The accelerator layout and polarimeter summary is shown in Figure 4.

Polarimeter	Target	P_{target}	I_{ave}	P_x	P_y	P_z
Injector Mott	ZA	N/A	2 μA	x	x	
Hall A Compton	γ	> 99%	70 μA			x
Hall A Moller	e	~ 8 %	2 μA	x		x
Hall B Moller	e	~ 8 %	10 nA		x	x
Hall C Moller	e	~ 8 %	1 μA			x

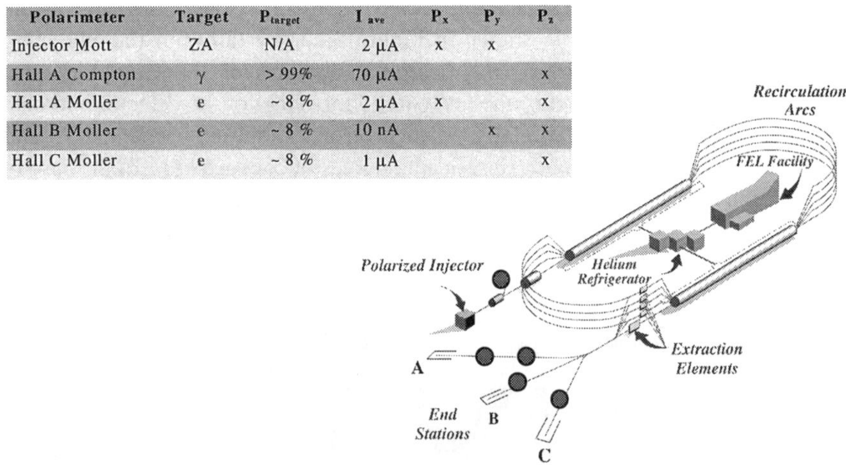

Figure 4. The CEBAF accelerator layout shows the position of the five JLab electron polarimeters.

4.1 *Jefferson Lab Polarized Source and Beam Transport*

The polarized electron beam at Jefferson Lab is produced by photoemission from a semiconductor strained GaAs photocathode [9-11] using polarized laser light (λ=850 nm). This experiment configured the accelerator for 5-pass recirculation (E=5.6 GeV) to the end stations. Although the electrons have an initial longitudinal spin orientation they precess while transporting through the magnetic fields of the accelerator. The net accelerator spin precession for this experiment is >10,000° and the spin direction is ultimately not oriented in the same direction at each polarimeter.

The capability for simultaneous beam delivery to multiple end stations has the advantage that polarimeter measurements can be done at the same time. However, considering spin precession, the measurable component of the beam polarization with respect to each polarimeter is generally not equal. The solution, depicted in Figure 5, is to perform the polarimetry in a way that does not rely upon only one measurement of a single component of the beam polarization. This is accomplished by adjusting the orientation of the beam polarization using a Wien filter spin rotator located at the polarized source. The measured experimental asymmetry is proportional to the component of the beam polarization along some analyzing component of a polarimeter. By varying the Wien angle the measurable component of the beam polarization will vary sinusoidally.

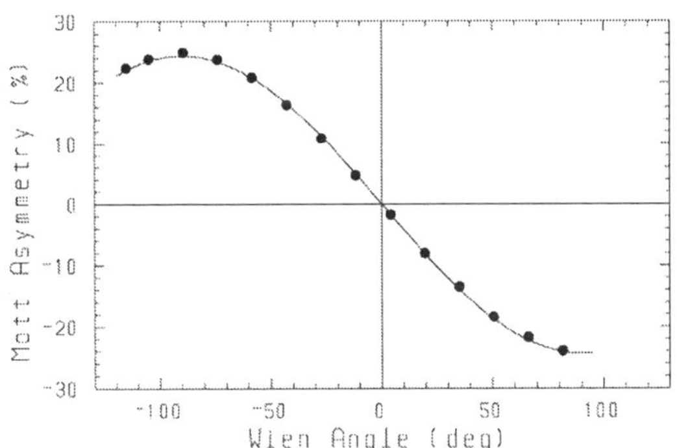

Figure 5. Plot of experimental asymmetry versus Wien angle traces a sinusoidal curve. The amplitude is proportional to the analyzing power and the phase is proportional to the spin precession.

4.2 Polarimeter Measurements and Experimental Results

Polarimeter data was collected at 12 unique spin angles (Wien angles) spanning $|\eta_{Wien}| < 110°$. The polarimeter data was then analyzed by each polarimeter group. The results are plotted together in Figure 6 against the common Wien angle.

4.3 Interpreting the Relative Analyzing Power

The amplitudes of the sinusoidal curves are proportional to the beam polarization. Taken together, they reflect the relative analyzing power of the five polarimeters, which is shown in Figure 7, normalized to the Mott. Inspecting the plot indicates that the three Moller polarimeters of different design differ by as much as 10%, yet polarimeters of three physics reactions (Mott, Moller, Compton) agree at the 2% level. In the latter case, this is within their quoted systematic uncertainties.

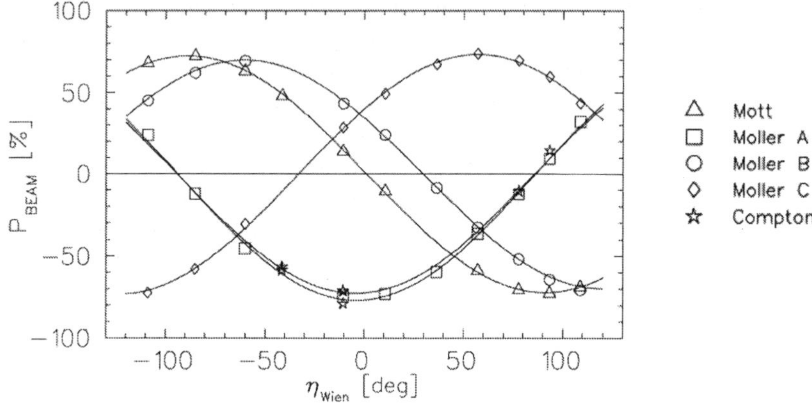

Figure 6. The polarimeter data versus Wien angle set point is shown for the *Spin Dance 2000* data.

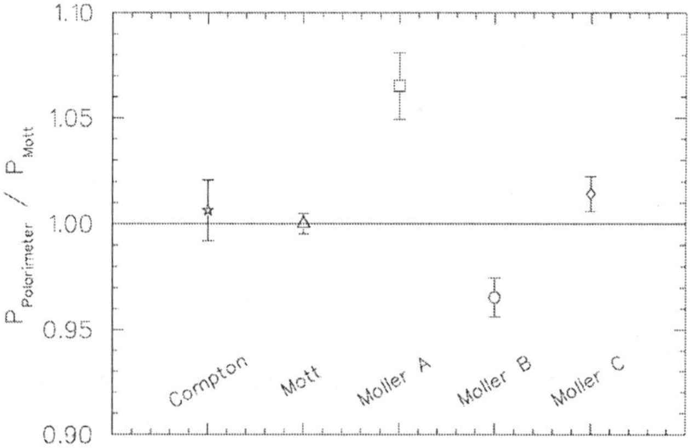

Figure 7. The relative analying powers for the five Jlab electron beam polarimeters, normalized to the Mott polarimeter for comparison, is shown.

4.4 Interpreting the Spin Precession Results

Although the beam energy is measured using conventional methods (magnetic spectrometer, elastic scattering) the spin precession results presented in Figure 6 yield two independent spin-based energy measurements.

The first method measures the precession difference between the source and end station polarimeters. This method yields a precision $\sim 10^{-4}$ due to the large total precession ($>10,000°$), but depends upon accelerator parameters such as the injector energy, linac gradients, and all magnetic transport sections. The gradients of the two linacs were not measured and therefore the results are described over a range of linac difference ($E_{12}=E_{linac1}-E_{linac2}$). However, a subsequent calibration measurement estimates the linac difference during the experiment at +4 MeV.

The second method measures only the relative precession difference between end station polarimeters after the beam is extracted from the accelerator. This method has a lower precision (10^{-3}) owing to the smaller net precession ($<500°$), but does not depend upon any of the accelerator parameters and therefore offers a separate independent spin-based measurement for comparison.

A summary plot of the two spin-based energy measurements for the polarimeters, along with a comparative energy measurement using a spectrometer [12], is shown in Figure 8. Altogether the results predict the beam energy to be 5648 MeV with variation less than 1.3%.

However, the results by both methods indicate a discrepancy associated with the Hall B transport arc beam deflection angle. The discrepancy is ±35 MeV and the sign is correlated with whether the second polarimeter is located in end station A or C. The discrepancy, using the two spin-based energy methods, indicate that the Hall B orbit was likely offset by -0.22°. Excluding the Hall B energy results the remaining 7 cases by 3 methods agree to better than 0.4%. Over this linac difference the total spin precession method has an uncertainty $<2 \times 10^{-4}$.

5   Conclusions

The *Spin Dance 2000* experiment provides a powerful tool to compare the relative analyzing powers of electron polarimeters. Jefferson Lab has the unique capability of having 5 electron polarimeters representing three physics reactions (Mott, Moller, Compton) to make such a comparison. The results indicate a nearly 10% disagreement between three Moller polarimeters, while at the same time there is less than 2% disagreement between a Compton, Mott, and Moller polarimeter. In the latter case, all claim systematic uncertainty near the level of the disagreement.

The two spin-based energy measurement methods are useful because of their sensitivity and resolution differences. The total precession method, combined with a measurement of the linac imbalance, yields an absolute uncertainty of 10^{-4} to 10^{-5}.

In conclusion, Jefferson Lab has the resources to test high precision (1%) absolute electron beam polarimetry. Yet, the polarimeter is often simply viewed as the tool. To reach the 1% absolute mark the polarimeter must be the experiment, not the tool.

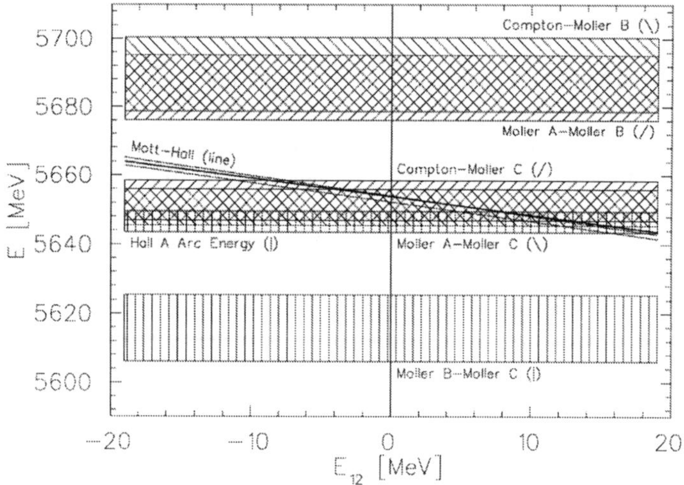

Figure 8. Graphical summary for all ten energy measurements in the range $|E_{12}| < 20$ MeV. Results that do not depend upon the linac imbalance are shown hatched and have constant bands of uncertainty. The lines which vary with E_{12} have an uncertainty $< 10^{-4}$ for $|E_{12}| < 12$ MeV.

6   Acknowledgements

Many people representing a variety of institutions have contributed significant effort to develop the electron polarimeters used at Jefferson Lab. It is important to acknowledge their efforts, and in particular, those that contributed to this experiment. These include, by institution, **Jefferson Lab** (E. Chudakov, H. Fenker, A. Freyberger, J. Hansknecht, J. Mitchell, M. Poelker, C. Sinclair, M. Steigerwald, M. Tiefenback), **CEA Saclay** (C. Cavata, S. Escoffier, F. Marie, T. Pussieux, P. Vernin), **Florida International University** (R. Nasseripour, B. Raue), **Karkov Institute** (V. Gorbenko), **Massachusetts Institute of Technology** (D. Higinbotham, R. Suleiman), **North Carolina Ag. and Tech. State University** (S. Danagoulian), **Old Dominion University** (V. Dharmawardane), **University of Virginia** (R. Fatemi, K. Joo, M. Zeier), **Universitaet Bonn** (T. Reichelt), **Vrije Universiteit** (B. Zihlmann).

References

1. N. Sherman, Phys. Rev. **103(6)**, 1601 (1956).
2. Ross *et al.*, Phys. Rev. A **38(12)**, 6055 (1988).
3. Ugincius, *et al.*, Nucl. Phys. **A158**, 418 (1970).
4. Steigerwald, *PST1999* Workshop, Erlangen, Germany (1999).

5. N. Falletto, *et al.*, Nucl. Instrum. Meth. A. **459**, 412 (2001).

6. E. Chudakov, *Jefferson Lab Hall A 1999 & 2000 Annual Reports* (2000).

7. R. Nasseripour, *et al.*, Bull. Am. Phys. Soc. **45**, 91 (2001).

8. M. Loppacher, Ph.D. Thesis, University of Basel, Switzerland (1996).

9. L.S. Cardman, Nucl. Phys. A **546**, 317c (1992).

10. C.K. Sinclair, 8th International Symposium on High Energy Spin Physics, AIP Conference Proceedings (1988).

11. M. Poelker, *et al.*, *PES2000* Workshop, Nagoya, Japan (2000).

12. J. Berthot and P. Vernin, Nucl. Phys. News **9**, 412 (1999).

OPAQUE SPIN FILTERS – A NEW TOOL FOR PRECISE NEUTRON POLARIMETRY

P. HAUTLE[1], W. HEIL[2], D. HOFMANN[2], H. HUMBLOT[3], T.M. MÜLLER[4],
O. ZIMMER[5,*]

[1] Paul Scherrer Institut, CH-5232 Villigen, Switzerland
[2] Institut für Physik, D-55099 Mainz, Germany
[3] Institut Laue-Langevin, Avenue des Martyrs, F-38042 Grenoble, France
[4] Physikalisches Institut, Philosophenweg 12, D-69120 Heidelberg, Germany
[5] Physik-Department E18, Technische Universität München, 85748 Garching, Germany
**Corresponding author, e-mail: oliver.zimmer@ph.tum.de*

Studies of angular correlations in free neutron decay offer various possibilities to test the standard model of electro-weak interaction at low energy. Prominent examples are a test of the unitarity of the weak-quark-mixing Cabibbo-Kobayashi-Maskawa (CKM) matrix and a search for right-handed charged currents in the weak interaction. High-precision neutron polarization analysis is an important tool in these studies. A new technique of cold neutron beam polarimetry employs a spin filter analyzer of either polarized ^{3}He or polarized protons with a neutron analyzing power close to unity. A recent experimental comparison of such opaque spin filters has demonstrated a hitherto unattained precision. The technique is briefly reviewed. In addition, results of a first test of a new method to measure the efficiency of neutron spin flipping devices are given.

1 Introduction

Polarized slow neutrons are an excellent tool to study fundamental symmetries and details of the weak force both in hadronic and semileptonic interactions (for a recent review see [1]). The β-decay of the free neutron, $n \rightarrow pe^-\overline{\nu}_e$, offers various experimental possibilities to investigate the properties of the electro-weak Hamiltonian. The absence of any nuclear environment, well-understood radiation corrections and experimental access to many different observables render this process favorable for precision studies. Such investigations have a longstanding history. They have provided precise values of several weak interaction parameters. In particular, the weak axial-vector coupling constant of the neutron could be determined much more precisely than by any other means [2]. This, in turn, allowed to derive important weak cross sections, *e.g.* that of the fusion reaction $pp \rightarrow de^+\nu_e$, which determines the hydrogen burning rate of the sun and other stars. Nowadays, testing the standard model of electro-weak interaction is in the focus of interest, although rapid progress in other fields will continue to raise requests for improved values of neutron decay observables for more practical needs. A present example is the neutron lifetime, which determines the abundance of primordial helium in the universe. In a recent calculation, this quantity crucial for big bang nucleosynthesis turned out to be not sufficiently well known [3]. The neutron lifetime plays a second important role as one of the ingredients to determine values of the nucleonic weak vector and axial-vector couplings independently from nuclear β-decay. For this purpose, it is combined with the value of the neutron β-asymmetry (neutron spin – electron angular correlation) and that is where polarized neutrons enter the game. Any non-perfect neutron polarization p_n will reduce the value of the measured asymmetry, which, among other corrections, then has to be divided by p_n to extract the true asymmetry. Hence, p_n has to be well known. In the most recent experiment, an accuracy of 0.7 % was attained [4]. The neutron polarimetry was performed according to the state-of-the-art at that time, using an analyzer which involves magnetic mirror polarizers [5]. With its precision of 0.3 %, polarimetry thus represented one of the main systematic uncertainties. This experiment, combined with the world average of the neutron lifetime, raises the question whether or not the weak-quark-mixing CKM matrix is indeed unitary [6]. Another case, where neutron polarization analysis has become limiting, is the neutrino asymmetry (neutron spin – neutrino angular correlation), a tool to search for right-handed charged currents in the frame of left-right symmetric models [7].

Obviously, experiments have reached a level of precision, which necessitates to sharpen the tool of neutron polarimetry. In this paper, we review a method based on a particular type of spin filter. In addition, we present first data on a new trick to measure the efficiency of a spin flipper, which completes the method.

2   Definition of the problem

Most experiments on angular correlations in polarized neutron decay are carried out using a cold neutron beam. Even at strong neutron sources, limited counting statistics is always a crucial point, which urges one to use a beam polarizer with highest possible efficiency. A popular device is the bender type supermirror polarizer [8], which is employed in many neutron laboratories. Future works may exploit the advantages of a spin filter of polarized ^{3}He, like low background and the precisely known neutron cross section, in particular at neutron spallation sources [9]. However, with respect to bare efficiency as a polarizer for a cold neutron beam, the ^{3}He spin filter still requires a further improvement of the nuclear polarization in order to become fully competitive for particle physics experiments with slow neutrons. A technique of neutron polarimetry based on neutron transmission measurements through ^{3}He as beam polarizer, even without using an analyzer, was proposed in [10] and has recently been demonstrated successfully [11].

The widespread bender type supermirror polarizer consists of a bent stack of thin glass plates coated with polarizing mirrors. It can polarize cold beams with large cross section of many tens of cm^2 at high transmission, and attaining high polarization values well above 95 %. However, although being very efficient, there

276

are serious disadvantages with respect to neutron polarimetry. The difficulties are caused by non-homogeneities of the neutron transmission T and the generated polarization p_n:

$$T = T(\overset{1}{r}, \Omega, \lambda) \quad \text{and} \quad p_n = p_n(\overset{1}{r}, \Omega, \lambda). \tag{1}$$

Any device used for polarization analysis of such a beam has to account for the dependence on the geometrical parameters $\overset{1}{r}$ and Ω, which denote a position within the polarized beam and the solid angle, respectively. In particular, using a non-homogeneous analyzer with large cross section may result in erroneous averaging of different partial beams with different polarization. Under such circumstances, the task of neutron polarimetry requires:

1. Selecting partial beams to be analyzed separately out of the polarized beam;
2. Measuring the neutron wavelength spectrum of the polarized partial beams;
3. Measuring the polarization of the beams as a function of the wavelength;
4. Calculating the wavelength averaged polarization of the total beam.

In past experiments on polarized neutron decay angular correlations, Stern-Gerlach and double supermirror analyzers have been used [12,5,13]. Stern-Gerlach devices have severely limited acceptance with respect to beam size and divergence, and the difficulties of supermirror polarizers already stated before require a rather complex experimental procedure, which renders neutron polarimetry a tedious and time consuming task.

3   Solution of the problem: the opaque spin filter

3.1   Basic idea

The idea of the opaque spin filter was introduced in [14] and shall briefly be sketched here. A neutron spin filter contains atoms with a spin-dependent cross section, which are homogeneously distributed within a slab intersected perpendicularly by the beam. To an excellent approximation, the transmission and analyzing power of a homogeneous device do not depend on $\overset{1}{r}$ and Ω,

$$T = T(\lambda) \quad \text{and} \quad A = A(\lambda) , \tag{2}$$

but only on the neutron wavelength. Homogeneity offers a great advantage for polarization analysis, because the total beam can be analyzed at once. Since inherently all partial beams are spatially weighted in the same way, the correct mean value of the polarization is thus obtained. The analyzing power is given as

$$A = \tanh(\sigma_p PNd) , \tag{3}$$

where P is the nuclear polarization, N the number density of polarized nuclei, d is the neutron path length through the target, and $\sigma_p(\lambda)$ is the polarization cross section. Thus, A can be calculated if one knows the parameters characterizing the spin filter. Moreover, the situation becomes particularly simple if the product of these quantities can be made sufficiently large, such that $A \to 1$. For instance, a value of $\sigma_p PNd = 3.8$ is easily attainable experimentally and corresponds to a near-to-ideal analyzer with $A = 0.999$. In this limit of what we have called *opaque spin filter*, A is insensitive to uncertainties of the filter parameters, and a modest knowledge of the latter is already sufficient to know the analyzing power very well. This in turn leads to high accuracy, reliability and simplicity of neutron polarimetry.

3.2   The polarized ^{3}He and proton spin filters

Practical implementations of a spin filter for neutrons are polarized gas of ^{3}He, and solid-state polarized proton targets. For ^{3}He, $\sigma_p = 5333$ barn for thermal neutrons (*i.e.* at $\lambda = 0.18$ nm), and the absorption dominated cross section increases in proportion to λ. Hence, A will rapidly increase with λ, which is shown in figure 1 for three filter cells with different neutron optical thickness (see also [15]).

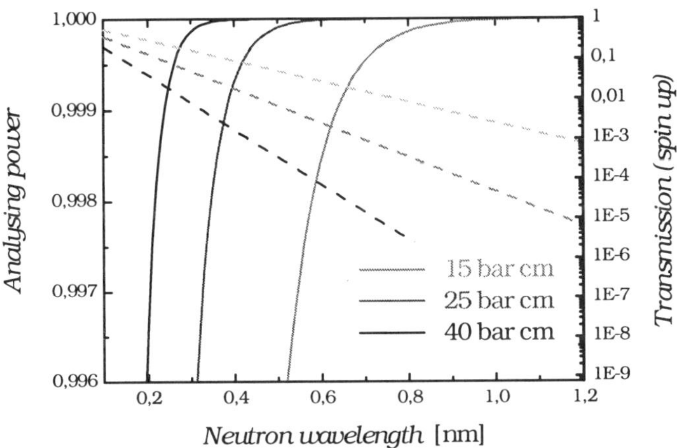

Figure 1. Solid lines: analyzing power of three different ^{3}He spin filter cells. Dashed lines: corresponding neutron transmission of the spin component with high transmission ("spin up"). The curves shown refer to a nuclear polarization of 50 %. The strong λ dependence is due to the cross section being proportional to λ. Each filter cell covers a range of neutron wavelengths where A is sufficiently close to unity and the transmission is still sufficiently large to provide a measurement with sufficient counting statistics.

The corresponding rapid decrease of the transmission of neutrons through the gas necessitates the use of several different filter cells if a white cold neutron beam with a broad wavelength spectrum shall be analyzed.

Figure 2 shows the polarization cross section of frozen 1,2-propanediole, doped with 1.1×10^{20} cm^{-3} Cr(V) paramagnetic centers (with a total hydrogen content of $N = 6.7\times10^{22}$ cm^{-3}) and measured for cold neutrons. The hydrogen protons were polarized via dynamical nuclear polarization (DNP) [16]. The data were obtained from neutron count rate ratios ("flipping ratios"), using a spin flipper between a supermirror polarizer and the polarized target, behind which a neutron detector was installed. The target was 3 mm thick and polarized to up to 58 % at temperature 1 K and a field of 3.5 T. A brief description of the facility and a comparison of nuclear polarization measurements via neutron transmission and NMR can be found in [17].

The wavelength dependence of σ_p was measured with a chopper via time-of-flight. For $\lambda > 1.2$ nm, σ_p is mainly due to incoherent neutron scattering (with a small contribution by absorption), $\sigma_p(\lambda > 1.2 \text{ nm}) = -(53.3 + 1.85 \text{ nm}^{-1}\lambda)$ barn.

For shorter wavelengths it is influenced by solid-state effects. The rapid rise at $\lambda < 0.2$ nm is caused by the effect of the chemical binding of the protons on the neutron scattering length.

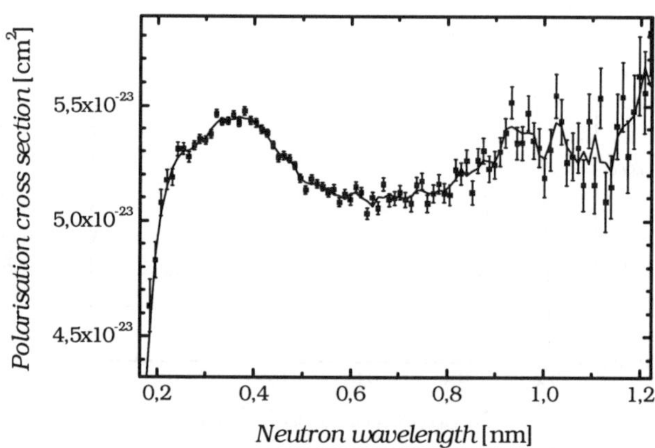

Figure 2. Polarization cross section σ_p of 1,2-propanediole, doped with 1.1×10^{20} cm^{-3} Cr(V).

These data show that, using equation 3, a rather modest value of $P = 55$ % is sufficient to attain $A \geq 0.999$ for $\lambda > 0.2$ nm with a block target of 2 cm thickness.

The cross section varies only weakly with λ. Thus, in contrast to the ^{3}He spin filter, only a single target is needed to analyze the whole λ range of cold neutrons.

As the main result of an experimental inter-comparison of these two different types of analyzer, the measured neutron polarization values agreed to 0.026(55) %, covering a large fraction of the entire cold neutron wavelength spectrum (0.45 nm $\leq \lambda \leq 1.2$ nm) [14]. This excellent agreement, obtained with two devices which differ strongly with respect to the neutron polarizing process (spin-dependent absorption, respectively scattering), has demonstrated the reliability of the opaque spin filter method.

4   Determination of the neutron polarization and the efficiency of a spin flipper from only four count rates

4.1   Two different flipping ratios

Normally the polarization of slow neutrons is defined relative to an external magnetic field, applied to the beam polarizer. Losses of polarization are avoided using a guide field on the whole neutron flight path after the polarizer. A smooth change of the spatial direction of the magnetic field along the neutron beam induces a corresponding change of the neutron polarization vector; if the adiabatic condition is fulfilled, the spins will stay parallel, respectively anti-parallel to the guide field.

In asymmetry experiments in polarized neutron decay, it is necessary to measure count rates of the charged decay products for two opposite directions of the neutron polarization. A simple and widespread procedure to do this involves a spin flipper, which inverts the neutron polarization with respect to the magnetic guide field. Since such devices perform only with efficiency smaller than unity, complete neutron polarimetry requires the measurement of the flipper efficiency as well as the polarization. Recently, a new method has been proposed [18]. It exploits the fact that, for dynamically polarized targets, one can operate the target with two opposite directions of the nuclear polarization with respect to the magnetic guide field.

The experimental setup of the method consists of the (supermirror) beam polarizer, a spin flipper and the opaque spin filter analyzer, followed by a neutron detector. A beam chopper is used to measure the dependence of the count rates on λ. We denote the analyzer states with the index $j = 0$ or 1 ($j = 0$, when both the beam polarizer and the analyzer have high transmission for the same neutron spin component). The state of the spin flipper is denoted by the index k, $k = 0$ (or 1) corresponding to the spin flipper being inactive (active). Combining these different settings, one can measure four different count rates N_{jk}. We define two "flipping ratios" with the analyzer held in a fixed state,

$$R_0 := \frac{N_{00}}{N_{01}} = \frac{1 + p_n A_0}{1 - f p_n A_0} \quad \text{and} \quad R_1 := \frac{N_{11}}{N_{10}} = \frac{1 + f p_n A_1}{1 - p_n A_1}. \tag{4}$$

The analyzing power in each analyzer state, A_0 respectively A_1, may be slightly different. However, in both states the spin filter shall be opaque, *i.e.* with the A_j close to unity and therefore well known. The expressions can be converted:

$$p_n = \frac{(R_1 - 1) R_0 A_0 - (R_0 - 1) A_1}{(R_0 R_1 - 1) A_0 A_1},$$

$$f = \frac{(R_1 - 1) A_0 - (R_0 - 1) R_1 A_1}{(R_0 - 1) A_1 - (R_1 - 1) R_0 A_0}. \tag{5}$$

Approximate expressions for the usual case of the flipper efficiency being close to unity were already given in [18]. The method is very sensitive to both p_n and f since, for $p_n A_j$ close to unity, the non-ideality of f has only weak influence on the flipping ratio R_1 but it strongly affects R_0.

4.2 First experimental demonstration

In the present study we have employed the opaque, gaseous ^{3}He spin filter analyzer. It was polarized in the state $j = 1$, using optical pumping of metastable ^{3}He at low pressure with subsequent compression [19]. The experiment was performed during a comparison of spin filter and supermirror analyzers at the Institut Laue-Langevin, Grenoble, using the setup mentioned before. More details can be found in [20]. The nuclear polarization was switched between the states $j = 0$ and 1 via adiabatic fast passage (AFP). For this purpose the ^{3}He filter cell was located in a homogeneous magnetic guide field with an additional coil creating a high frequency field perpendicular to the static field. The frequency was ramped manually through the resonance at a speed chosen to fulfill the adiabatic condition for the ^{3}He-spins [21]. Figure 3 shows the resulting flipping ratios, obtained with a filter cell with 11 bar cm at an initial polarization of 52 %. A typical spin relaxation times T_1 of the ILL filter cells is 100 s. By flipping the neutron spin each 5 s one gets rid of this effect. Using equation 5, the measured flipping ratios determine the flipper efficiency f. The neutron polarization was determined from R_1 with $f = 1$, which is a good approximation here, since, as said before, R_1 is rather insensitive to f. $p_n(\lambda)$ at the present conditions was already given in figure 3 of [20]. The result for f is presented in figure 4. The second data set for the shorter wavelengths was obtained with another filter cell with 32 bar cm and initial polarization $P = 54$ %. Using equation 4 for R_1 (again with $f = 1$) we obtain $A(\lambda)$ before and after two times AFP. From a fit to these data we conclude that the efficiency of each single manual AFP sweep was about 96 %.

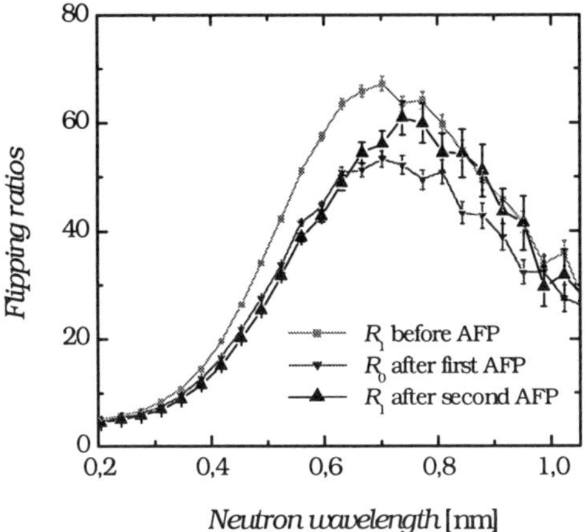

Figure 3. Flipping ratios measured with a filter cell of polarized ^{3}He as analyzer. The condition $A \to 1$ of the opaque spin filter is fulfilled in the wavelength range above 0.7 nm. After a first AFP sweep through the nuclear resonance the flipping ratio is decreased in the whole wavelength range shown. After the second sweep, the values of R_1 are smaller than before only in the range $\lambda < 0.7$ nm, which is due to loss of nuclear polarization during the sweeps. For $\lambda > 0.7$ nm, where the spin filter stays opaque even after the slight decrease of nuclear polarization, the original flipping ratios are recovered. This behavior demonstrates the flipping ratio R_0 being suppressed by the effect of the flipper efficiency in equation 4.

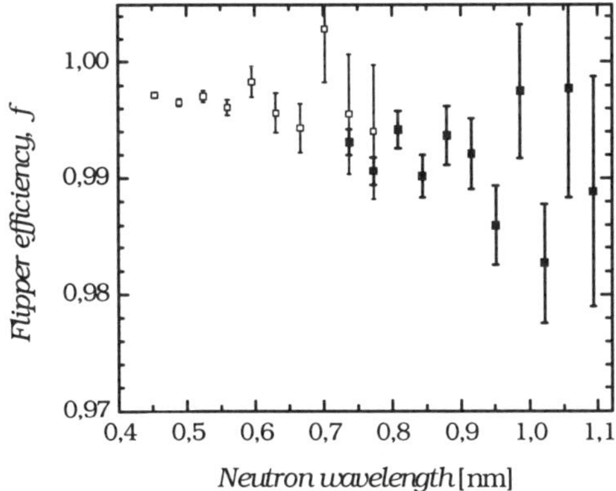

Figure 4. Flipper efficiency calculated from equation 5, using measured values of the flipping ratios R_0 and R_1. The values of the solid symbols were derived from the data shown in fig. 3, and the open symbols from a measurement with a ^{3}He spin filter at a larger value of the product PNd of filter parameters. The smaller error bars there are due to the strong increase of neutron intensity towards smaller wavelengths.

This first test clearly demonstrates the feasibility of the method. In our experiment, where the AFP was done manually (turning a knob of the hf source at appropriate speed), the loss of nuclear polarization during inversion of the nuclear polarization was 4 % (relative). This can be reduced at least to 0.03 %, which was demonstrated in [21]. Note, however, that such a high efficiency of AFP is actually not mandatory for the purpose of neutron polarimetry.

In the case of the polarized proton analyzer, polarization reversal by AFP is not efficient enough [22]. It therefore has to be performed by DNP, which takes about half an hour at the conditions chosen in our experiment. Thus, a polarized proton target is a good alternative to the ^{3}He spin filter in a laboratory where the latter is not available, which, however, is somewhat more convenient in practical use at the neutron beam.

References

1. *Particle physics with slow neutrons*, Proceedings of Int. Workshop held at Institut Laue-Langevin, Grenoble, France October 22-24, 1998, ed. by Zimmer O., Butterworth J., Nesvizhevsky V. and Korobkina E., *Nucl. Instr. Meth. A* **440** (2000) pp. 471–790.
2. Dubbers D., *Progr. Part. Nucl. Phys.* **26** (1991) p. 173.
3. Lopez R.E., Turner M.S., *Phys. Rev. D* **59** 103502 (1999) pp. 1-14.
4. Reich J., Abele H., Astruc Hoffmann M., Baeßler S., v. Bülow P., Dubbers D., Nesvizhevsky V., Peschke U., Zimmer O., in [1] pp. 535-538.
5. Serebrov A., Aldushchenkov A., Lasakov M., Kuznetsov I., Stepanenko I., *Nucl. Instr. Meth. A* **357** (1995) p. 503.
6. Abele H., Astruc Hoffmann M., Baeßler S., Dubbers D., Glück F., Müller U., Nesvizhevsky V., Zimmer O., Is the Unitarity of the quark-mixing-CKM-matrix violated in neutron β-decay? submitted to *Phys. Rev. Lett.*
7. Serebrov A.P., Kuznetsov I.A., Stepanenko I.V., Aldushenkov A.V., Lasakov M.S., Mostovoi Yu.A., Erozolimskii B.G., Dewey M.S., Wietfeldt F.E., Zimmer O., Börner H.G., *J. Exp. Th. Phys. JETP* **86** (1998) p. 1074.
8. Schärpf O., *Physica B* **156 & 157** (1989) p. 639.
9. Snow W.M., Bazhenov A., Blessinger C.S., Bowman J.D. Chupp T.E., Coulter K.P., Freedman S.J. *et al.*, in [1] pp. 729-735.
10. Greene G.L., Thompson A.K., Dewey, M.S., *Nucl. Instr. Meth. A* **356** (1995) pp. 177-180.
11. Rich D.R., Bowman J.D., Crawford B.E., Delheij P.P.J. *et al.*, A measurement of the absolute neutron beam polarization produced by an optically-pumped ^{3}He neutron spin filter, accepted for publication in *Nucl. Instr. Meth. A*.
12. Yerozolimsky B.G., *Nucl. Instr. Meth. A* **420** (1999) pp. 232-242.

13. Nastoll H., Schreckenbach K., Baglin C., Bussière A., Guillaud J.P., Kossakowski R., Liaud P., *Nucl. Instr. Meth. A* **306** (1991) p. 65.

14. Zimmer O., Müller T .M ., Hautle P., Heil W., Humblot H., *Phys. Lett. B* **455** (1999) pp. 62-68.

15. Surkau R., Becker J., Ebert M., Grossmann T., Heil W., Hofmann D., Humblot H., Leduc M.,Otten E.W., Rohe D., Siemensmeyer K., Steiner M., Tasset F., Trautmann N., *Nucl. Instr. Meth. A* **384** (1997) p. 444.

16. Abragam A., Goldman M., Nuclear Magnetism: Order and Disorder, *Clarendon Press, Oxford,* (1982).

17. Zimmer O., Müller T.M., Dubbers D., Hautle P., Bunyatova E.I., Korobkina E.I., Nesvizhevsky V.V., An experiment for the measurement of the spin dependence of slow neutron absorption by protons, Proceedings of the V International Seminar on Interaction of Neutrons with Nuclei, Dubna 1997.

18. Zimmer O., *Phys. Lett. B* **461** (1999) pp. 307-314.

19. Heil W., Anderson K., Hofmann D., Humblot H., Kulda J., Lelievre-Berna E., Schärpf O., Tasset F., *Physica B* **241-243** (1998) p. 56.

20. Zimmer O., Hautle P., Heil W., Hofmann D., Humblot H., Krasnoschekova I., Lasakov M., Müller T .M ., Nesvizhevsky V., Reich J., Serebrov A., Sobolev Yu. Vassilev A., *Nucl. Instr. Meth. A* **440** (2000) pp. 764-771.

21. Wilms E., Ebert M., Heil W., Surkau R., *Nucl. Instr. Meth. A* **401** (1997) p. 491.

22. Hautle P., van den Brandt B., Konter J.A., Mango S., *Nucl. Instr. Meth. A* **356** (1995) pp. 108-110.

A HIGH-ENERGY PHOTON POLARIMETER BASED ON PAIR PRODUCTION

M. H. WOOD[1], B. WOJTSEKHOWSKI[2], D. TEDESCHI[1], B. VLAHOVIC[3],
D. ABBOTT[2], J. ASAI[4], G. FELDMAN[5], T. HOTTA[6], M. KHANDAKER[7],
H. KOHRI[6], T. MATSUMURA[6], T. MIBE[6], T. NAKANO[6], G. ORIELLY[5],
A. RUDGE[8], P. WEILHAMMER[8], T. YORITA[6], AND R. ZEGERS[9]

(1) University of South Carolina, USA
(2) Thomas Jefferson National Accelerator Facility, USA
(3) North Carolina Central University, USA
(4) Canadian Light Source, Canada
(5) George Washington University, USA
(6) RCNP, Osaka, Japan
(7) Norfolk State University, USA
(8) CERN, Switzerland
(9) Japan Atomic Energy Research Institute, Mikazuki, Japan

With the recent installation of the Hall B coherent bremsstrahlung source at the Thomas Jefferson National Accelerator Facility, it is desirable to have an experimental determination of the photon beam polarization. We have built a polarimeter which detects electron-positron pairs created from the linearly-polarized beam. The device consists of multiple silicon microstrip detectors for determining the relative orientation of the pair as well as the distance separating the e^+ and e^-. The polarimeter was calibrated at the SPring-8 facility using a compton-backscattered photon beam in the energy range of 1.5 GeV $\leq E\gamma \leq$ 2.4 GeV. This measurement was the first made for this process at these energies.

1 Introduction

The production of e^+e^- pairs by linearly polarized photon beam was proposed by Yang [1] and Berlin and Madansky [2] as a method for measuring the photon beam polarization. L. Maximon and H. Olsen [3] performed an accurate calculation of the process using QED. The technique was successfully used for beam polarization measurement at beam energies less than 300 MeV [4]. At higher energies, the method is limited by the small angle separating the electron and positron.

By using silicon microstrip detectors (MSDs), the particles can be detected with small distances ($\sim$100-μm) separating them and with good position resolution ($\sim$ few μm). This technology is employed in out polarimeter which has been built to measure the polarization of a linearly polarized photon beam in the few GeV energy range [5]. Another improvement was to detect both particles in the pair and measure the asymmetry with respect to the

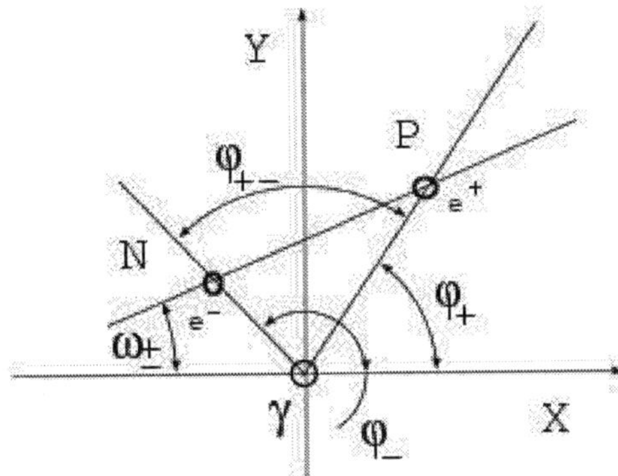

Figure 1. This drawing displays the coordinate system of e^+e^- pair detected by a MSD. The angle ω is given between the x-axis and the vector connecting the e^+ (P) and the e^- (N).

angle ω (see Figure 1). Following the analysis of L. Maximon and H. Olsen [3], the theoretical asymmetry for this experimental setup is ~ 0.25. Following the analysis of L. Maximon and H. Olsen [3], the theoretical asymmetry for this experimental setup is ~ 0.25. This paper will discuss the calibration and testing of the polarimeter.

2 Experiment

The polarimeter was calibrated at the SPring-8 facility which is a third-generation synchrotron radiation facility operated by Japan Synchrotron Radiation Research Institute. The laboratory is an 8 GeV electron storage ring. By compton-backscattering polarized laser light off of the electron beam, a photon beam was produced with energies from 1.5 GeV to 2.4 GeV. The beam polarization axis was set either parallel or perpendicular to the beam direction and was switched between the two orientations several times throughout the duration of the experiment. Table 1 lists important parameters of the experiment.

There were a number of components to the polarimeter assembly. The photon beam was incident first on a multi-wire proportional chamber(MWPC) which was a veto detector for charged-particle background. After the veto detector, the electron-positron pairs were produced by the photon beam in 0.1 mm carbon foil. Immediately after the carbon converter was a second MWPC used to give a coincidence signal with the downstream detectors. The

Table 1. Experimental parameters for polarimeter calibration at SPring-8

Electron beam intensity	100 mA
Photon energy range	1.5-2.4 GeV
Laser photon polarization	98-99%
Photon polarization	73-92%
Photon intensity	130-230 kHz

two particles traveled through an 80-cm vacuum pipe onto four MSDs stacked in series. No magnetic fields were used to increase the angle separating the e^+e^- pair. The position and separation of the two particles was measured with the MSDs. Each detector contains 384 strips with the thickness of each strip being 50 μm. The active area of each detector is about 1 cm^2. The second, third, and fourth detectors were oriented so that the strips were rotated by 90°, 45°, and −45°, respectively, with respect to the first detector. Four detectors were used to recover events when the both particles in the pair struck a single strip in one detector. The position information for each particle could be retrieved from the remaining three detectors. After the MSDs, the separation of the e^+e^- pair was increased by a 4-kGauss permenant magnet. The positions and speeds of the particles were measured once more by two drift chambers and a time-of-flight (TOF) wall after the magnet.

3   Analysis

Figure 2 shows the raw ADC output from one MSD for a single event. Figure 2 is a plot of the same detector and event after pedestal subtraction. The electron and positron signals are clearly visible above the electronic noise. For every event, the position of the particle was determined by the strip with the largest ADC value out of the cluster of strips that were activated.

With the particle position determined, the distance between the pair was calculated as well as the angle ω. Figure 2 contains plots of the ω and distance distributions. These results are from events with only two particles observed in all four detectors. Further analysis will include the cases with only one particle detected and with more than 2 particles detected. For this paper, the cleanest case is considered.

With the ω information for every event, the asymmetry was calculated. The asymmetry is defined as

$$A(\omega) = \frac{1}{P_\gamma} \frac{\sigma_\parallel - \sigma_\perp}{\sigma_\parallel + \sigma_\perp} \tag{1}$$

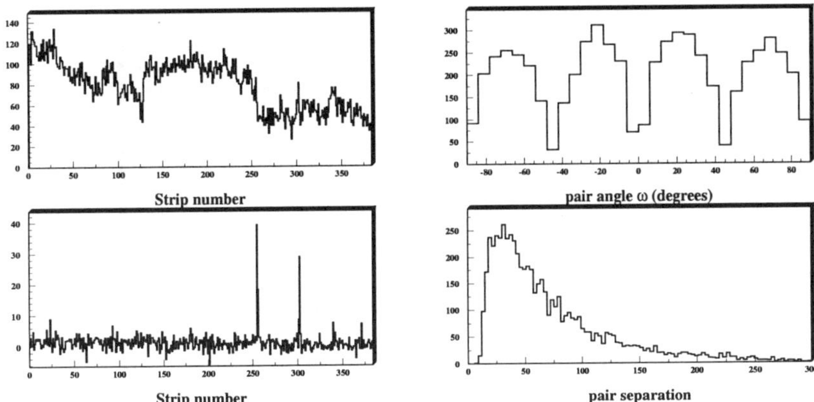

Figure 2. The graphs on the left are: (top) raw ADC value for each strip of a single MSD from one event, (bottom) the same event after pedestal subtraction. The plots on the right are: (top) distribution of pairs with repect to ω, (bottom) distance separating the pair (units are the number of strips).

where P_γ is the average photon beam polarization, and $\sigma_\parallel$ ($\sigma_\perp$) is the yields when the beam polarization axis was the parallel(perpendicular) to the beam direction. Both $\sigma_\parallel$ and $\sigma_\perp$ are normalized to their respective integrated beam currents. A plot of $A(\omega)$ versus ω is given in Figure 3. Fitting a $\cos(2\omega)$ to the points produces an asymmetry of 0.20 ± 0.01.

4 Conclusions

The measured asymmetry is close to the theoretical expectation of 0.25. Refinements to the analysis should bring the measured value into closer agreement with the prediction. One refinement will be a cut on the sharing energy of the e^+e^- pair. Another improvement to the analysis will be to incorporate the position and timing information from the downstream drift chambers and the TOF paddles.

With this preliminary analysis, the asymmetry is confirmed to be substantial. This first measurement of the asymmetry at these high beam energies establishes pair production as a strong technique for photon polarimetry. A polarimeter with silicon MSDs is being designed for Hall B at the Thomas Jefferson National Accelerator Facility. The installation of the device is scheduled for the end of 2001.

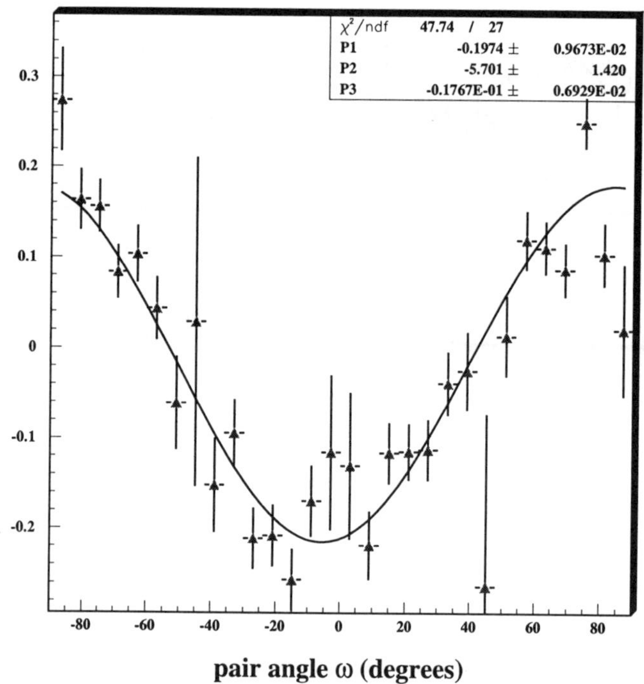

pair angle ω (degrees)

Figure 3. The measured asymmetry versus the angle ω.

Acknowledgments

The authors would like to thank the staff of the SPring8 facility. The experiment ran very smoothly and made this measurement possible.

References

1. C. N. Yang, *Phys. Rev.* **77**, 722 (1950).
2. J. H. Berlin and L. Madansky, *Phys. Rev.* **78**, 623 (1950).
3. L. C. Maximon and H. Olsen, *Phys. Rev.* **126**, 310 (1962).
4. G. Barbiellini *et al*, *Phys. Rev. Lett.* **9**, 396 (1962).
5. B. B. Wojtsekhowshi *et al*, proceedings from JLAB/USC/NCCU/GWU workshop on "Polarized Photon Polarimetry", June 2-3, 1998, p.257.

A HIGH PRECISION SCANNING POLARIMETER FOR THE TRIUMF PROTON-PROTON PARITY VIOLATION EXPERIMENT

W.D. RAMSAY, J. BIRCHALL, D.A. HORNING, L.LEE, S.A. PAGE, A.A. RAUF, G. RUTLEDGE, W.T.H. VAN OERS

Department of Physics and Astronomy, University of Manitoba, Winnipeg, MB, R3T 2N2, Canada

P.W. GREEN, G. ROY

Centre for Subatomic Research, University of Alberta, Edmonton, AB, Canada T6G 2N5

A.A. HAMIAN

University of Washington, Nuclear Physics Laboratory, Box 354290, Seattle, WA 98195-4290, USA

C.A. DAVIS, C.D.P. LEVY

TRIUMF, 4004 Wesbrook Mall, Vancouver, BC, V6T 2A3, Canada

N.A. TITOV, A.N. ZELENSKI

Institute for Nuclear Research, Academy of Sciences of Russia, 60th October Anniversary Prospect 7a, Moscow, 117312, Russia

J.D. BOWMAN, R.E. MISCHKE

Physics Division, Los Alamos National Laboratory, PO Box 1663, Los Alamos, NM 87545, USA

N.T. OKUMUSOGLU

Karadeniz Technical University, 61080 Trabzon, Turkey

TRIUMF experiment 497 has measured the parity violating longitudinal analyzing power, A_z, in $\vec{p}p$ scattering at 221 MeV incident proton energy. Corrections comparable to the size of the result were required for first moments of transverse polarization. The experiment used counting mode scanning polarimeters based on pp elastic scattering from hydrogen in scanning CH_2 targets. At 200 nA beam current, the first moment precision was approximately $\pm 6\mu m$ for a one hour run. A new experiment at TRIUMF, E761, has now upgraded these polarimeters by adding large solid angle (40 msr) detectors operating in current mode and responding predominantly to elastic scattering from carbon in the CH_2. The new detectors may be operated at the same time as, and calibrated against, the full coincidence, counting, detectors. In test runs, an effective analyzing power of ~ 0.8 was obtained and first moments of transverse polarization were determined to $\pm 0.34\mu m$ in a one hour run.

1 Introduction

TRIUMF experiment 497 has recently measured the parity violating longitudinal analyzing power, A_z, in $\vec{p}p$ scattering at 221 MeV incident proton energy[1]. In this experiment, a longitudinally polarized beam of protons was passed through a 40 cm liquid hydrogen target. The helicity of the incident beam was flipped from right handed to left handed and the change in transmission through the target was measured. Small "helicity correlated" changes in beam properties other than helicity introduced false parity violating signals which had to be corrected for. One of the largest, and hardest to control, of these systematic errors was that arising from first moments of transverse polarization. First moments were measured by two Polarization Profile Monitors (PPMs) separated by 1.8m. A new parity experiment at TRIUMF, E761, has now upgraded these devices and greatly improved their performance.

2 First Moments of Transverse Polarization

The parity *allowed* analyzing power, $A_y \simeq 0.3$ in $\vec{p}p$ scattering at 221 MeV, is very large compared to the parity *violating* analyzing power, $A_z \simeq 10^{-7}$, and even small transverse polarization components in the beam are a potential source of systematic error. The most problematic effect of transverse polarization is from so-called *intrinsic* first moments, which arise from a distribution of transverse polarization across the beam. They can be reduced by cyclotron tuning, but are not usually less than a few μm.

The *sensitivity* to first moments of transverse polarization was measured by scanning a horizontally polarized beam vertically and a vertically polarized beam horizontally. For the TRIUMF parity setup, the false A_z from $\langle yP_x \rangle$ was found to be $1.9 \times 10^{-7}/\mu m$ at PPM1 and $-2.5 \times 10^{-7}/\mu m$ at PPM2. From $\langle xP_y \rangle$, it was $-1.5 \times 10^{-7}/\mu m$ at PPM1 and $2.0 \times 10^{-7}/\mu m$ at PPM2. From these numbers, one sees that if the first moment at PPM2 is approximately 75% of that at PPM1, then the false A_z from the two will cancel, equivalent to reducing the *effective* sensitivity to zero.

3 The Polarization Profile Monitors (PPMs)

As described in detail by Berdoz *et al.*[2], each of the two PPMs has four rotating CH_2 blades, two scanning vertically and two scanning horizontally. Each blade is 5 mm along the beam and 1.6 mm transverse to the beam. The original PPMs operate in a full coincidence mode. Two forward and two recoil scintillators for each of the left, right, bottom, and top branches detect

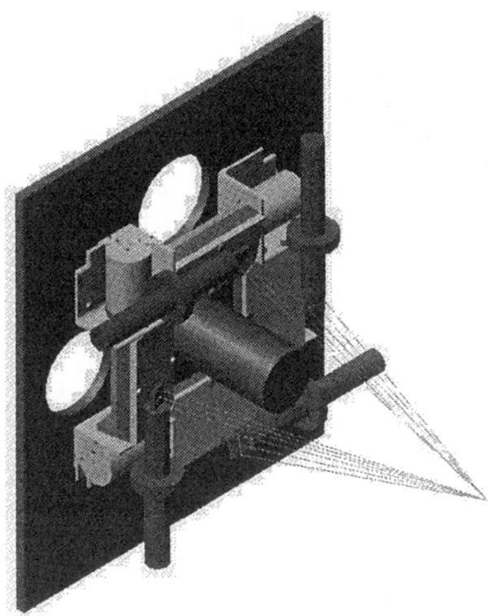

Figure 1. Location of the new current mode detectors between the small counting mode detectors and the beampipe. The new detectors are 88 mm x 200 mm x 12.7 mm paddles of BC412 scintillator coupled to vacuum photodiodes. Absorbers in front of the new detectors range out inelastic events.

only pp elastic scattering events from hydrogen in the CH_2 blades. Although these PPMs have a stable, well known analyzing power, are hardware corrected for cross section variation across a scan, and are relatively insensitive to room background, they take approximately an hour to measure the average first moments to a statistical precision of $\pm 6\mu m$. This precision cannot be improved significantly, as individual detectors are close to their count rate limits.

4 Current Mode PPMs

To improve the speed of the PPMs, large solid angle (40 msr) scintillators were added between the existing forward arms and the beampipe as shown in Figure 1. The new detectors are paddles of BC412 scintillator measuring 88 mm x 200 mm x 12.7 mm, and covering scattering angles of 8.9^o to 16.3^o, just inside the 16.6^o to 18.4^o range of the old counting mode scintillators, and centered at the analyzing power maximum for pp elastic scattering from carbon[3,4]. Because of the large solid angle and large $(p,^{12}C)$ cross section, individual particles cannot be counted with the new detectors and they are operated in current mode. Each scintillator is coupled to a Hamamatsu R2046PT 76 mm diameter

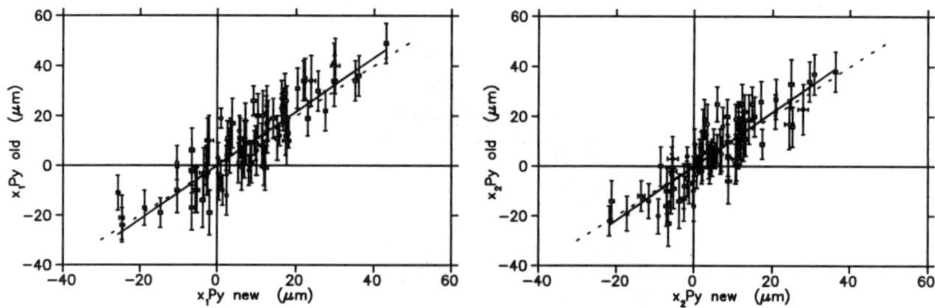

Figure 2. First moments as measured simultaneously by the new and old PPMs. The errors on the old PPMs are a factor of approximately 17 worse than the new PPMs. The dotted line is a slope of unity (old = new) and the solid line is a fit to the data.

vacuum photodiode. The photocathode current reaches approximately 40 nA when a scanning target passes through the 200 nA beam. Absorbers are placed in front of the current mode detectors to range out inelastic events.

In addition to 40 times larger solid angle, the new detectors gain the advantage of approximately an order of magnitude greater cross section as well as a larger analyzing power.

4.1 Performance of the current mode PPMs

The polarization is calculated by first taking the ratio $r = \sqrt{\frac{L^+ R^-}{L^- R^+}}$, where the $+$ and $-$ refer to the two spin states and the L and R refer to left and right detectors. This cancels the effect of differences in detector efficiency, cross section, solid angle, and $+/-$ current difference. The polarization may then be found from $P_y A_y = \frac{r-1}{r+1}$.

The optimum thickness of the absorbers was found empirically by adjusting their thickness for maximum left-right asymmetry with vertical polarization. Calibrated against the known $A_y = 0.31$ of the full coincidence polarimeter, 38 mm of steel plus 13 mm of copper gave a current mode effective analyzing power of ~ 0.8.

Figure 2 shows first moments of transverse polarization measured simultaneously with the old and new polarimeters using longitudinal polarization under a wide range of cyclotron and beamline conditions. The plots assume $A_y = 0.31$ for the old PPMs and $A_y = 0.80$ for the new PPMs. The solid lines are best fit straight lines through the origin. For PPM1 the fitted slope is 1.08 ± 0.06 and for PPM2 it is 1.09 ± 0.07, consistent with the assertion that old and new are measuring the same thing. Also note that the errors on

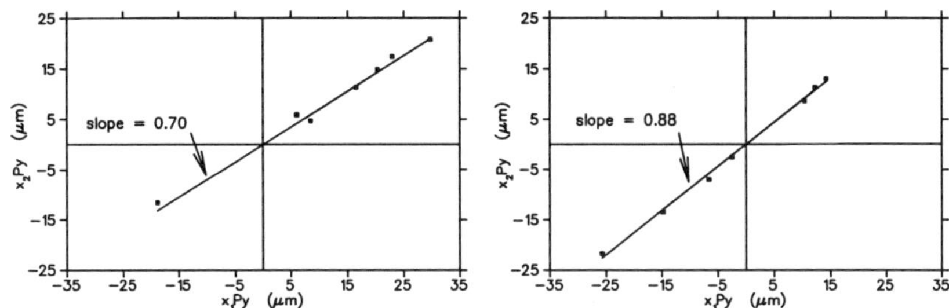

Figure 3. First moment ratios. Shown are groups of first moment readings obtained under two different beamline tunes. The data on the left were taken with a more strongly converged tune. One sees that, for a fixed tune, the absolute value of the first moments can vary dramatically, but the PPM2/PPM1 ratio remains relatively constant.

the old PPM are approximately 17 times larger than the small errors on the new PPM; we can now measure in a minute what used to take hours.

As an example of what one can do with such a precision device, Figure 3 shows the first moments at PPM1 and PPM2 under a wide variety of cyclotron tunes, but separated into two groups of data, each group sharing the same beamline magnet settings. The tune for the left panel had a stronger geometric convergence than that for the right panel. One sees that the *ratio* of first moments, as indicated by the slope, is reasonably constant for a given beam tune, whereas the absolute value of the moments can vary over a wide range. The ability to make such ratio measurements in a reasonable amount of time will permit us to tune for very low effective sensitivity to first moments as described in section 2.

Acknowledgments

This work was supported in part by the Natural Sciences and Engineering Research Council of Canada; TRIUMF receives federal funding via a contribution agreement through the National Research Council of Canada.

References

1. TRIUMF preprint TRI-PP-01-11 (2001); nucl-ex/0107014 .
2. A.R. Berdoz, *et al.*, *Nucl. Instrum. Methods* A **457**, 288 (2001).
3. M.W. McNaughton, *et al.*, *Nucl. Instrum. Methods* A **241**, 485 (1985)
4. E. Aprile-Giboni, *et al.*, *Nucl. Instrum. Methods* **215**, 147 (1983)

A POLARIMETER FOR PROTON BEAMS WITH A VERY LOW DUTY FACTOR

E.J. STEPHENSON, S. GOLBECK, B. VON PRZEWOSKI, T. RINCKEL, AND P. SCHWANDT

Indiana University Cyclotron Facility, Bloomington, IN 47408 USA

We report on a polarimeter that operates on the injection beam line for the IUCF Cooler where the duty factor is about 10^{-8}. Using a thick carbon target and photomultiplier tubes without scintillators, scattered proton current pulses are recorded and used to calculate asymmetries for each pulse (0.8 Hz). The analyzing power varies from 0.3 at 120 MeV to 0.7 at 200 MeV. This polarimeter has been used to scan the $G\gamma=2$ imperfection resonance in the Cooler Injector Synchrotron and, with a 50% transmitting target, to monitor pulses injected into the Cooler ring for spin dynamics studies.

1 Introduction

Proton beams at IUCF begin at the H^- polarized ion source, then proceed through pre-acceleration to 7 MeV in an RFQ/DTL, stripping accumulation and energy ramping in the Injector Synchrotron (CIS), and injection into the electron-cooled storage ring (Cooler). The polarization may be measured at 7 MeV using p+^{4}He elastic scattering or with one of the detector assemblies in the Cooler (based on p+p or p+C elastic scattering). If, at the end of beam preparation for an experiment, the Cooler beam appears to have a low polarization, a diagnostic polarimeter on the injection beam line would be useful.

However, injection line beam pulses are short (~20 ns) and infrequent (0.8 Hz), so the usual particle counting techniques would not yield a result in a reasonable time because of the low duty factor (~10^{-8}). This report summarizes our experience as we dealt with this challenge.

2 Polarimeter Development

One solution to the particle counting conundrum is to use a thick target and to treat the scattered flux as a current. This could work since the inclusive proton spectrum from a carbon target is known to have an analyzing power that peaks above 0.6 for angles between $12°$ and $13°$ [1]. The short duration of the beam pulse from CIS would make such a current appear as a single pulse. This creates two requirements for any detector system: (1) the response must be <u>linear</u>, and (2) there should be <u>no background</u>.

Using equipment available at IUCF, we tried various combinations of targets, scintillator detectors, and ionization chambers in the CIS vault. That environment

"

proved to be too harsh. There was a large flash of neutrons and gammas from beam hitting the Lambertson extraction magnet, and all of the detectors were sensitive to RF pickup from the electric field kicker used to send the beam out of CIS. These problems became more manageable as soon as we moved to a location on the Cooler side of the radiation shielding, away for CIS. There tests showed that there was really no signal in time with the beam whenever beam pulses were absent, thus satisfying one of our initial requirements.

Scintillator and photomultiplier tube (PMT) detectors showed the most promise because of their fast time response. But with carbon targets 5 mm in thickness, the number of scattered protons led to signals that saturated the detectors. Each pulse from CIS contained approximately 10^9 protons, and about 10^5 of those were scattered toward the active part of the PMT. Eventually we settled on PMT tubes with no scintillator as a good choice. In this case, electrons are produced by protons passing through the photocathode surface.

To check linearity, we compared the PMT response with signals derived from a "wall gap beam monitor," a device that registers the image charge of the beam produced on a metal cylinder that surrounds the passing beam. Large pulse-to-pulse variations were seen in the extracted CIS beam, and larger beam pulses showed evidence for some pulse height saturation. Nevertheless, the correlation between PMT and beam monitor responses was very narrow. So we continued to pursue this scheme despite the non-linear response. The next step was to surround the target by four PMTs at a laboratory angle of $15.9°$ so that we could search for evidence of either vertical or horizontal polarization components in the ratios of PMT pulse heights.

3   Results

Initial results with a 200-MeV proton beam extracted from CIS showed that the PMT pulse heights differed by about a factor of 2 between spin up and down. If correlations with the size of the pulse, as measured with the beam monitor, are included, these correlations remain tight. Figure 1 shows part of a page taken from the online logbook. The horizontal axis in all cases is the beam monitor signal. The left and middle panels show the pulse heights in the left and right detectors, respectively. The difference between the strong bands is the spin-dependent effect.

It is important to note that each event represents a measurement of the polarization of that CIS pulse. Thus one can calculate the asymmetry, $\varepsilon = (L-R)/(L+R)$ from the L and R current pulses alone. A scatterplot of asymmetry as a function of the beam monitor is shown in the right-hand panel where the y-axis is 100ε. The two strong groups represent spin up and down, and are widely and cleanly separated.

The vertical width of these strong bands is an indication of the statistics of the asymmetry measurement. A Gaussian width of 0.01 to 0.02 indicates, based on statistics alone, that there are several thousand particles (either primary protons or secondary PMT electrons) associated with each CIS pulse.

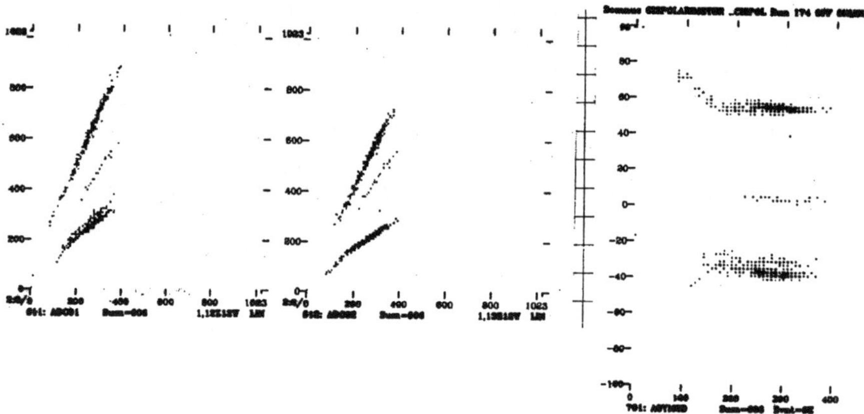

Fig. 1: Polarimeter on-line data. The horizontal axis is the beam intensity. The vertical axes are left detector pulse height (left panel), right detector pulse height (middle panel), and asymmetry (right panel).

The PMT gain behaves as solid angle, thus a change of PMT voltage can unbalance the asymmetry measurement. So no significance for the beam polarization can be attached to the fact that the average asymmetry is greater than zero. Only the difference (provided the unbalance is not extreme) carries useful information on the difference of the polarization between the up and down states.

The pulse-by-pulse measurement of asymmetry opens a new dimension in the diagnostics of the beam polarization. Once calculated, we arranged that the asymmetry was fed from the computer to a DAC so that the individual pulse polarizations could be watched on a meter in real time.

This allows various malfunctions to be observed that otherwise would be lost in the average polarization of a longer run. Two weaker bands visible in the right-hand asymmetry panel of Fig. 1 show that some pulses arrive with either a slightly reduced or a nearly vanishing polarization. Although the cause in this case was not investigated, such behavior could arise if the RF for a particular transition unit does not reach full power in the cavity that changes the population of the hyperfine states in the polarized ion source. Turning the RF power on and off is the mechanism used to switch from one polarization state to another. Other problems related to timing might also be an issue.

Data of the sort shown in Fig. 1 were useful in locating other problems. The upper correlations in the left and middle panels of Fig. 1 are straight. In other runs, some downward curvature for the largest pulse heights indicated that the PMTs involved were becoming saturated. In Fig. 1, there are a few events at low beam monitor signal values that appear to break away from the horizontal correlation in the asymmetry plot and head toward larger asymmetry magnitudes. Such behavior was observed for a number of runs and could be mapped by using a defocusing lens upstream of the RFQ to modulate the beam intensity over a broader range. The systematics of such studies indicated that there is some mechanism that can depolarize the 200-MeV proton beam near injection into CIS so that the horizontal bands represent an attuated asymmetry. The mechanism for such a depolarization is not known. It has not been observed at 120 MeV, the other energy at which this polarimeter has seen significant use.

Such depolarization effects make it difficult to obtain a calibration for the analyzing power associated with the polarimeter. Tests with steel absorbers placed in front of the PMTs show that an arrangement in which the elastically scattered protons nearly reach the end of their range at the photocathode surface can enhance the analyzing power by as much as 20% of its value. Using such absorbers and taking asymmetries for lower beam pulse intensities yields analyzing powers of about 0.3 at 120 MeV and 0.7 at 200 MeV. The actual beam polarization is assumed to be that measured at 7 MeV in the CIS injection beam line.

4　Applications

4.1　Continuous Polarization Monitor

In the switch from cyclotron to CIS injection of beam into the Cooler, one feature that was lost was the ability to continuously monitor the beam polarization coming in. This was resolved by creating a target, shown in Fig. 2, that is a 50% transmitting grid of holes 1 mm in diameter. The beam is many hole diameters in width. In this case, the polarimeter flux is reduced by only a factor of two, not enough to complicate a polarization measurement.

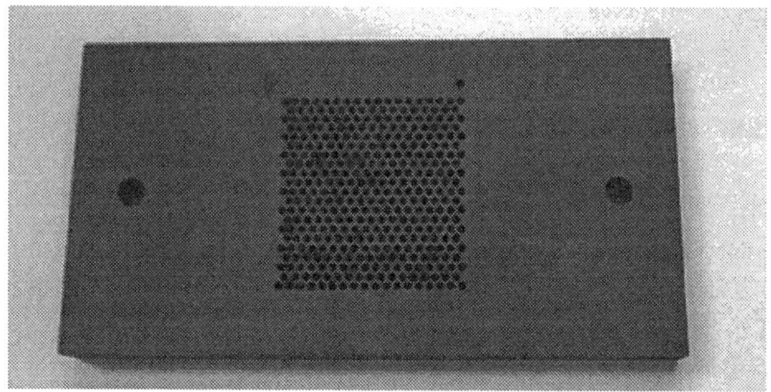

Fig. 2: 50% transmitting carbon sieve target used to provide an online polarization monitor.

4.2   The $G\gamma = 2$ Depolarizing Resonance in CIS

As a summer project for a student in the Research Experience for Undergraduates program (SG), we used this polarimeter to track the precession of the proton spin as it ramped through the $G\gamma = 2$ imperfection resonance in CIS. For this project, we needed to be able to monitor both the vertical and horizontal components of the polarization. The center energy of the resonance is at 108.4 MeV.

The effects of the resonance are controlled by placing a partial Siberian snake (longitudinal solenoid) into the CIS ring. Even a small current saturates the resonance so that the polarization of the proton beam reverses as CIS ramps in energy through 108.4 MeV. A map of the CIS polarization direction was obtained by kicking the beam out at a number of energies on either side of the resonance and using the four polarimeter PMTs to measure the two transverse beam polarization components. The results are shown in Fig. 3.

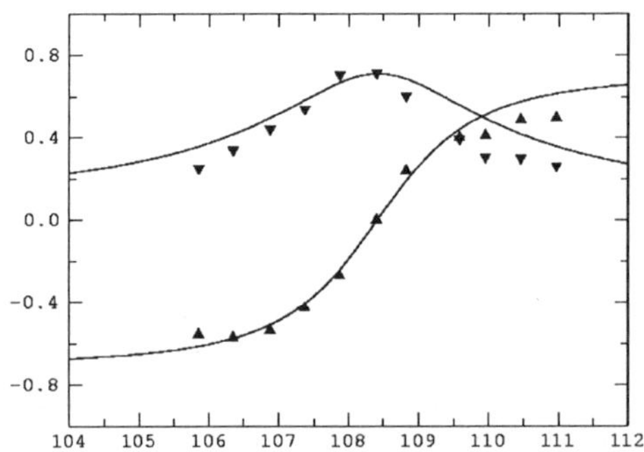

Fig. 3:   The horizontal (down-pointing triangles) and vertical (up-pointing triangles) beam polarization components as a function of the CIS extraction energy (MeV). The curves are a prediction based on the strength of the correction solenoid field and the measured beam polarization injected into CIS.

5   Conclusions

We have demonstrated the operation of a polarimeter capable of yielding pulse-by-pulse polarization values for a proton beam with a very low duty factor.

The authors acknowledge the support of the US National Science Foundation under grant NSF-PHY-9602872.

References

1. M.W. McNaughton *et al.*, Nucl. Instrum. and Methods **A241**, 435 (1985).

RHIC pC CNI POLARIMETER, CURRENT STATUS AND FUTURE PLANS

I.G. ALEKSEEV[1], M. BAI[2], B. BASSALLECK[2], G. BUNCE[2,3],
A. DESHPANDE[5], S. DHAWAN[5], J. DOSKOW[6], S. EILERTS[4], D.E. FIELDS[4],
Y. GOTO[7], H. HUANG[2], V. HUGHES[5], K. IMAI[8,3], M. ISHIHARA[7,3],
V.P. KANAVETS[1], K. KURITA[7,3], K. KWAITOWSKI[6], B. LEWIS[4],
B. LOZOWSKI[6], W. MACKAY[2], Y. MAKDISI[2], H.O. MEYER[6],
B.V. MOROZOV[1], M. NAKAMURA[8], B.V. PRZEWOSKI[6], T. RINCKEL[6,3],
T. ROSER[2], A. RUSEK[2], N. SAITO[7,3], B. SMITH[4], H. SPINKA[9],
D.N. SVIRIDA[1], M. SYPHERS[2], A. TAKETANI[7], T.L. THOMAS[4], J. TOJO[8],
D. UNDERWOOD[9], D. WOLFE[4], K. YAMAMOTO[8], L. ZHU[8]

*(1) Institute for Theoretical and Experimental Physics, B. Cheremushkinskaya 25,
Moscow 117259, Russia*
(2) Brookhaven National Laboratory, PO Box 5000, Upton, NY 11973-5000, USA
(3) RIKEN BNL Research Center, PO Box 5000, Upton, NY 11973-5000, USA
(4) University of New Mexico, 800 Yale, Albuquerque, NM 87131, USA
(5) Yale University, 217 Prospect, New Haven, CT 06511, USA
*(6) Indiana University Cyclotron Facility, 2401 Milo B. Sampson Lane,
Bloomington, IN 47408, USA*
*(7) Institute of Chemical and Physical Research RIKEN, 2-1 Hirosawa, Wako-shi,
Saitama, 351-0198, Japan*
(8) Kyoto University, Kitashirakawa-Oiwakecho, Sakyo-ku, Kyoto 606-8502, Japan
(9) Argonne National Laboratory, 9700 S. Cass Ave, Argonne, IL 60439, USA

Acceleration of polarized protons at RHIC requires fast and reliable measurements of proton beam polarization. There are not many physical processes which could be used to measure polarization of protons with momentum larger than 10 GeV. Elastic scattering with very small momentum transferred is one of these processes. Its advantages are large cross section, which gives large figure of merit, and weak energy dependence of the analyzing power. CNI polarimeter with ultra thin carbon ribbon target and silicon detectors for recoiled nuclei was successfully tested with internal polarized beam at AGS. Then the polarimeter consisting of 4 silicon detectors was used for RHIC polarized beam commissioning in the blue ring. This year polarimeters of 6 silicon detectors each are to work in both rings to ensure RHIC polarized beam collisions.

1 Introduction

Experimental study of polarized proton collisions at energies up to 250 GeV/c opens a new horizon in high energy polarization physics. RHIC experimental program will cover such topics as spin structure of proton, including gluon and sea quark polarization, parity violation in W and Z-boson production, spin effects in direct photon production and so on[1,2]. Realization of this program requires to measure beam polarization, which is not easy for 25–250 GeV/c beams. Process of elastic scattering of protons on an ultra thin carbon target in a Coulomb nuclear interference (CNI)[3,4,5] range gives several advantages: large cross section, weak energy dependence of the analyzing power, constant kinematics in the whole energy range, simple equipment. As drawbacks of the method one should count small (several percent) analyzing power and absence of precise theoretical calculations. Nevertheless large cross section provides good figure of merit and fast measurements.

In this paper we would like to present our experience of pC CNI polarization measurements at AGS and RHIC as well as plans for this year RHIC running.

2   Experimental conditions

The idea of detection of elastic proton-carbon scattering in our experiment is in the selection of carbon nucleus scattered to 90 degrees. The carbon produced in inelastic reactions has considerable forward momentum and can't hit the detector. The carbon nucleus is selected by the energy to velocity correlation.

AGS test setup was mounted in the AGS tunnel in the internal proton beam with energy 21.7 GeV/c^2. A proton hit ultra thin carbon ribbon target[6] (3.7 μg/cm^2 × 6 μm) and recoiled carbon was detected by one of the two arms. Each arm consisted of a very thin foil with a microchannel plate and a silicon strip detector. The microchannel plate was used to achieve time resolution better than bunch length which for AGS was about 25 ns. AGS beam polarization was controlled by two elastic polarimeters one on the internal beam[7] and the other on the extracted beam[8] and was $(44 \pm 4(stat) \pm 4(sys))\%$. Each spill the sign of the polarization was changed.

RHIC year-2000 polarimeter was mounted in blue ring in the warm straight section near the 12 o'clock interaction point. It consists of 4 silicon detectors placed at $\pm 45^o$ in the vertical plane. This allowed to measure both vertical and horizontal polarization. During the proton run RHIC was working in 6 bunches mode.

In both AGS test and RHIC year-2000 runs DAQ based on LeCroy 4300 FERA ADC/TDC[9] was used with LeCroy 2367 universal logic module (ULM) as a large FERA memory. The data was first stored in the ULM and then read-out to PC in large portions. The collected statistics was about $2 \cdot 10^7$ events for the whole AGS test and $\approx 5 \cdot 10^6$ events per one RHIC measurement.

3 Data processing and event selection

The kinetic energy (E) of the carbon nucleus is connected to the time of flight (t) by the non-relativistic equation:

$$E = \frac{ML^2}{2t^2}, \tag{1}$$

where M is carbon mass and L is the flight distance. The distribution of RHIC events in the $t - E$ plane is shown in fig. 1a. Carbon locus is easily seen. And under it one can see a locus, corresponding to α-particles. For event selection we also used the same data ploted in $t - 1/\sqrt{E}$ plane, where carbon locus is a straight line (fig. 1b). Distributions over the recoiled particle mass (in atomic units) and 4-momentum transferred are shown in fig. 1c and d.

Physical asymmetry is given by equation:

$$A = \frac{\sqrt{N_{UL}N_{DR}} - \sqrt{N_{UR}N_{DL}}}{\sqrt{N_{UL}N_{DR}} + \sqrt{N_{UR}N_{DL}}},$$

where N_{ij} are numbers of events collected for beam polarization **Up** and **Down** by the **Left** and **Right** arms correspondingly. Integrated raw asymmetry measured at AGS was 0.00533 ± 0.00027, which showed good perspectives of the method. During RHIC running the beam polarization was measured a number of times with as good as 10% statistical precision[10]. Mean RHIC polarization at the injection energy (24.3 GeV/c^2) during year-2000 run was $(18.0 \pm 1.4 stat \pm 3.1 sys)\%$. The main sources of the systematic error was uncertainty of AGS beam polarization during AGS run and uncertainty in recoiled carbon energy measurement. The systematical error of the measurement itself was rather small.

4 DAQ with wave form digitizers

Switching from the commissioning to the real work required much faster DAQ than that decribed above. For this purpose DAQ based on wave form digitizers (WFD) was developed[11]. Each WFD CAMAC module contains 4 independent channels. Each channel includes 420 MHz 8-bit ADC and a Xilinx Virtex FPGA chip[12] for data processing. The very first test of the WFD was done during RHIC year-2000 run, where FPGA were configured just to record the waveform when external trigger occures. The resulting time versus amplitude distribution obtained in the off-line analysis is shown in fig. 2.

For this year run a new code for WFD was developed and tested with pulse generator signal. With this code each WFD channel is selftriggered and completely independent from others. In the main running mode FPGA code for each bunch (1) calculates and substracts pedestal; (2) finds amplitude

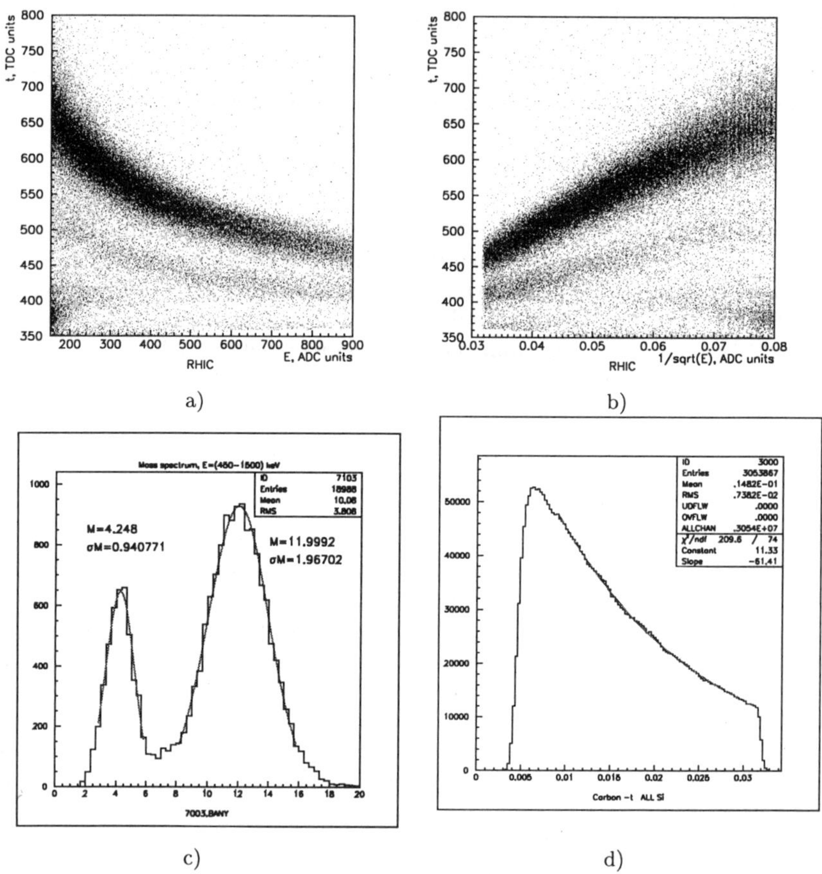

Figure 1. Event distributions: in the plane TDC - ADC values (a), in the plane TDC - $1/\sqrt{\text{ADC}}$ (b), as function of recoiled particle mass (c) and momentum transferred (d).

and integral of the signal; (3) finds time when the signal reaches 1/4 of its maximum value; (4) checks the time, amplitude and integral correlation; (5) if the event is correct increments proper scalers. There are 512 24-bit scalers in every channel, which gives histograms of good events distribution over bunch number and particle energy. This design allows deadtimeless very fast data accumulation: you load time, amplitude, integral correlation for the events you like to collect; put the target in and have some run; then readout scalers.

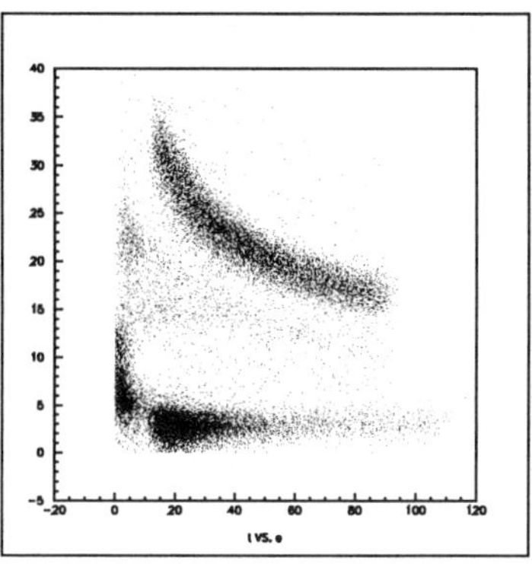

Figure 2. Time versus amplitude event distribution from WFD test (arbitary units).

No dataflow on CAMAC backplane is assumed while running. Other modes included to check algoritms and performance, with deadtime.

For this year of RHIC running we have 6 silicon detector (at 90° and $\pm45^o$) in each of two (blue and yellow) polarimeters. All the detectors and preamplifiers are tested and installed in the RHIC tunnel. We are going to have 12 WFD modules (6 per each polarimeter) and work in 60 bunches mode. This will be the first year of real polarized proton collisions.

References

1. D. Underwood et al., *Part. World.* **3**, 1 (1992).
2. C. Bourrely and J. Soffer, *Phys. Lett.* **B 314**, 132 (1993).
3. B.Z. Kopeliovich and L.I. Lapidus, *Yad. Phys.* **19**, 218 (1974).
4. N.H. Buttimore et al., *Phys. Rev.* **D 18**, 694 (1978).
5. B. Kopeliovich, *High Energy Polarimetry at RHIC* **hep-ph**/9801414.
6. W.R. Lozowski and J.D. Hudson, *Nucl. Instr. Meth.* **A 303**, 34 (1991).
7. H. Huang et al., *Phys. Rev. Lett.* **28**, 2982, (1994).
8. H. Spinka et al., *Nucl. Instr. Methods* **211**, 239, (1983);
   K. Krueger et al., *Phys. Lett* **B 459**, 412, (1997).
9. **http://www.lecroy.com/lrs/dsheets/dslib.htm**
   http://www.lecroy.com/lrs/dsheets/dsheets.htm
10. H. Huang et al., *Commissioning of RHIC p-carbon CNI Polarimeter*, Proceedings of the 14th International Spin Physics Symposium, Spin 2000, Osaka, Japan, AIP Conference Proceedings **570**, 795 (2000).
11. S. Dhawan, private communication.
12. **http://www.xilinx.com**

DEVELOPMENT OF A POLARIZED DEUTERON BEAM FOR THE IUCF COOLER

E.J. STEPHENSON, C.E. ALLGOWER, A.D. BACHER, V. DERENCHUK, H. NANN, P. PANCELLA, AND T. RINCKEL

Indiana University Cyclotron Facility, Bloomington, IN 47408 USA

We describe the operation of a polarized D⁻ source that generates beam for the IUCF Cooler. Two polarimeters operating in the CIS injection line at 4 MeV and in the Cooler at 231 MeV were used to measure the deuteron beam polarization.

1   Introduction

Vector and tensor polarized deuteron beams are needed in the IUCF electron-cooled synchrotron (Cooler) for experiments that will investigate three-body forces and search for the isospin-forbidden $dd \rightarrow \alpha\pi^0$ reaction. The pulsed, polarized beam is generated in an atomic beam source whose features are outlined in the next section. The beam's first acceleration is through an RFQ whose output energy is 4 MeV. At that energy, the ^{3}He$(d,p)^4$He reaction is used for a first measurement of the polarizations. The beam is accumulated in the Injector Synchrotron where it is ramped to 90 MeV and transferred to the Cooler. For these tests, the Cooler was ramped to 231 MeV, the energy for the $dd \rightarrow \alpha\pi^0$ search. There the PINTEX detector system was used to observe the polarizations using d+p elastic scattering. The analyzing powers for the polarimeters at 4 and 231 MeV were taken from the literature [1-6].

2   Polarized Ion Source

The pulsed atomic-beam source (CIPIOS) now being used at IUCF contains two stages, each consisting of a separator sextupole magnet followed by RF transition units. The first stage is equipped with a medium-field unit; the second a combination of strong and weak-field units. This allows, at some expense in beam intensity, a greater variety of deuteron polarization states, including states consisting of a single projection of the deuteron's magnetic quantum number.

We used four polarized states, as shown in Table 1, as well as an unpolarized state for the tests described here. Table 1 gives the ideal polarizations for these states. Normally, background and inefficiencies in the separators and RF transitions units will lead to inefficiencies that reduce the actual polarization to about 90% of the ideal value. (During these tests, the WF transition unit had a low efficiency, so the V− state, as defined below, produced a lower polarization.)

Table 1: Ideal Polarization States of CIPIOS

state	p_y	p_{yy}	first RF	second RF
vector + (V+)	1	1	MF 3↔4	SF 2↔6
vector − (V−)	−1	1	MF 1↔4	WF
tensor + (T+)	0	1	MF 1↔4	SF 2↔6
tensor − (T−)	0	−2	MF 1↔4	SF 3↔5

Prior to these tests, an extensive search had been completed using the RF transition unit parameters to obtain the largest possible polarization for each state.

3   4-MeV Polarimeter

As shown in Fig. 1, the polarimeter at the end of the RFQ consists of a gas cell containing ^{3}He at a pressure of two atmospheres. Protons from the ^{3}He(d,p)^{4}He reaction are collimated ahead of plastic scintillation detectors. The large, positive reaction Q-value produces clean proton peaks, as shown in the right side of Fig. 1. This makes it possible to scale all events above an energy threshold rather than to digitize each event for computer processing, an advantage for pulsed beam.

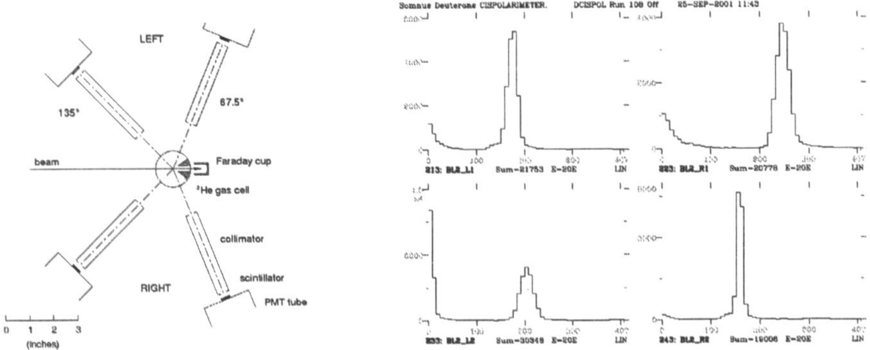

Fig. 1: [left] Layout of the 4-MeV polarimeter. [right] Sample pulse height spectra from the 135° (top) and 67.5° (bottom) detectors on the left and right sides of the beam (shown in left and right panels).

The analyzing powers used for this polarimeter are a smooth interpolation through the measurements of Gruebler [1], Clark [2], and Bittcher [3] to the energy given by energy-loss calculations for the beam at the center of the gas cell. These values are A_y = 0.008 and 0.150, and A_{yy} = −0.463 and 0.420, at 67.5° and 135° respectively. The low values of A_y make the vector polarization poorly determined statistically; a steep energy dependence adds systematic uncertainty. So one goal of the calibration was to check these A_y values against the Cooler. Other low energy reactions have a larger A_y, but not the large Q-value needed for easy observation.

4   Cooler Polarization Measurement

We used the PINTEX gas target and detector system to measure the polarization of the circulating deuteron beam in the Cooler, having made sure first that an intrinsic depolarizing resonance was avoided during the energy ramp. The gas target consisted of a storage tube along the beam that confines H_2 gas before it is pumped away. The large area detectors completely surrounded the beam. The stack consisted of a thin trigger scintillator, an XY multiwire chamber, a UV multiwire chamber (rotated $45°$ with respect to XY), and a final thick scintillator. The scintillators were segmented azimuthally into 4 pieces. A trigger required that there be two signals (from both deuteron and proton) in the thick detector scintillator on opposite sides of the beam and in coincidence with the front scintillator.

It was necessary to separate elastic scattering events from the large flux of deuteron breakup events. The additional tests applied in analysis included:

- quality of track reconstruction in the multiwire chambers (good χ^2 fit),
- traceback of the tracks to a position within the gas cell target,
- coplanarity (tracks separated by $180°$ in ϕ),
- correct correlation of thick scintillator energy with track angle,
- location of the events on the correct kinematic locus in θ_d by θ_p.

The remaining breakup background was subtracted based on event rates close to the kinematic locus.

Events along a part of the span of proton angle were divided among 9 bins that ranged from $\theta_{c.m.}=87.7°$ to $132.7°$ in $5°$-wide bins. For each bin, the vector and tensor analyzing powers were interpolated linearly between the measurements made at 270 MeV [4,5] and 200 MeV [6]. Polarizations calculated from the data in each bin should be consistent. (Initial problems for the vector analyzing powers led to a correction in the laboratory angle of a little over one degree.)

For analysis, the events in each scattering angle bin were plotted according to the azimuthal angle ϕ for later Fourier decomposition. To make these data smoother, each polarized state was divided by the unpolarized state to remove notch inefficiencies at scintillator segment boundaries. The results are shown in Fig. 2.

5   Results

Each spin state in the Cooler measurement can yield the vector and tensor polarizations from the Fourier components of its azimuthal distribution alone. These are:

$$F = \int_0^{2\pi} \sigma(\phi)d\phi = 2\pi\sigma_0[1 - \sqrt{\tfrac{1}{8}}T_{20}\,p_{yy}]$$

$$G = \int_0^{2\pi} \sigma(\phi)\cos\phi\,d\phi = \sqrt{3}\,\pi\sigma_0 iT_{11}\,p_y$$

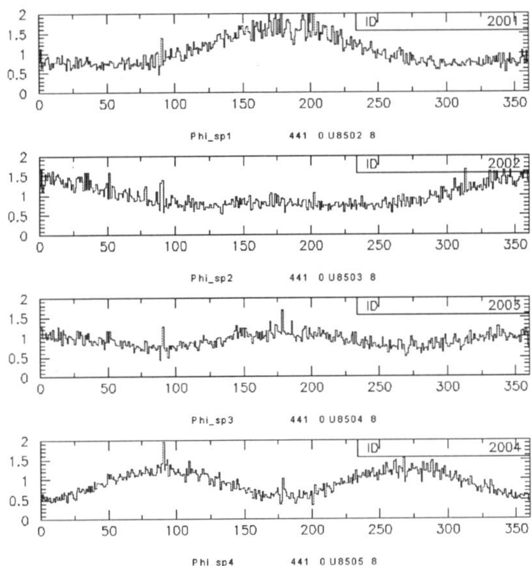

Fig. 2: Azimuthal distributions of d+p scattering for the 4 polarization states of Table 1.

$$G = \int_0^{2\pi} \sigma(\phi)\cos 2\phi\, d\phi = -\sqrt{\tfrac{3}{4}}\,\pi\sigma_0 T_{22}\, p_{yy}$$

With these coefficients, the vector and tensor polarizations become:

$$[\pi\sigma_0] = \frac{F}{2} - \frac{T_{20} H}{\sqrt{6}\,T_{22}}$$

$$p_y = \frac{G}{\sqrt{3}\,[\pi\sigma_0]i T_{11}}$$

$$p_{yy} = \frac{2H}{\sqrt{3}\,[\pi\sigma_0]T_{22}}$$

Because all of the analyzing powers are known, it is possible to obtain both the vector and tensor polarization from a single azimuthal distribution alone (even without prior division by the unpolarized state).

The vector and tensor polarizations for the 9 bins are shown in Figs. 3 and 4. In all cases the polarization values are consistent across angles. The averages for each are compared in Table 2 with the measurements made at 4 MeV at the end of the RFQ. The uncertainties for the vector analyzing power are large at 4 MeV; so we regard these results as consistent. The tensor polarization is systematically 7% larger in the Cooler measurement. Since the Cooler values exceed the polarization expected from CIPIOS, we conclude that the T_{22} analyzing powers [4-6] are low. Part of this may arise from the linear interpolation made to 231 MeV.

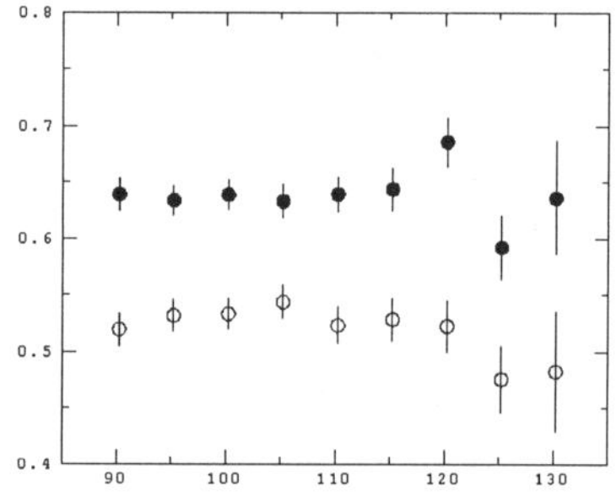

Fig. 3: Vector polarization p_y values for the V+ (dots) and V− (circles) states vs. $\theta_{c.m.}(p)$.

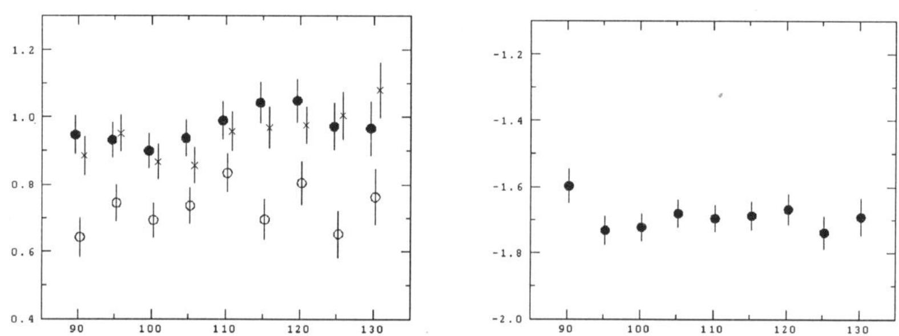

Fig. 4: Tensor polarization p_{yy} values [left] for the V+ (dots), V− (circles), and T+ (x) states, and [right] for the T− state as a function of the center of mass proton scattering angle, $\theta_{c.m.}(p)$.

Table 2: Average polarizations (at each energy in MeV)

state	vector(4)	vector(231)	tensor(4)	tensor(231)
V+	0.909∓0.031	0.852∓0.008	0.891∓0.013	0.961∓0.020
V−	−0.684∓0.030	−0.702∓0.008	0.695∓0.014	0.728∓0.020
T+			0.875∓0.013	0.949∓0.020
T−			−1.579∓0.013	−1.693∓0.015

The authors acknowledge support under NSF grant NSF-PHY-9602872.

References

1. H. Gruebler *et al.*, Nucl. Phys. **A176**, 631 (1971).
2. Harvey W. Clark, Ph.D. thesis, Ohio State University, 1981.
3. M. Bittcher *et al.*, Few-Body Systems **9**, 165 (1990).
4. N. Sakamoto *et al.*, Phys. Lett. B **367**, 60 (1996).
5. H. Sakai *et al.*, Phys. Rev. Lett. **84**, 5288 (2000).
6. H. Sakai and K. Sekiguchi, private communication.

VII. Applications and New Techniques

INTERACTION OF NUCLEAR SPIN POLARIZED ATOMS WITH SOLID SURFACES

D. FICK

*Philipps-Universität, Fachbereich Physik and Zentrum für Materialwissenschaften,
D-35032 Marburg, Germany*

and

*University of Wisconsin-Madison, Department of Physics,
Madison, WI 53706, USA*

E-mail: dieter.fick@physik.uni-marburg.de

Since the use of surface ionizers in polarized heavy ion sources in the seventies and also due to the use of storage cell targets in storage rings increasing interest grew on the interaction of nuclear spin polarized atoms with solid surfaces. Out of it a new technique to perform NMR on surfaces emerged. We report on various NMR experiments on Li adsorbed on metal and semiconductor surfaces, the results of which might be of interest for the PST community.

1 Introduction

The surface ionizers have been one of the most challenging devices of the first polarized heavy ion sources [1,2]. They consisted of an oxygen covered tungsten strip heated up to 1800 K, from which only positive ions desorb thermally. The high temperatures were chosen in order to keep the mean residence time as short as possible and thus depolarization eventually small. One of the most intriguing questions at that time was concerned with the nuclear polarization of the desorbing ions. The predictions ranged from complete depolarization to complete conservation. Happily enough, none of the extreme scenarios were valid. Fig. 1 displays the polarization of the desorbing ions as a function of ionizer temperature as determined in a steady state experiment through beam foil spectroscopy [1]. At the highest surface temperature tested, the polarization of both ^{6}Li and ^{7}Li was fully conserved, whereas at the lowest temperature investigated, ^{7}Li was completely depolarized. This difference in depolarization bore the real surprise. It could be retraced to the different electric quadrupole moments of both isotopes, the depolarization being caused by their interaction with electric field gradients (EFG) fluctuating in time due to the Li diffusion in the oxygen adlayer. The Fourier components around the Larmor frequency

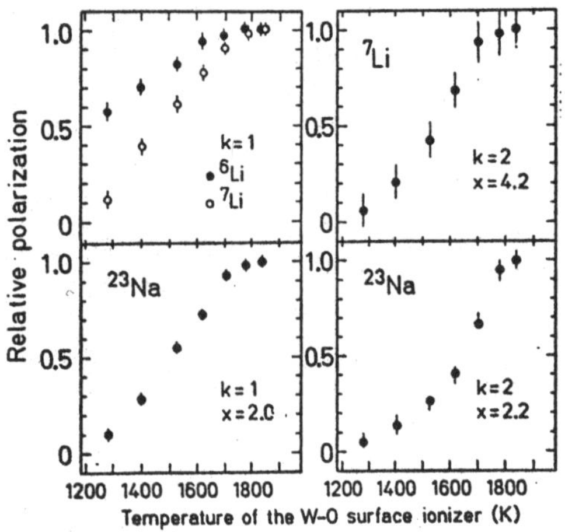

Figure 1. Polarization of the ionized alkali beams as a function of ionizer temperature T for various polarization tensor moments k. The data are normalized to unity for each isotope and moment at the highest temperature investigated.

$\omega_L = \mu \cdot B / \hbar$ drive the transition in between the nuclear spin sublevels causing eventually their complete equilibrium (depolarization). It had been this very result which pointed to the feasibility of nuclear magnetic resonance (NMR) experiments on surfaces. Because of its inherent insensitivity, even though highly desirable, it is a topic unthinkable for conventional NMR, still now.

To proceed further, we will now leave the historical path and rather jump immediately to one of the present frontiers in this field, β–NMR with adsorbed ^{8}Li. In these experiments the sensitivity as compared to conventional NMR has been increased by about thirteen orders of magnitude. Now, as compared to conventional solid state NMR, for which at least 10^{17} equivalent nuclear spins are necessary, as little as 10^4 equivalent nuclear spins are sufficient to perform an NMR experiment on a single crystal surface.

2   Experimental

Opposite to solid state NMR experiments, in which polarized recoil nuclei from nuclear reactions can be used, β–NMR on surfaces requires a gentle landing of the adsorbates and thus thermal nuclear spin polarized ^{8}Li atoms (nuclear spin I = 2). Since the half life of ^{8}Li of about 0.8 s the experiment has to be

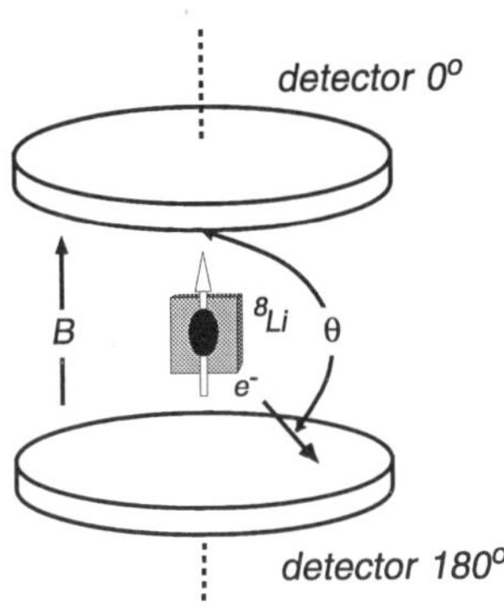

Figure 2. Principle of the measurement of the directional asymmetry ϵ of β–decay electrons.

situated at an accelerator, here the MP–Tandem accelerator in Heidelberg. The experimental setup consists of three essential parts: a bakeable source to produce a thermàl nuclear spin polarized ^{8}Li beam, an NMR region including detectors for the ^{8}Li decay electrons, and a conventional UHV chamber with standard analysis tools. The setup has been described in great detail recently [3].

Since ^{8}Li is a β–decaying nucleus, spin polarization of the adsorbed ^{8}Li can be detected via the directional asymmetry ϵ of its β–decay electron (Fig. 2).

$$\epsilon = \frac{N(0°) - N(180°)}{N(0°) + N(180°)} \tag{1}$$

In the experiment, the thermal ^{8}Li atoms are accumulated on the surface for 0.5 s followed by the detection of the β–electron asymmetry ϵ as function of time for 4 s. These steps are repeated many times [3].

The observed exponential time dependence of $\epsilon(t)$ bears the information on the nuclear spin relaxation rate α or its inverse, the T_1–time, well known from NMR literature [4,5]:

$$\epsilon(t) = \epsilon(0)e^{-\alpha t} = \epsilon(0)e^{-t/T_1}. \tag{2}$$

Contrary to conventional NMR experiments, the determination of the T_1 time does not require the application of resonant rf–fields, since the nuclear spin polarization in the thermal equilibrium (of the order of 10^{-6}) can be neglected as compared to the initial polarization of the ^{8}Li ensemble of 0.8 to 0.9 after adsorption.

314

3 Fluctuating interactions in time and depolarization

In order to describe the frequency spectrum of an fluctuating interaction, time correlation functions $G(t)$ and their Fourier transformed, the spectral density $j(\omega)$ are used [4,5,6]. Only in rare cases they might be obtained rigorously. Thus parametrizations are used, out of which the BPP one, named according to the initials of their authors, is the most widely used.

$$G(t) = G(0)\, e^{-t/\tau_c} \tag{3}$$

Here G(0) denotes the square of the fluctuating, time averaged interaction and τ_c a proper correlation time, typical for the processes encountered. For simplicity it is assumed that G(0) does not depend on the m–quantum numbers of the nuclear spin sublevels between which the transitions occur. This assumption is equivalent to the fact that the Zeeman interaction is by far the largest one [7].

The Fourier spectrum of this correlation function at the Larmor frequency ω_L is given by

$$j(\omega_L) = \frac{2}{\pi}\, G(0)\, \frac{\tau_c}{1 + (\omega_L \tau_c)^2}. \tag{4}$$

Relaxation rates are related to $j(\omega_L)$ within a factor of unity by

$$\alpha = 1/T_1 \propto \frac{1}{\hbar^2}\, j(\omega_L). \tag{5}$$

(In practice the exact expression has to be calculated for each polarization tensor moment [7].) The following educating experimental examples, which deal mainly with the adsorption of Li on a Ru(001) single crystal surface will demonstrate the properties of these expressions.

For Li adsorption on a well prepared and characterized Ru(001) single crystal surface Fig. 3 shows $\alpha(T)$ data taken at B = 0.15 T over a wide temperature range [6]. To getter any type of defects, which are dressed by adsorbed Li at first, the surface was always precovered with about 2 % of a monolayer (ML) of stable Li. The data above 1000 K were obtained without any precoverage similar to the ones in Fig. 1, using now, however, laser induced fluorescence to detect the polarization of the desorbing ^{6}Li atoms [8]. Two main features stick out from the data in Fig. 3: There is a) a contribution linear in temperature underlying b) a resonance–like structure. Whereas the latter is due to diffusion of the Li adsorbates over the surface, the linear one is caused by the fluctuating spins of the conduction electrons. In s–state dominated systems, as here, the main electronic relaxation is caused by the Fermi contact interaction in between nuclear and electronic spins. The necessary mutual spin

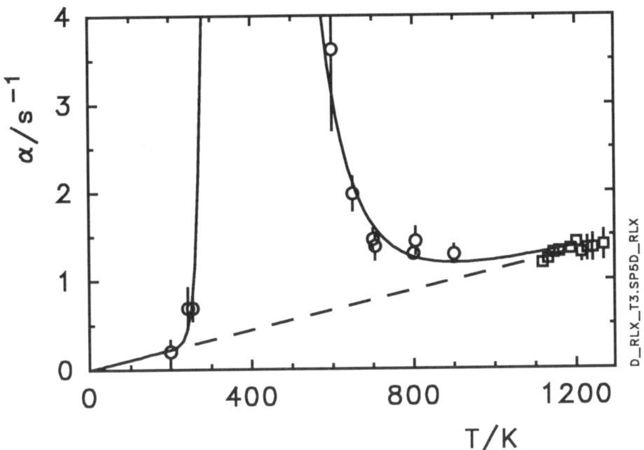

Figure 3. The nuclear spin lattice relaxation rate α as a function of temperature for Li adsorption on a Ru(001) surface. The solid line gives the sum of diffusional and electronic relaxation. The latter part is given by the dashed line.

flips of the the nucleus and the electron can only occur around Fermi energy, since only here the electrons can scatter into empty states. This explains the linearity of the relaxation rate with surface temperature [4,5]. Typical correlation times for metals (eq. 3), now called τ_e, are given by [9]

$$\tau_e = \hbar \, DOS(\epsilon_F) \simeq \hbar/\epsilon_F \qquad (6)$$

and are around 10^{-15} s. ($DOS(\epsilon_F)$: density of states at Fermi energy ϵ_F). Since they are much, much smaller than the Larmor frequency ω_L of the nuclear spin in the applied external magnetic fields (around 10^6/s) the electronic relaxation will be magnetic field independent ($\omega_L\tau_c \ll 1$, eq. 4), as observed indeed in the experiments (not shown). For this particular situation the relaxation rates called α_K can be calculated rigorously to be

$$\alpha_K = \frac{1}{T_1} = \frac{256\pi^3}{9\hbar} \, \mu_e^2 \left(\frac{\mu(^8\mathrm{Li})}{I}\right)^2 |<|\psi(0)|^2>|^2 \, DOS(\epsilon_F) \cdot \tau_e \cdot \frac{kT}{\hbar}. \qquad (7)$$

($|<|\psi(0)|^2>|^2$: probability to find an electron at the nucleus) We adopt here for later purposes a more general formulation [10]. One yields the usual one found in textbooks for 'free–electron metals' by inserting eq. 6 into the above eq. A fit to the data yields for the local density of states at Fermi energy [6]

$$LDOS(\epsilon_F) = |<|\Psi(0)|^2>| \, DOS(\epsilon_F) = (0.13 \pm 0.01) \, \mathrm{eV}^{-1}\mathring{A}^{-3}, \qquad (8)$$

316

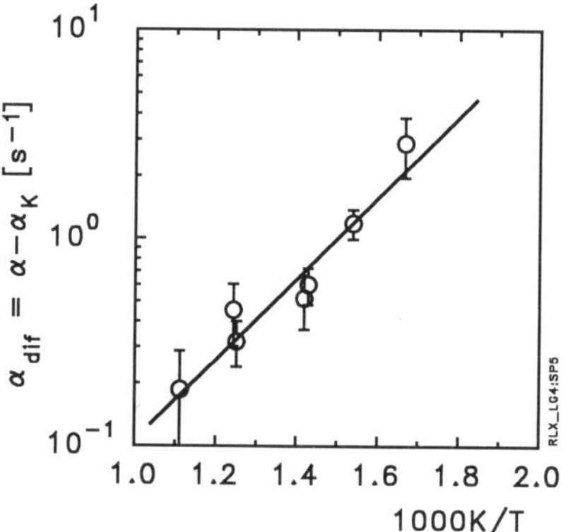

Figure 4. $\alpha_{\rm dif}$ as taken from Fig.3, however drawn into an Arrhenius plot.

in agreement with ab initio all electron calculations [11].

The resonance like structure in Fig. 3 is due to thermally activated diffusion. It generates a fluctuating interaction of the nuclear quadrupole moments with EFG's caused by charge distributions of less than cubic symmetry. They are always present on surfaces. Since the experimental data indicate a maxiumum of α with T, the correlation times

$$\tau_c = \tau_0 \, e^{E_{\rm dif}/kT} \tag{9}$$

($E_{\rm dif}$: diffusion energy) have now to be around ω_L^{-1} because the spectral density (eq. 4) adopts a maximum only there ($\omega_L \tau_c = 1$). The solid line is a fit to

$$\alpha = \alpha_K + \alpha_{\rm dif}, \tag{10}$$

whereby α_K is given by eqs. 7 and 6 and $\alpha_{\rm dif}$ by inserting eqs. 4 and 9 into eq. 5. Only because of the very different correlation times involved, both contributions can be added incoherently. Taking into account the result of eq. 8, the fit to the data yields $E_{\rm dif} = (0.45 \pm 0.11)$ eV [6].

Beyond the maximum, at higher temperatures, diffusion becomes faster and $\omega_L \tau_c \ll 1$. Then the spectral density becomes magnetic field independent (not shown) and proportional to τ_c (eq. 4). Plotting therefore $\alpha - \alpha_K = \alpha_{\rm dif}$ in an Arrhenius–plot, as done in Fig. 4, one obtains a straight line, the slope of which determines $E_{\rm dif}$ in agreement with the result above.

For low enough temperatures ($\omega_L \tau_c \gg 1$) diffusional relaxation will be proportional to $1/B^2$ (eq. 4). Fig. 5 shows such data for T = 120 K, taken at

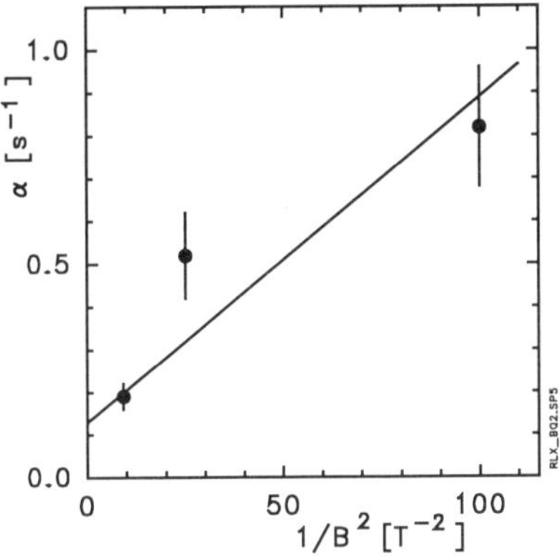

Figure 5. Spin lattice relaxation $\alpha = 1/T_1$ as a function of the inverse magnetic field squared taken at a substrate temperature of 120 K and a coverage of $\Theta = 0.12$ ML.

a coverage of 0.12 ML (dosing with stable Li) [11]. Now the main source for the EFG is the adsorbed Li itself, through an induced electric dipole at its site. From the slope together with data at 200 K (not shown) [6] again a diffusion energy can be extracted, which amounts now to be $E_{\mathrm{dif}} = 0.176$ eV [11]. This is considerably smaller than the one found above at much smaller coverages. We attribute this difference to different diffusion processes: diffusion over steps and terrace diffusion, respectively [6,11].

4   Li adsorption on the Si(111)-(7×7) surface

Due to its dangling bonds the (7×7) reconstruction of the Si(111) surface has around room temperature many hall marks of a metallic surface [12]. Fig. 6 displays data for $\alpha = 1/T_1$ as function of temperature, taken at a coverage as low as 10^{-4} ML [13]. They are independent of magnetic field strength and follow quite well a linear temperature dependence. Literally, following the discussion of the previous section, these data may be taken as a sign of the metallicity of the (7×7) reconstruction of the Si(111) surface. However, the relaxation rates exceed by about 70 % those found for ^{8}Li adsorbed on the metal surface Ru(001) (Fig. 3). The recently found narrow band around ϵ_F with a band width in the 100 meV regime points to considerably enlarged correlation times ($\tau_e = 10^{-14}$ s) as compared to 'real' metals. According to eq. 7 they enlarge the relaxation rates! Photon electron spectroscopy on an alkali–metal (K) covered Si(111)-(7×7) surface displays with increasing coverage vanishing

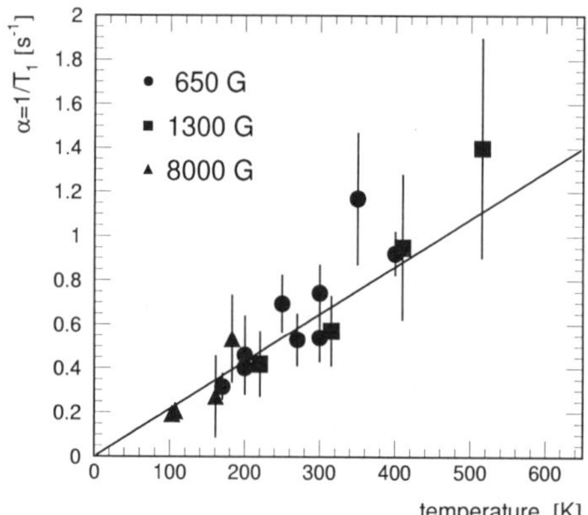

Figure 6. Nuclear spin relaxation rates $\alpha = 1/T_1$ as a function of substrate temperature for three different magnetic fields for ^{8}Li adsorbed on a clean Si(111)-(7×7) surface.

emission from states near the Fermi energy [14]. That is commonly interpreted as transition from a metallic to a semiconducting surface. In order to investigate this result further, the coverage dependence of $\alpha = 1/T_1$ (Fig. 7) has been measured at a fixed surface temperature (110 K) and high magnetic field (0.8 T) in order to suppress diffusional relaxation (Sect. 3) [15]. For each data point the surface has been freshly coveraged with Li from a getter source. The relaxation rates were then observed for ^{8}Li atoms adsorbed at a very low additional coverage of about 10^{-4} ML. Starting at extremely low coverages the relaxation rate firstly increases with coverage to more than twice its value. Around $\Theta \approx 0.15$ it passes a maximum. At about this coverage all empty dangling bonds of the surface are occupied by one unpaired electron. Beyond that coverage, dangling bonds are filled further, until at monolayer coverage they are occupied completely by electron pairs. They hybridize in a binding and antibinding state. The binding state moves well below Fermi energy generating an isolator (semiconductor) with vanishing $\mathrm{DOS}(\epsilon_F)$ [14]. That is why the T_1-times become now very long, more than 30 s! Beyond a coverage of 1 ML the relaxation rates increase again. The Li overlayer becomes then fully metallic again. Thus, from extremely low coverage to high ones a metal–isolator–metal transition exists on the Si(111)–surface.

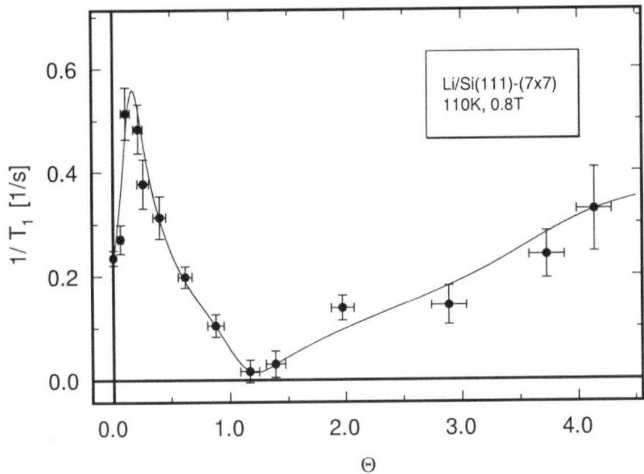

Figure 7. At a surface temperature of 110 K and a magnetic field of 0.8 T nuclear spin relaxation rates as function of Li–coverage Θ.

5 Summary

Obviously NMR experiments on solid surfaces using "hyperpolarized" species give detailed information on the fluctuating electromagnetic fields which cause the depolarization (relaxation) of the nuclear spin. In this report, it should finally be mentioned that their origin is by no means restricted to electronic and diffusional processes. It is just the opposite: There are usually too many to be identified by one experiment alone, as the subject index of the seminal work of Abragam indicates easily [4]: Nuclear spin relaxation caused by paramagnetic impurities, spin diffusion, spin–phonon coupling, two photon Raman–scattering, etc., to mention just a few. A good example for this situation, as explained by Tom Wise in his invited talk, is the spin relaxation of ortho H_2 molecules because of the fluctuations of the molecule axes due to wall collisions in a storage cell [16].

Acknowledgments

Support under various contracts from both the Bundesministerium für Bildung und Forschung (BMBF) and the Deutsche Forschungsgemeinschaft (DFG) is kindly acknowledged.

References

1. E. Steffens. *Nucl. Instrum. Methods*, 184:173–185, 1981.

2. D. Krämer, K. Becker, K. Blatt, R. Caplar, D. Fick, H. Gemmeke, W. Haeberli, H. Jänsch, O. Karban, I. Koenig, L. Luh, K.-H. Möbius, V. Necas, W. Ott, M. Tanaka, G. Tungate, I. M. Turkiewicz, A. Weller, and E. Steffens. *Nucl. Instrum. Methods*, 220:123–132, 1984.

3. H. J. Jänsch, G. Kirchner, O. Kühlert, M. Lisowski, J. J. Paggel, R. Platzer, R. Schillinger, H. Tilsner, C. Weindel, H. Winnefeld, and D. Fick. *Nucl. Instrum. Meth. B*, 171:537–550, 2000.

4. A. Abragam. *Principles of Nuclear Magnetism.* University Press, Oxford, 1978.

5. G. Schatz and A. Weidinger. *Nuclear Condensed Matter Physics.* Wiley, Chichester, 1996.

6. H. D. Ebinger, H. Arnolds, C. Polenz, B. Polivka, W. Preyß, R. Veith, D. Fick, and H. J. Jänsch. *Surf. Sci.*, 412/413:586–615, 1998.

7. M. Riehl-Chudoba, U. Memmert, and D. Fick. *Surf. Sci.*, 245:180–190, 1991.

8. M. Kaack and D. Fick. *Phys. Rev. B*, 51:17902–17909, 1995.

9. N. W. Ashcroft and N. D. Mermin. *Solid State Physics.* CBS Publishing, Asia, 1987.

10. W. W. Warren, Jr. *Phys. Rev. B*, 3:3708–3724, 1971.

11. G. Kirchner, M. Czanta, G. Dellemann, H. J. Jänsch, W. Mannstadt, J. J. Paggel, R. Platzer, C. Weindel, H. Winnefeld, and D. Fick. *Surf. Sci., to be published*, 2001.

12. R. Losio, K. N. Altmann, and F. J. Himpsel. *Phys. Rev. B*, 61:10845–10853, 2000.

13. H. Winnefeld, Ph.D. thesis, Philipps–Universität, Marburg, 2000.

14. H. H. Weitering, J. Chen, N. J. DiNardo, and E. W. Plummer. *Phys. Rev. B*, 48:8119–8135, 1993.

15. C. Weindel, Ph.D. thesis, Philipps–Universität, Marburg, 2000.

16. T. Wise, W. Haeberli, B. Lorentz, P. A. Quin, F. Rathmann, B. Schwartz, T. G. Walker, A. Wellinghausen, J. T. Balewski, J. Doskow, H. O. Meyer, R. E. Pollock, B. v. Przewoski, T. Rinckel, S. K. Saha, and P. V. Pancella. *Phys. Rev. Lett.*, 87:042701-1–042701-4, 2001.

BIOMEDICAL MRI WITH LASER POLARIZED GASES

TIMOTHY CHUPP

University of Michigan
Departments of Physics and Biomedical Engineering

The field of medical imaging with polarized rare gases, spawned by the remarkable developments of polarized ^{3}He targets, has brought optical scientists together with medical researchers to perfect techniques and pursue new opportunities for biomedical research. This brief overview aims to present the field from several perspectivesfrom historical to physical, to biomedical. The biomedical perspective describes work to date and the potential applications of imaging in medicine and research.

1   Introduction

In 1994, a short article was published in the journal **Nature** showing Magnetic Resonance Images of ^{129}Xe gas that had filled the airways of an excised mouse lung[1]. The images were acquired at SUNY, Stony Brook, on Long Island, NY. But the gas, prepared by laser optical pumping methods, in Princeton, New Jersey, was transported over 100 km by car in a small glass cell immersed in a cup of liquid nitrogen. (The gas was "polarized" in Princeton to provide 10,000 times greater NMR signal per atom than produced by "brute force." This compensated for the 10,000 times lower concentration of gas.) Reading the Nature paper led many in the field of laser optical pumping to turn their attention to the new possibilities, and it led many radiologists to seek laser physicists as collaborators to help develop the potential biomedical applications. Now physicists, radiologists, neuroscientists, medical researchers, and clinicians are working together in teams around the world.

Figures 1 and 2 show magnetic resonance images produced with laser polarized ^{3}He and ^{129}Xe. Figure 1 shows a comparison of ^{133}Xe scintigraphy and ^{3}He lung imaging and indicates the remarkable improvement in detail and resolution possible. Figure 2, shows images of ^{129}Xe gas in the lung and dissolved in tissue and blood of a rat that had been breathing a polarized ^{129}Xe – oxygen mixture. In contrast to helium, xenon crosses the blood–gas barrier in the lungs, dissolves in blood, and is carried to distal organs where it diffuses into tissue as the blood flows through capillaries from artery to vein. The NMR frequencies of ^{129}Xe depend on the tissue in which it is dissolved and it is possible to separetely image the airspace, blood, and tissues.

The development of techniques of laser enhanced nuclear polarization (or hyperpolarization), has been most strongly motivated by nuclear and particle physics. Targets of polarized ^{3}He are used in accelerator experiments such as those that probe the elementary particle, short range structure of the neutron [2]. Polarized ^{3}He is also used to polarize neutrons for nuclear physics and neutron scattering research [3]. The requirements of these experiments have pushed us to understand the physics and technical limitations of optical pumping at high densities. Now, we can produce liters (at STP) of ^{3}He, polarized to 50% or more, and similar quantities of ^{129}Xe. We conclude emphasizing some of the potential applications and future promise of this new technique that provides a wonderful example of transfer of technology motivated by fundamental physics research.

The isotope ^{3}He was accidently discovered by Alvarez at the Berkeley cyclotron in a test run preliminary to the intended use of the Cyclotron as a mass spectrometer to detect ^{3}He produced in nuclear reactions [4]. (It had been assumed, following the suggestion of Hans Bethe, that ^{3}He was unstable.) After the experiment, the magnet was ramped down with the RF on, revealing an ion with $Z/A = 2/3$. Once the Cyclotron magnet was shimmed properly for this mass, the discovery of ^{3}He was confirmed. One surprising consequence of that discovery was that ^{3}He and not ^{3}H is the stable A=3 isotope. With two protons and one neutron, the ^{3}He nucleus must have half integer spin, and naive consideration of the Pauli principle suggests that the protons' spins pair in a singlet $l = 0$ state. In that case the entire angular momentum and magnetic moment of the ^{3}He nucleus would be due to the neutron.

^{3}He is rare, and this is a problem. Primordial abundance of ^{3}He produced in Big Bang nucleosynthesis is $[^3\text{He}]/[^4\text{He}]=0.00004$. Additional ^{3}He has been produced in stellar burning, in the atmosphere due to cosmic ray interactions, and underground due to natural radioactivity. Cosmic ray production of ^{3}He on the moon, which does not have atmospheric shielding from cosmic rays, has left much greater abundances imbedded in lunar rocks [5], though mining the moon may remain impractically expensive. Most of the stored ^{3}He reserve, less than 1000 kg, has come from the decay of tritium (^{3}H) produced for thermonuclear weapons. The isotope, ^{3}He, has long been extremely important in nuclear particle physics. Calculation of the magnetic moment has clarified the role of meson exchange corrections. Nuclear reactions induced by ^{3}He and ^{3}H allow study of isospin symmetry and isospin dependence in a unique way because the A=1 isodoublet is much more difficult experimentally – accelerated neutron beams are not feasible. Perhaps most important is the fact that the neutron polarization in a polarized ^{3}He nucleus is $\approx 87\%$. This has allowed important short range properties of the neutron to be measured including the neutron electric charge distribution – the electric form factor G_E^n – and the neutron deep inelastic scattering spin structure functions

$g_n(x)$. Therefore beams and targets of polarized ^{3}He have been sought since at least the 1950's. It was the late 1980's before the basic physical processes and technology came together to foresee practical polarized ^{3}He targets.

Another feature of ^{3}He has motivated work to develop polarization techniques. ^{3}He turns out to be a potentially perfect spin filter for polarization of neutrons. The strong neutron absorption reaction n+^{3}He $\rightarrow$ p + ^{3}H is nearly 100% polarization dependent, due to an unbound $J = 0$ resonance in ^{4}He. With the neutron and ^{3}He spins opposite, the absorption cross section is $\sigma_a(v) = 5230\,\frac{v_0}{v}$ barns (v_0 =2200 m/s is the rms velocity of thermal neutrons). Other n+^{3}He interactions are negligible. Passing a neutron beam through a filter of polarized ^{3}He produces a beam polarized parallel to the ^{3}He, though reduced in flux. Polarized neutron beams are widely sought for condensed matter and materials science research and for studies of the nuclear interactions of scattered, absorbed, and decaying neutrons. The decay of polarized neutrons provides the opportunity to study the weak interactions [6], and weaker interactions [7] such as those that emerge from extensions of the Standard Model of elementary particle interactions including SuperSymmetry.

The two methods for polarizing ^{3}He, discovered in the early 1960s, became competing techniques in the 1980s. Metastability exchange was pursued by Leduc, Schaerrer and co–workers who teamed up with Otten and Heil at Mainz [8] and with Woodward, McKeown and Milner [9]. This led to the neutron spin filter program at ILL, Grenoble [10], and quasielastic scattering measurements at Mainz [8,11], both using a two stage, titanium pump compressor to increase the ^{3}He pressure from 1 torr to $\approx$ 1 bar. The metastability exchange technique has also been used to pump ^{3}He into a cooled cell in a quasielastic scattering experiment [12] and to fill a "storage cell" that is coaxial with the circulating 30 GeV positron beam in the HERA ring at DESY, Hamburg, Germany [13]. Spin exchange has been most successful in producing high density polarized ^{3}He that is essential for targets used in extracted beam experiments such as the SLAC End Station A deep inelastic scattering program [14].

While the technical advances in polarized ^{3}He have been largely motivated by work on polarized targets for nuclear and high energy physics, ^{129}Xe polarization was advanced in the optical pumping studies led by Happer and co–workers [15]. Early in the 1980s they began extensive investigations of spin exchange between noble gases and optically pumped alkali-metal vapors [16]. They studied many processes involved in optical pumping of alkali–metal vapors in the presence of buffer gases providing extensive data on the xenon–Rb system [15]. ^{129}Xe polarization of nearly 100% in small $\approx$ cm^3 volumes was produced.

A variety of experimenters have since used laser polarized ^{3}He and other noble gases to enhance NMR measurements. The low temperature work at Ecole Normal Superieur has used NMR to measure polarization and probe such phenomena as spin waves, and ^{3}He–^{4}He mixtures [17]. Geometric phases have been measured with ^{129}Xe [18]. Measurement of the NMR splittings of ^{129}Xe and ^{3}He in the presence of an electric field is used to search for T–violation [19]. Conventional NMR research with ^{129}Xe (*i.e.* not laser enhanced) has focused on variety of problems including cross polarization, molecular dynamics, xenon molecules (*e.g.* XeF$_5$), diffusion in porous media, polymers, and liquid crystals. ^{131}Xe has been used to study quadrupolar relaxation on surfaces, in macromolecules, and porous media. Xenon has been extremely important because it is normally gaseous, can be easily frozen or liquified, is relatively soluble, and is characterized by large NMR chemical shifts up to 500 ppm between the gas and dissolved phases. Alex Pines and his group recognized that many of these applications of NMR research could be enhanced with laser polarized ^{129}Xe [20], and inspired the original pursuit of MRI with laser polarized noble gases [21].

2　Nuclear Polarization Techniques

Optical polarization employing either hyperfine spin exchange with an alkali-metal vapor [22] for ^{3}He and ^{129}Xe or optical pumping of metastable helium atoms, for ^{3}He [23] are now used to produce liter quantities (at STP) with polarization $P_3 > 50\%$ that are used for neutron polarization, polarized targets, and for medical imaging. For this to work, relaxation of nuclear spin must be balanced by polarization rates. (Note that nuclear spin relaxation in a biological environment *in vivo* or *in vitro* is completely different from relaxation in a carefully prepared polarization system.) Rare gas nuclear spin relaxation occurs by bulk collisions with impurities, dipole–dipole interactions in the bulk, magnetic field gradients, and surface wall interactions. The most important impurity in a biological system is paramagnetic O$_2$. Relaxation rates are proportional to the oxygen impurity level with rate constants $k(\text{O}_2$-$^{129}\text{Xe}) \approx 0.3/\text{s}^{-1}/\text{amagat}$ [24] and $k(\text{O}_2$-$^3\text{He}) \approx 0.45~\text{s}^{-1}/\text{amagat}$ [25] at 14.1 kG and at room temperature with temperature dependence $\approx T^{-1/}$.

Several devices combine optical pumping and polarization with delivery of the polarized gas to a subject or a storage container. For ^{3}He, the basic designs used for polarized targets are applied for both metastability exchange and spin exchange pumped systems. The metastability exchange systems have a valved port that connects to a transport container. Gentile has recently presented a relatively compact and inexpensive compressor that may see wide use [26]. For spin exchange systems an additional valve of the appropriate material is straight forward. With ^{3}He polarization relaxation times of several days typical in glass containers, transport almost anywhere can be contemplated.

For ^{129}Xe, the high rate of Rb electron spin depolarization in spin–rotation collisions limits the rate of ^{129}Xe production, and a method of accumulation is essential. Cates, Happer, and co–workers have shown that frozen and liquid xenon provide very long nuclear spin relaxation times for ^{129}Xe [27,28], and that freezing is an ideal accumulation method [29]. Relaxation times are on the order of an hour at liquid N_2 temperatures and days at liquid He temperatures [30].

For human studies, it is sufficient to collect the polarized gas in a plastic bag, where it is held for several minutes before inhalation and a breath–hold. For animal studies, voluntary breathing is not possible, and delivery to the animal requires a polarized gas ventilator. There are many technical difficulties, and very few such ventilators have been constructed [31,32].

3 Low Field Imaging

NMR with nuclei polarized by laser optical pumping is less dependent on large magnetic fields than conventional NMR, and the potential of low–field imaging has emerged. The signal to noise ratio (SNR) is the important parameter. For conventional, NMR, the signal S due to a nuclear spin $I = \hbar/2$ (for ^{1}H, ^{3}H, and ^{129}Xe) with concentration $[I]$ is proportional to the product of precession frequency (ω) and magnetization: $S \propto \omega \mu_I [I] P$ where $I\omega = \mu_I B$ and the brute–force polarization is $P_I \approx \frac{\mu_I B}{kT}$. Thus $S_{brute} \propto \mu_I^3 B^2 [I]$. In contrast, for NMR with laser polarized nuclei, P_I is independent of field, and $S_{laser} \propto \mu_I^2 B[I]$. The most important MRI noise sources are Johnson noise due to the pick–up coil resistance, R_c, amplifier noise, and dissipation in the sample due to loading characterized by R_s. Skin depth effects generally increase the coil resistance so that $R_c \propto \sqrt{B}$. The signal to noise ratios for brute–force and laser polarization for fixed bandwidth are

$$SNR_{brute} \propto \frac{B}{\sqrt{1 + \alpha B^{-3/2}}} \qquad SNR_{laser} \propto \frac{1}{\sqrt{1 + \alpha B^{-3/2}}} \qquad (1)$$

where $\alpha \approx (0.2 \text{ T})^{-3/2}$ [33]. This shows that above about 0.2 T, the signal to noise ratio for laser polarized NMR and MRI increases very little, *i.e.* is approximately independent of B.

There are many advantages that may be gained from NMR and MRI at lower fields. The cost of magnets is less, open geometry permanent and conventional magnets may provide more friendly NMR scanners (important for pediatrics), high field effects such as susceptibility dependence may be less. Low field work has been most effectively pursued by [34]. They have shown

326

that the combination of 0.1 T magnet and a low–polarization metastability pumped ^{3}He polarizer can produce lung images with resolution comparable to standard ^{133}Xe nuclear medicine techniques. One-dimensional images of polarized ^{3}He have been to study diffusion effects [25]. Very low field imaging, at 0.003 T has been demonstrated by Walsworth and collaborators [35].

4 Imaging Polarized ^{129}Xe and ^{3}He Gas

Though either ^{3}He and ^{129}Xe may be used for gas imaging, the majority of lung ventilation imaging studies, have used ^{3}He. Helium has a number of advantages over xenon for creation of high-resolution gas images: the magnetic moment of ^{3}He is nearly three times larger than that of xenon, and it has generally been easier to create high magnetization with ^{3}He. ^{3}He polarizations are generally of 20%– 50% whereas typical ^{129}Xe polarizations are currently 5%. A recent study imaging both gases concluded that in general helium is approximately 10 times more sensitive than xenon for MRI studies [36].

Helium also has fewer biological effects than xenon. Helium is biologically inert and the only consequence of helium inhalation (apart from the well known change in voice pitch) is the risk of lowering the blood oxygen content due to oxygen being removed from the inhaled gas. Xenon on the other hand, is anesthetic at concentrations greater than 35%. These effects are well known and have been addressed in CT studies where xenon is used to measure regional cerebral blood flow (rCBF) by monitoring the spatial and temporal attenuation of x-rays

Though helium provides greater signal strength and fewer medical complications, a major concern for widespread clinical studies with helium will be the limited supply of ^{3}He discussed earlier. Most studies are performed with ^{3}He gas with an isotopic concentration of approximately 99% at a cost of approximately 100-150 USD/liter. Xenon is present in the air at a concentration of approximately 0.04%. The abundance of the spin 1/2 isotope, ^{129}Xe, is 26.44%. Naturally abundant xenon can be purchased for approximately 10 USD/liter. ^{129}Xe enriched to approximately 75% can be purchased for about 300 USD/liter. This price is determined primarily by demand and could drop dramatically if specific clinical uses are identified.

Lung ventilation imaging is currently based on nuclear medicine scintigraphy of either ^{133}Xe or aerosol sprays with Tc. Laser–polarized noble–gas imaging research with animals and human subjects has already shown that tomographic (slice selected) high resolution images can be produced. A comparison of ^{133}Xe scintigraphy and laser–polarized ^{3}He images is shown in figure 1. The first human ventilation studies with ^{3}He were performed in Mainz [37] and at Duke. The group at Mainz has continued with more clinical studies of volunteers with diagnosed lung diseases [38,37,39]. Other studies have looked at helium images of from the lungs of smokers [40] and find ventilation defects in the few cases studies. In fact even apparently healthy, active, volunteers have ventilation defects that are revealed in the high resolution laser polarized ^{3}He MRIs [41]. A study of subjects with chronic asthma suggests that ventilation defects may allow a measure of the progression and treatment of the disease[42]. Though it will be some time before the utility of high resolution lung images is clarified, it is clear that they provide new information and raise new questions.: for example, what are the mechanisms of signal destruction in diseased lungs [43]. The lungs are not the only organ amenable to gas imaging. The sinus cavities [44] and bowel [45] can also be imaged by using laser-polarized ^{3}He or ^{129}Xe.

In contrast to ^{3}He, which is most useful for imaging air spaces such as the lungs and colon, ^{129}Xe is soluble in blood with $\approx 17\%$ solubility and tissue with varying solubility [46]. Many of the biological properties of xenon have been established through research with radioactive isotopes, particularly ^{133}Xe. Xenon freely diffuses across biological membranes including the blood gas barrier in the lungs and capillary walls between blood and tissue. Xenon is metabolically inert, and is carried to distant organs where it accumulates in tissue. The NMR spectrum of ^{129}Xe *in vivo* depends on the tissue enviromnemt as shown in figure 3, which is from the body and head of a rat that had been breathing a mixture of ^{129}Xe and oxygen gas [47]. Similar spectra have been observed in humans after a single breath–hold of laser polarized ^{129}Xe [48]. The peaks in the rat body spectrum have been identified on the basis of the time dependence of the peaks, the location of each resonance determined by imaging (see figure 2), and the chemical shifts revealed in *in vitro* experiments.

The images of ^{129}Xe in dissolved phases shown in figure 2 demonstrate some potential medical applications that may emerge in the coming years. Images of the lungs in the gas phase (figures 2A and 2D) show the region of ventilation. In a healthy lung, xenon crosses the blood–gas barrier, appearing also in tissue (figures 2B and 2E) and blood phase images (figures 2C and 2F). We discuss further analysis of lung function in the next section. The blood carries the ^{129}Xe magnetization from the lungs to the left side of the heart. In the heart, the blood phase signal is dominated by pooled blood in the left heart chambers. Perfusion in the healthy heart is indicated by the appearance of ^{129}Xe magnetization in the dissolved tissue and fat phases in the heart also shown in figures 2B and 2E. Restricted blood flow (ischemia) and unperfused regions (infarction) would be revealed by the absence of the dissolved tissue phase in that region.

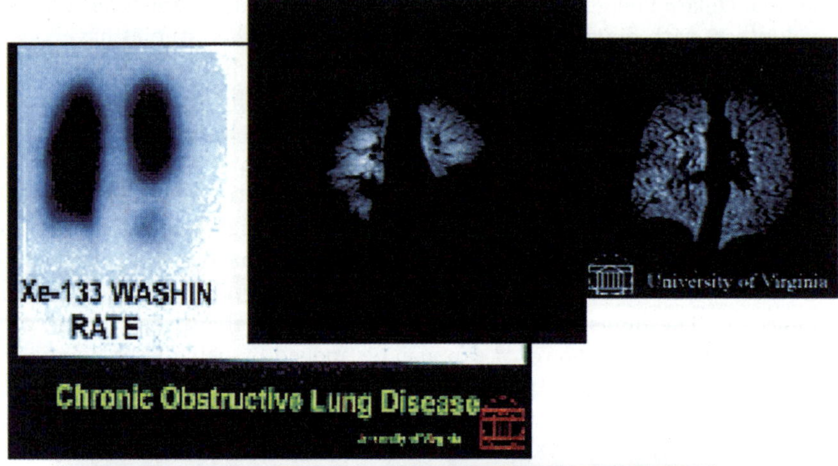

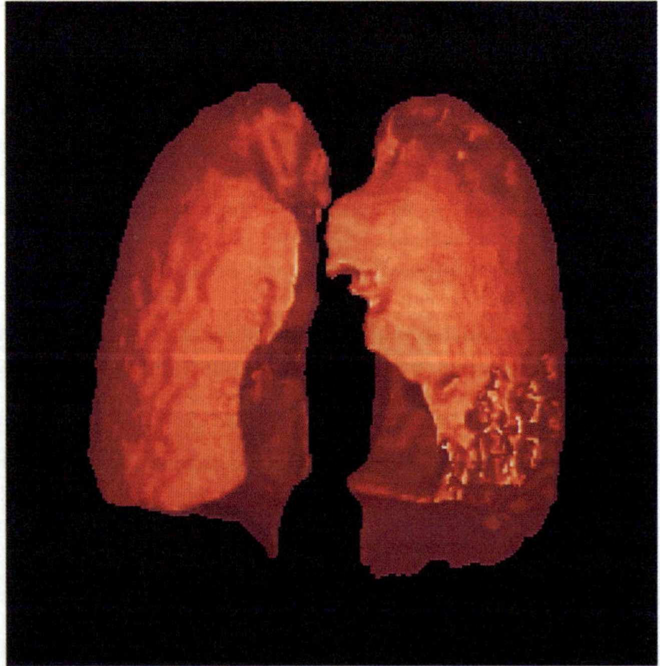

Figure 1. Top: Three lung images: Left: A ^{133}Xe nuclear medicine scan of a patient with chronic obstructive pulmonary disease (COPD). Center: A laser polarized ^{3}He MRI of the same patient. Right: Laser polarized ^{3}He MRI of a healthy volunteer. Bottom: a surface rendering of the lung produced with ^{3}He MRI. *Images Courtesy of the University of Virginia. Used with permission.*

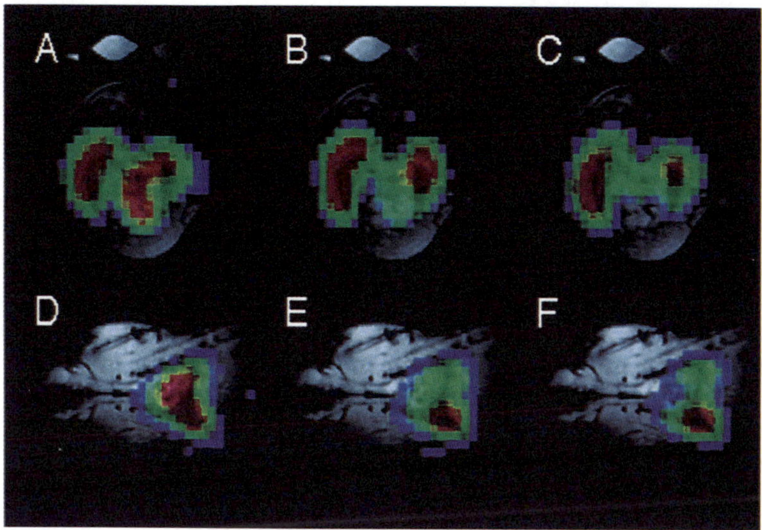

Figure 2. Magnetic Resonance Images of ^{129}Xe in the lungs and dissolved in the blood and tissue of a rat. The gray–scale images are conventional proton MRI (spin–echo) images that show the animal's anatomy. Superimposed images show the concentration of ^{129}Xe magnetization for each of three spectral features corresponding to xenon in the gas phase (C and F), dissolved in tissue (B and E) and dissolved in blood (A and D). Panels (A through C) are called axial images across the body, and (D through F) are coronal images through the body.

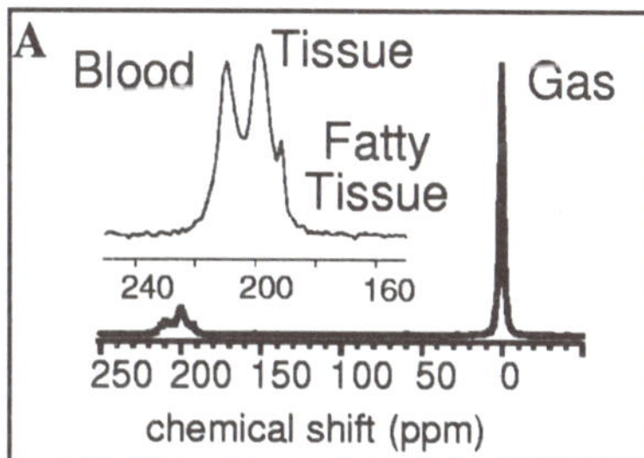

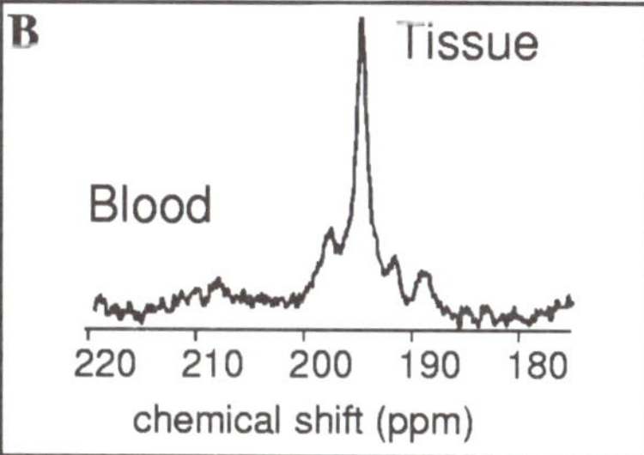

Figure 3. Spectra from the body and head of a rat breathing a ^{129}Xe–O$_2$ mixture. Left: the body spectrum showing signals from gas in the lungs (at 0 ppm), and dissolved in the blood (210 ppm) and tissue (192 ppm and 199 ppm). Right, the spectrum from the head showing several tissue peaks and possibly the blood peak near 210 ppm.

5 Future Possibilities

The future is exceptionally bright for research in biomedicine, neuroscience, and materials science with laser polarized rare gas imaging. The scientific problems relating to polarization techniques, delivery of polarized gas with devices and in solutions are challenging, but progress continues. Ventilation images of animals and humans in the US and Europe provide unprecedented resolution and are likely to provide new information, as is often the case when we can look at something with greater sensitivity, precision, and resolution. Figure 1 provides a stunning example. The new techniques possible with ^{129}Xe provide resolution in chemical shift frequency and time that promise to develop into ways of measuring perfusion of specific tissues as well as organs, complementing PET. The potential for a completely quantitative measure of perfusion promises broad application. All of these possibilities have been discussed in this review.

However, research with a new imaging modality does not ensure its application as a medical diagnostic procedure. Among the potential applications of high resolution lung ventilation imaging, colonoscopy, lung function assessment, and perfusion measurement, MRI with laser polarized gases must pass the tests of

1. sensitivity to disease or injury

2. specificity for a unique diagnosis

3. effectiveness based on cost and risk

With the promise of these and a host of other potential applications, pre–clinical efforts are underway in the US and Europe. In the US, the efforts are organized by the commercial interest that would produce the polarized gas in regional centers and ship it, overnight, to medical facilities. In Europe, a collaboration of industry, academic, and hospital based researchers is developing the pre–clinical program. The goals of both these groups include regulatory approval for administration of polarized gas as a contrast agent and its use for medical diagnosis. Interestingly, the final step in regulatory approval, following demonstration of safety and other issues, is demonstration of efficacy: the sensitivity and specificity for diagnosis of specific maladies that physicians would be likely to use.

Acknowledgments

The author is grateful to several colleagues for discussions and advice and for scientific inspiration and guidance. They are: Scott Swanson, Bernie Agranoff, Jim Brookeman, Gordon Cates, Kevin Coulter, Tom Chenevert, Will Happer, Bob Koeppe, Pierre–Jean Nacher, Eduardo Oteiza, Matt Rosen, Brian Saam, Ron Walsworth, Robert Welsh, and Jon Zerger. Images were provided by Brian Saam, Jim Brookeman, Tom Chenevert, Hans–Ulrich Kauczor, and Pierre–Jean Nacher.

References

1. M. S. Albert, G. D. Cates, B. Driehuys, W. Happer, B. Saam, C. S. Springer, and A. Wishnia. *Nature*, 370(6486):199–201, 1994.
2. T. E. Chupp, R. J. Holt, and R. G. Milner. *Annu. Rev. Nucl. Part. Sci.*, 44:373–411, 1994.
3. K. P. Coulter, A. B. McDonald, W. Happer, T. E. Chupp, and M. E. Wagshul. *Nucl. Instrum. Methods Phys. Res. Sect. A-Accel. Spectrom. Dect. Assoc. Equip.*, 270(1):90–94, 1988.
4. L. Alvarez. *Discovering Alvarez*. University of Chicago Press, 1987.
5. L. J. Wittenberg, J. F. Santarius, and G. L. Kulcinski. *Fusion Technol.*, 10(2):167–178, 1986.
6. J. D.Jackson, S. B.Trieman, and H. W. Wyld. *Phys. Rev.*, 106:517–521, 1957.
7. P. Herczeg. *Progress in Nuclear Physics, ed. W.-Y.P. Hwang*. Elsevier, 1991.
8. J. Becker, W. Heil, B. Krug, M. Leduc, M. Meyerhoff, P. J. Nacher, E. W. Otten, T. Prokscha, L. D. Schearer, and R. Surkau. *Nucl. Instrum. Methods Phys. Res. A*, 346(1-2):45–51, 1994.
9. C. L. Bohler, L. D. Schearer, M. Leduc, P. J. Nacher, L. Zachorowski, R. G. Milner, R. D. McKeown, and C. E. Woodward. *J. Appl. Phys.*, 63(8):2497–2501, 1988.
10. W. Heil, K. Andersen, D. Hoffman, H. Humbolt, J. Kulda, E. Lelievre-Berna, O. Schärpf, and F. Tasset. *Physica*, B267-268:328–35, 1998.
11. J. Becker, J. Bermuth, M. Ebert, T. Grossmann, W. Heil, D. Hofmann, H. Humblot, M. Leduc, E. W. Otten, D. Rohe, et al. *Nucl. Instrum. Methods Phys. Res. A*, 402(2-3):327–336, 1998.
12. C. E. Woodward, E. J. Beise, J. E. Belz, R. W. Carr, B. W. Filippone, W. B. Lorenzon, R. D. McKeown, B. Mueller, T. G. Oneill, G. Dodson, et al. *Phys. Rev. Lett.*, 65(6):698–700, 1990.

13. K. Ackerstaff et al. *Phys. Lett. B*, 404(3-4):383–389, 1997.

14. P. L. Anthony et al. *Phys. Rev. D*, 54(11):6620–6650, 1996.

15. X. Zeng, Z. Wu, T. Call, E. Miron, D. Schreiber, and W. Happer. *Phys. Rev. A*, 31(1):260–278, 1985.

16. W. Happer, E. Miron, S. Schaefer, D. Schreiber, W. A. Vanwijngaarden, and X. Zeng. *Phys. Rev. A*, 29(6):3092–3110, 1984.

17. P.J. Nacher and E. Stolz. *J. Low Temp. Phys.*, 101:6170–6177, 1995.

18. S. Appelt, G. Wackerle, and M. Mehring. *Phys. Lett. A*, 204(3-4):210–216, 1995.

19. M. A. Rosenberry and T.E. Chupp, *Phys. Rev. Lett.*, 86: 22-25 (2001). PhD thesis, University of Michigan, 1999.

20. D. Raftery, H. Long, T. Meersmann, P. J. Grandinetti, L. Reven, and A. Pines. *Phys. Rev. Lett.*, 66(5):584–587, 1991.

21. Y. Q. Song, B. M. Goodson, and A. Pines. *Spectroscopy*, 14:26–33, 1999.

22. M. A. Bouchiat, T. R. Carver, and C. M. Varnum. *Phys. Rev. Lett*, 5:373–375, 1960.

23. F. D. Colgrove, L. D. Schaerer, and G. K. Walters. *Phys. Rev.*, 132, 1963.

24. C. J. Jameson, A. K. Jameson, and J. K. Hwang. *J. Chem. Phys.*, 89(7):4074–4081, 1988.

25. B. Saam, N. Drukker, and W. Happer. *Chem. Phys. Lett.*, 263(3-4):481–487, 1996.

26. T. R. Gentile, G. L. Jones, A. K. Thompson, R.R. Rizi, D. Lipson, D.A. Roberts, M.D. Schnall, J.R. Brokeman, and J.P. Mugler III. *European Radiology*, 9:B17, 1999.

27. G. D. Cates, D. R. Benton, M. Gatzke, W. Happer, K. C. Hasson, and N. R. Newbury. *Phys. Rev. Lett.*, 65(20):2591–2594, 1990.

28. K. L. Sauer, R. J. Fitzgerald, and W. Happer. *Phys. Rev. A*, 59(3):R1746–R1749, 1999.

29. B. Driehuys, G. D. Cates, E. Miron, K. Sauer, D. K. Walter, and W. Happer. *Appl. Phys. Lett.*, 69(12):1668–1670, 1996.

30. M. Gatzke, G. D. Cates, B. Driehuys, D. Fox, W. Happer, and B. Saam. *Phys. Rev. Lett.*, 70(5):690–693, 1993.

31. L. W. Hedlund, M. S. Chawla, X. J. Chen, G. P. Cofer, H. E. Moller, and G. A. Johnson. *European Radiology*, 9:B32, 1999.

32. M. S. Rosen, T. E. Chupp, K. P. Coulter, R. C. Welsh, and S. D. Swanson. *Rev. Sci. Instrum.*, 70(2):1546–1552, 1999.

33. W. A. Edelstein, G. H. Glover, C. J. Hardy, and R. W. Redington. *Magn.Reson.Med.*, 3(4):604–618, 1986.

34. L. Darrasse, E. Durand, G. Guillot, P.J. Nacher, and G. Tastevin. *Magn.*

Res. Mat. in Phys, Biol. and Med., Suppl. 1 to Vol. 6:101, 1998.

35. C. H. Tseng, G. P. Wong, V. R. Pomeroy, R. W. Mair, D. P. Hinton, D. Hoffmann, R. E. Stoner, F. W. Hersman, D. G. Cory, and R. L. Walsworth. *Phys. Rev. Lett.*, 81(17):3785–3788, 1998.

36. H. E. Moller, M. S. Chawla, X. J. Chen, B. Driehuys, L. W. Hedlund, C. T. Wheeler, and G. A. Johnson. *Magn. Reson. Med.*, 41(5):1058–1064, 1999.

37. M. Ebert, T. Grossmann, W. Heil, W. E. Otten, R. Surkau, M. Leduc, P. Bachert, M. V. Knopp, L. R. Schad, and M. Thelen. *Lancet*, 347(9011):1297–1299, 1996.

38. P. Bachert, L. R. Schad, M. Bock, M. V. Knopp, M. Ebert, T. Grossmann, W. Heil, D. Hofmann, R. Surkau, and E. W. Otten. *Magn.Reson.Med.*, 36(2):192–196, 1996.

39. H. U. Kauczor, M. Ebert, K. F. Kreitner, H. Nilgens, R. Surkau, W. Heil, D. Hofmann, E. W. Otten, and M. Thelen. *JMRI-J. Magn. Reson. Imaging*, 7(3):538–543, 1997.

40. E. E. de Lange, J. P. Mugler, J. R. Brookeman, J. Knight-Scott, J. D. Truwit, C. D. Teates, T. M. Daniel, P. L. Bogorad, and G. D. Cates. *Radiology*, 210(3):851–857, 1999.

41. J. P. Mugler, B. Driehuys, J. R. Brookeman, G. D. Cates, S. S. Berr, R. G. Bryant, T. M. Daniel, E. E. deLange, J. H. Downs, C. J. Erickson, et al. *Magn.Reson.Med.*, 37(6):809–815, 1997.

42. T. A. Altes, P. L. Powers, J. Knight-Scott, E. E. de Lange, B. A. Alford, and J. R. Brookeman. In *Abstracts of the 85th Annual Meeting of the Radiological Society of North America*, 1999.

43. H. U. Kauczor, R. Surkau, and T. Roberts. *Eur. Radiol.*, 8(5):820–827, 1998.

44. R. R. Rizi, I. E. Dimitrov, A. Thompson, G. Jones, T. R. Gentile, M. Ishii, R. Reddy, M. D. Schnall, and J. S. Leigh. *Magn.Reson.Med.*, 39(6):865–868, 1998.

45. K.D. Hagspiel, T.A. Altes, J.P. Mugler III, M.J. Spellman, J.F. Mata, N.J. Tustison, R.T. Rudy, and J.R. Brookeman. *European Radiology*, 9:B31, 1999.

46. R.Y.Z. Chen, F.C. Fan, S. Kim, K.M. Jan, S. Usami, and S. Chien. *J. Appl. Physiol.*, 49:179–181, 1980.

47. S. D. Swanson, M. S. Rosen, K. P. Coulter, R. C. Welsh, and T. E. Chupp. *Magn. Reson. Med.*, 42(6):1137–1145, 1999.

48. J.R. Brookeman. In Jan den Hollander, editor, *Advanced Spectroscopy*, pages 505–512. I.S.M.R.M., 1998.

A HIGHLY POLARIZED ^{8}LI$^+$ ION BEAM AT ISAC

C.D.P. LEVY[1], R. BAARTMAN[1], J.A. BEHR[1], A. HATAKEYAMA[1,2],
Y. HIRAYAMA[2], R.F. KIEFL[1,3], G.D. MORRIS[1,4], R. NUSSBAUMER[1],
R. POUTISSOU[1] AND G.W. WIGHT[1]

(1)TRIUMF, 4004 Wesbrook Mall, Vancouver, BC, Canada V6T 2A3

*(2)Department of Physics, Graduate School of Science, Osaka University,
Osaka 560-0043, Japan*

*(3)Department of Physics and Astronomy, 6224 Agriculture Road, University of
British Columbia, Vancouver, BC, Canada V6T 1Z1*

(4)Los Alamos National Laboratory, MST-10, Los Alamos, NM 87545 USA

A nuclear-spin vector polarization of 0.69 has been achieved in a ^{8}Li$^+$ ion beam at
the ISAC radioactive beam facility at TRIUMF. The incident 30 keV unpolarized
^{8}Li$^+$ ion beam is neutralized in a Na vapour cell, and the resulting fast atomic
beam is polarized by collinear optical pumping. The polarized beam is then re-
ionized with high efficiency in a He gas cell and directed to a polarimeter or β-NMR
experimental facility. A novel standing wave Ti:sapphire laser produces the optical
pumping. It operates at 673 nm, on two cavity modes separated by the ground
state hyperfine splitting in ^{8}Li. The laser modes are broadened by two electro-optic
modulators in series.

1   Introduction

Beams of short lived, polarized radioactive atoms have applications in condensed matter and nuclear physics experiments. A nuclear-spin polarized ^{8}Li$^+$
ion beam line has recently been commissioned at the radioactive beam facility
ISAC at TRIUMF, where β-NMR studies of thin structures and superconduc-
tors have begun. The ^{8}Li (I=2) is polarized using fast beam collinear optical
pumping, a technique that has been employed in several guises at ISOLDE
at CERN for many years [1]. We have a unique facility at ISAC, in that the
polarized ^{8}Li beam is efficiently re-ionized to form an intense polarized ion
beam. We can vary the implantation depth of the ions in a target by adjusting the target bias voltage, and in principle supply 3 or more experiments
simultaneously. The polarized beam is directed to a polarimeter, a β-NMR
high-voltage experimental platform, and a second experimental beam line.

Figure 1 shows details of the polarizer section. The incident beam, up to
10^8 ions/s at an energy of 30 keV, is neutralized with up to 90% efficiency
in a Na vapour jet, longitudinally polarized in a 1.9 m long optical pumping
region, and then re-ionized with 70% efficiency in a cold He gas target [2]. Ion
beam transmission through the polarizer is 82% when the cells are off. The

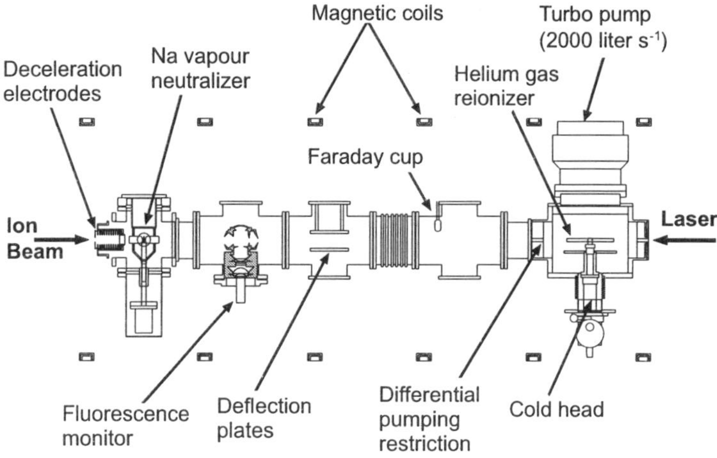

Figure 1. The collinear optically pumped polarizer at ISAC

counter-propagating laser beam is Doppler shifted into resonance with the D_1 absorption transitions of ^{8}Li. The laser frequency is fixed, and deceleration electrodes allow fine tuning of the Doppler shift. Laser-induced fluorescence (LIF) from the atomic beam is observable with a photon counter. Deflection plates remove ions from the atomic beam, and magnetic field coils in a Helmholtz configuration maintain a 1.0 mT field along the polarizer axis.

2 Laser System

The beam is polarized by optical pumping of ^{8}Li atoms on the D_1 transition with circularly polarized light. Both ground state hyperfine levels, split by 382 MHz, must be pumped in order to achieve high polarization.

The optical pumping laser is a modified Spectra-Physics 3900S standing wave Ti:sapphire (TiS) laser pumped by an 18 W argon-ion laser. Lasing at 673 nm is achieved with a rear cavity mirror optimized for that wavelength and an R = 99% output coupler. The cavity length is shortened to increase the longitudinal cavity mode spacing to 381 MHz, measured by observing the beat frequency in a fast detector. A 1 mm thick, R = 10% intra-cavity etalon ensures operation on only two modes. The crystal gain medium is in the centre of the cavity, where the nodes of one mode overlap the anti-nodes of the other, minimizing gain competition. Both modes lase continually, observed using a spectrum analyzer as a filter. Typical laser power is ∼100 mW on

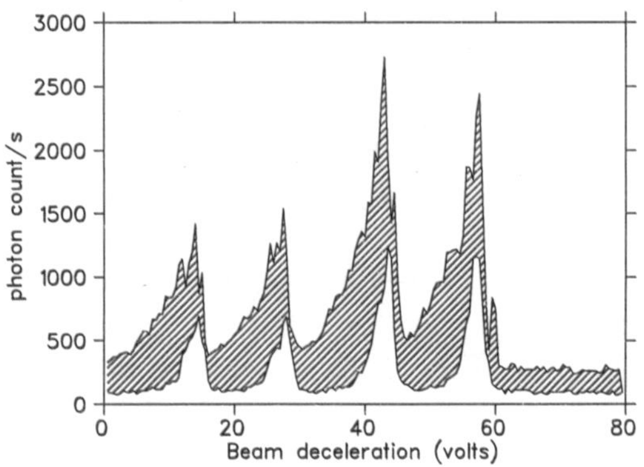

Figure 2. LIF from ^{7}Li observed with the Na cell at 435 °C (upper) and 385 °C (lower), with estimated neutralization efficiencies of 80% and 30%, respectively.

each mode. The instantaneous laser linewidth of $\sim$1 MHz is broadened by acoustic vibrations to $\pm$20 MHz on a 1 s timescale.

It is important to know the absorption bandwidth of the fast atomic beam. Figure 2 shows LIF spectra of a stable 20 keV ^{7}Li beam, produced by the same ISAC beam production target and ion source used to produce radioactive beams. 1 mW of linearly polarized light was used to probe the beam in a zero holding field. The D_2 line was chosen because of its narrow $^2P_{3/2}$ state hyperfine structure. The spectra show 4 peaks, due to scanning 2 hyperfine levels across the 2 laser modes. The upper spectrum shows unresolved low energy satellites caused by Na-beam interactions [3]. The linewidth is equivalent to $\sim$100 MHz in ^{8}Li at 30 keV energy. The lower spectrum, where the Na jet is much thinner, shows less broadening, mainly from the inherent $\sim$2 eV energy spread at the ion source. Because of the broadening, the Na cell is operated at a temperature of 400 – 420 °C, where the neutralization efficiency is $\sim$50 – 70%. The laser power must be spread over the absorption linewidth during the 2 μs transit time of the atomic beam. Resonant electro-optic modulators (EOMs) at 19 MHz and 28 MHz in series, driven by low power signal generators, produce laser sidebands spaced at about 10 MHz intervals. That is dense enough to pump all atoms effectively [4].

The TiS frequency is prevented from drifting by a feedback system based on a single-mode frequency-stabilized He-Ne laser. Both laser beams are directed into a single 2 GHz free-spectral-range spectrum analyzer. The frequency interval between the He-Ne mode and one of the TiS modes, modulo 2 GHz, is measured by computer and averaged. Any change is automatically corrected within 2 s by adjusting the cavity length with the piezo-mounted rear TiS mirror. The average-frequency stability is ± 10 MHz. The system also monitors the number of TiS modes. If a third small mode appears, a correction applied to the etalon tilt eliminates it.

3 Polarimeter results

The polarimeter beam line axis is perpendicular to the polarizer — connected by two 45° electrostatic bends — and hence the original longitudinal spin becomes transverse. The polarimeter measures the β-decay asymmetry of the $^8\mathrm{Li}^+$ implanted into a Pd foil target, using left and right β detectors outside the vacuum chamber. Each detector consists of two plastic scintillator sheets connected to two photomultipliers on coincidence. 50 μm stainless steel windows cover a total solid angle of 16% of 4π. A magnetic field of 15 mT in the same direction as the spin decreases the polarization relaxation rate $1/\mathrm{T}_1$.

For tuning the deceleration voltage, the β counts are integrated for typically 10 s at each voltage. Three peaks are seen (see Fig. 3), due to scanning 2 hyperfine lines across 2 equally spaced laser modes. The central peaks correspond to pumping of both hyperfine levels. The integrated (steady-state) asymmetry is reduced to $1/(1+\tau/\mathrm{T}_1)$ times the beam asymmetry, where $\tau=1.21$ s is the $^8\mathrm{Li}$ lifetime. Tuning is also possible using LIF, but the polarimeter will be essential for low current beams such as $^{11}\mathrm{Li}$.

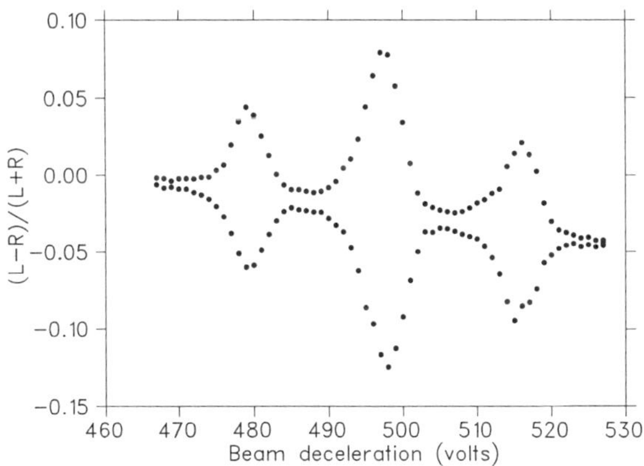

Figure 3. β-decay asymmetry for both light helicities, Na cell at 370 °C (estimated neutralization efficiency 25%), EOMs off. The statistical error bars are too small to show. The variable offset is due to the beam shifting position on the foil.

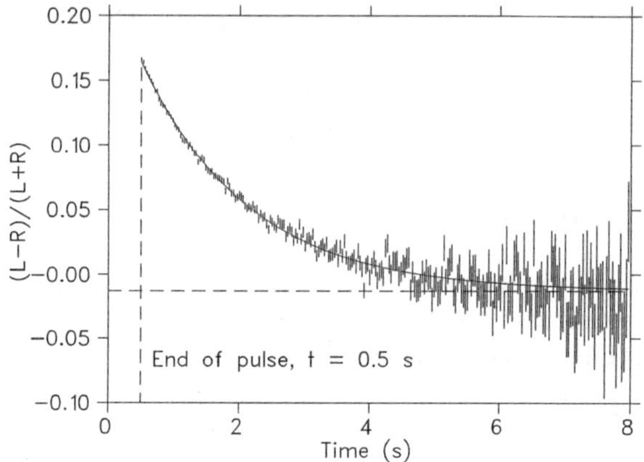

Figure 4. β-decay asymmetry in Pd, Na cell at 400 °C, recorded for $\sim$10 min at one helicity. The baseline is nonzero, due to unequal detector efficiencies.

T_1 is found by pulsing the beam on for 0.5 s every 10 s, and recording the time-resolved asymmetry while the beam is off (Fig. 4 shows typical data). Knowing T_1, a beam polarization of 0.69$\pm$0.02 under optimal conditions was calculated from integrated asymmetry data [4].

Acknowledgments

TRIUMF receives federal funding via a contribution agreement through the National Research Council of Canada. Y. Hirayama and A. Hatakeyama are partly supported by a Grant-in-Aid in Scientific Research of the Japanese Ministry of Education, Science and Culture.

References

1. For example: M. Keim *et al.*,*Eur. Phys. J.* A 8, 31 (2000); P. Lievens *et al.*,*Nucl. Instrum. Meth.* B 70, 532 (1992); E.W. Otten *Treatise on Heavy Ion Physics*, ed. D.A. Bromley, vol. 8 (Plenum, New York, 1988) p. 517; E. Arnold *et al.*, *Phys. Lett.* B 311 (1987).
2. C.D.P. Levy *et al.*, *Proc. Fifth Radioactive Nuclear Beams Conf.*, (Divonne, 2000), to be published in *Nucl. Phys.* A.
3. N. Bendali *et al.*, *J. Phys. B: At. Mol. Phys.* 19, 233 (1986).
4. Y. Hirayama *et al.*, these Proceedings.

OPTICAL PUMPING AT THE ISAC POLARIZER

A. Hatakeyama[1,2†], Y. Hirayama[1], J. A. Behr[2], H. Izumi[1],
C. D. P. Levy[2], D. Melconian[2] and T. Shimoda[1]

*(1) Department of Physics, Graduate School of Science,
Osaka University, Osaka 560-0043, Japan*
(2) TRIUMF, 4004 Wesbrook Mall, Vancouver, BC, Canada V6T 2A3
†*E-mail: hatakeya@triumf.ca*

We have carried out detailed simulations of optical pumping that is being performed at the ISAC radioactive beam facility at TRIUMF, using both density
matrix and rate equation formalisms. The former is a rigorous quantum mechanical method and the latter is an approximation that neglects coherences. We found
that rate equations describe the optical pumping process well under our experimental conditions. The rate equation formalism predicts a significant increase in
nuclear polarization, if the laser bandwidth is in effect broadened with electro-optic
modulators to cover the Doppler broadening of the ^{8}Li beam to be polarized. Following this prediction, we performed the optical pumping experiment and achieved
about 70% polarization under optimal conditions.

1   Introduction

The polarized beam line at the ISAC radioactive beam facility at TRIUMF
is intended to produce highly nuclear-polarized radioactive beams. The first
species to be polarized in this beam line is ^{8}Li (nuclear spin I=2, lifetime
$\tau = 1.21$ s) for β-NMR studies of condensed matter. The optical pumping is
performed on a fast atomic ^{8}Li beam at an energy of 30 keV, obtained after
neutralization of the ion beam by charge transfer collisions with a Na vapour.
The polarized ^{8}Li atomic beam is then reionized by collisions with a He gas
and delivered to experimental sections [1,2].

High polarization of alkali-metal atomic beams is achieved using collinear
optical pumping with a two-frequency laser beam that pumps the atoms from
both ground state hyperfine levels. However, relatively low laser power (up to
about 200 mW) and a short optical pumping time (about 2 μs) require a careful
optimization of the optical pumping conditions with the help of theoretical
calculations. We therefore performed detailed calculations to simulate the
optical pumping process. Calculation methods we used are discussed below,
followed by a comparison with experimental results.

2 Calculation methods

We first simulated the optical pumping process with the density matrix formalism, a rigorous quantum mechanical method in which the Liouville equation is solved. We solved it for all matrix elements, following Tremblay and Jacques[3]. This is straightforward in principle, but complicated in practice because of the large number of differential equations. We found that for D_1 pumping with two frequencies tuned to the two ground state hyperfine levels of alkali atoms, a phenomenon called "coherent population trapping" can occur when the frequency difference precisely equals the hyperfine splitting. Coherently trapped atoms in superpositions of hyperfine levels cannot be excited by laser light and therefore the pumping efficiency is greatly reduced. We note that coherent population trapping has been studied in various systems, including the same D_1 pumping scheme with two circularly polarized laser beams[4,a]. However this phenomenon seems to be overlooked in optical pumping experiments and needs to be kept in mind, especially when one uses high power, narrow bandwidth laser light for a short optical pumping time.

We can easily avoid coherent trapping. The magnetic field applied for optical pumping ($\sim$1.0 mT in our experiment) causes Zeeman splitting of the hyperfine levels, resulting in detuning of the hyperfine splittings from the laser frequency difference. Also, a laser bandwidth of about 1 MHz for each independently-oscillating laser mode in our laser reduces the hyperfine coherence[3]. With the coherence suppressed in this way under typical experimental conditions, it was found that the nuclear polarization calculated using the density matrix formalism agreed well with that using the rate equation formalism, which is an approximation neglecting coherence (see Fig. 1). Accordingly, we decided to use the simpler rate equation formalism to predict the polarization. We assumed a laser bandwidth of 1 MHz and a Gaussian energy broadening of the ^{8}Li beam. We also assumed all ^{8}Li atoms interact with a laser beam with a diameter of typically 8 mm at the entrance of the optical pumping region and a beam divergence of typically 2 mrad. The laser intensity was set uniform on the cross section.

3 Experimental results and comparison with calculations

The key to achieving high polarization is to match the optical pumping light bandwidth to the energy (Doppler) broadening of the ^{8}Li beam. The typical broadening is about 100 MHz, much greater than the laser bandwidth

---

[a]Coherent population trapping in the D_1 pumping scheme has recently been observed also in the TRINAT laboratory at TRIUMF using laser-cooled atoms.

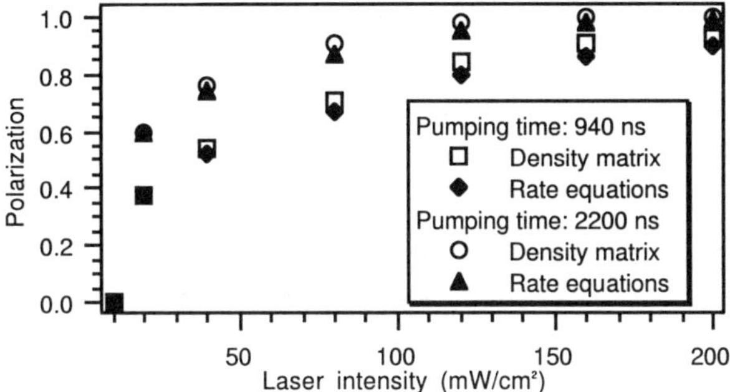

Figure 1: Nuclear polarizations calculated for two optical pumping times using the density matrix formalism and the rate equation formalism, assuming a laser bandwidth of 1 MHz, no laser beam divergence, no Doppler broadening of the ^{8}Li beam and a magnetic field of 1.0 mT.

of 1 MHz. It is caused by multiple collisions with Na atoms in the neutralizer [2], and cannot be avoided if a high neutralization efficiency is required. To overcome this problem, we use resonant electro-optic modulators (EOMs) to produce laser sidebands and broaden the effective laser bandwidth. We performed calculations using the rate equation formalism to determine what resonant frequencies of EOMs would be effective. We found that a combination of a 19 MHz EOM and a 28 MHz EOM worked effectively in terms of both the total width and the sideband spacing of about 10 MHz. More closely spaced sidebands gave a minor increase in the polarization.

Figure 2 shows schematic examples of the line profile of a single laser frequency modulated by the two EOMs, denoted by mode 1, 2, 3, together with the calculated polarizations for each EOM mode under typical experimental conditions. One sees a significant increase (about 30%) with EOM mode 3.

Encouraged by these calculated results, we performed the experiment with the two EOMs. The nuclear polarization was measured by β-decay asymmetry of ^{8}Li$^+$ implanted continuously into a Pd foil stopper using right and left detectors outside the vacuum chamber [2]. The effective asymmetry factor A_{eff} times the nuclear polarization P can be determined from the β-ray count in the steady-state at each detector for each polarization direction (which was changed by laser helicity flipping) as,

$$A_{eff}P = \frac{\sqrt{R}-1}{\sqrt{R}+1} \qquad \left(R = \frac{N_R^+}{N_L^+}\bigg/\frac{N_R^-}{N_L^-}\right), \tag{1}$$

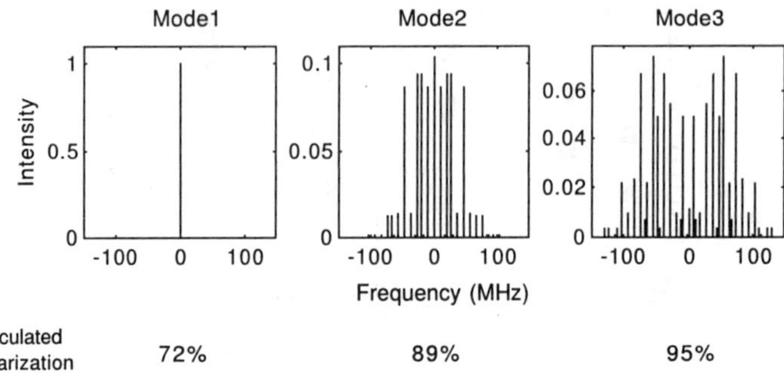

Figure 2: Schematic figures showing the sideband distribution generated by 19 MHz and 28 MHz EOMs in series. The intensities are normalized to the value for mode 1, in which the laser is not modulated. The optical phase shifts caused by the EOMs are 1.4 rad and 1.4 rad (mode 2) and 1.4 rad and 2.8 rad (mode 3) for the 19 MHz and 28 MHz EOMs, respectively. The nuclear polarizations are calculated assuming a Doppler broadening of the ^{8}Li beam of 125 MHz (FWHM) and a laser power of 200 mW.

where $N_{R(L)}^{+(-)}$ is the β-ray count at the right (left) counter for $\sigma^{+(-)}$ laser light. The effective asymmetry factor A_{eff} is expressed by the product of three correction factors as $A_{eff} = A_{solid}A_{relax}A_{asym}$. $A_{solid} = 0.92$ is a solid angle correction factor, and $A_{asym} = -0.33$ is the β-decay asymmetry parameter for ^{8}Li. $A_{relax} = 1/(1 + \tau/T_1) = 0.62$ is a correction for the spin relaxation in the Pd foil. The spin relaxation time T_1 in the Pd foil is 2.0 s, which was measured by pulsing the ^{8}Li beam and observing the polarization decay[2].

Table 1 shows the dependence of the polarization on the Na temperature and the EOM mode. One sees that at all temperatures the broadening of the laser bandwidth by the EOMs gives higher polarization. At higher Na temperatures (with higher vapour density and hence more energy-broadening collisions) the EOM effect is more prominent, about a 50% increase in polarization at 430°C. The highest polarization of about 70% is obtained for the smallest Doppler broadening at 370°C with EOM mode 3.

Table 1: Nuclear polarizations measured at three Na temperatures with three EOM modes. The laser power is estimated at 140 mW at the entrance of the optical pumping region.

Na temperature	370°C	400°C	430°C
EOM mode 1	53.3±1.6%	49.8±1.5%	39.8±1.2%
EOM mode 2	66.4±2.0%	65.0±2.0%	54.8±1.6%
EOM mode 3	68.7±2.1%	68.1±2.0%	60.6±1.8%

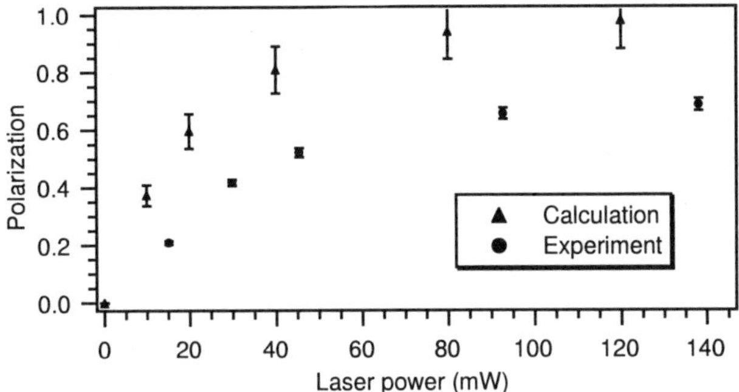

Figure 3: The dependence of nuclear polarization on the laser power, at a Na temperature of 370°C. The calculation assumes a Doppler broadening of the ^{8}Li beam of 60 MHz (FWHM). The errors for the calculated values originate from a Monte Carlo simulation of the Doppler broadening.

The calculation did not agree quantitatively with the experiment. One example is seen in Fig. 3, which shows the laser power dependence of polarization. The measured polarization seems to saturate at about 70%, while the calculated values approach 100%. This may indicate that a subset of ^{8}Li atoms is not properly pumped, or the polarization is partially destroyed after the optical pumping region, such as during reionization by collisions with a He gas. Further investigations are required.

Acknowledgments

This work is partly supported by a Grant-in-Aid in Scientific Research of the Japanese Ministry of Education, Science and Culture. TRIUMF receives federal funding via a contribution agreement through the National Research Council of Canada.

References

1. C.D.P. Levy *et al.*, *Proc. Fifth Radioactive Nuclear Beams Conf.*, (Divonne, 2000), to be published in *Nucl. Phys. A.*
2. C.D.P. Levy *et al.*, these Proceedings.
3. P. Tremblay and C. Jacques, *Phys. Rev. A* **41**, 4989 (1990).
4. E.S. Fry *et al.*, *Phys. Rev. Lett.* **70**, 3235 (1993).

$\vec{n}\text{-}\vec{d}$ SCATTERING MEASUREMENTS WITH THE TUNL DYNAMICALLY POLARIZED DEUTERON TARGET

D.M. MARKOFF[1,4], R.D. FOSTER[1,4], C.R. GOULD[1,4], D.G. HAASE[1,4], J.O. POOLE [1,4], B.W. RAICHLE[3,4] AND W. TORNOW[2,4]

(1) Physics Department, NC State University, Raleigh, NC 27695-8202, USA
(2) Physics Department, Duke University, Durham, NC 27708-0308, USA
(3) Appalachian State University, Boone, NC 28608, USA
(4) Triangle Universities Nuclear Laboratory, Durham, NC 27708, USA

We have completed initial measurements of the spin-dependent transmission asymmetry of a longitudinally polarized neutron beam (5-12 MeV) through a polarized deuterium target. The nd scattering observable is $\Delta\sigma_L$, the longitudinal cross-section difference. Our current data indicate that the three-nucleon calculations, with and without the inclusion of three-nucleon force terms to the dominant nucleon-nucleon potential, underpredict the magnitude of $\Delta\sigma_L$. This paper outlines the experiment with a description of the polarized neutron beam production and the polarized deuterium target production. Improvements to the apparatus currently in progress will also be presented.

1 Introduction

Polarized beams of protons and deuterons and polarized targets have enabled the measurements of a number of nucleon-nucleon (NN) spin-dependent scattering observables. With these data, theorists have developed and fine-tuned NN potential models that describe quite well the two-nucleon strong interaction. However, the underbinding of the triton and other few-nucleon systems, indicates an incomplete description of the tensor and central components of the NN interaction.

To determine the strength of the tensor component in NN interactions, a comprehensive program at TUNL was carried out to measure the spin-dependent total cross-section difference, $\Delta\sigma$ in the $\vec{n}\vec{p}$ system. This observable is defined as the difference in total cross section for scattering when the beam and target are antialigned and the total cross section for the aligned configuration. For the longitudinal spin case, we have

$$\Delta\sigma_L \equiv \sigma(\rightleftarrows) - \sigma(\rightrightarrows), \tag{1}$$

where σ is the total cross section and the arrows indicate the beam and target relative spin configurations. Results from these measurements indicate that the high-precision NN potential models underpredict the strength of the tensor force in the low-energy two-nucleon interaction[1].

344

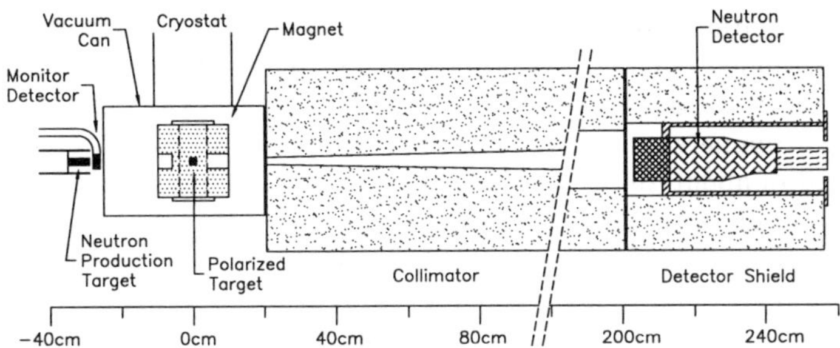

Figure 1. Schematic diagram of the $\vec{n}$-$\vec{d}$ transmission asymmetry experiment from the neutron production cell to the 0° detector.

The TUNL polarized target group is now pursuing $\Delta\sigma$ measurements in the $\vec{n}\vec{d}$ system. These observables are sensitive to the same three-nucleon force (3NF) terms that are traditionally added to the NN potential models to reproduce the triton binding energy. Therefore, a measurement of these observables will provide an independent test of the 3NF approach used in bound-state calculations.

2 Experiment Technique

The cross section difference is linearly proportional to the neutron transmission asymmetry, ϵ_n, which is the difference between zero-degree detector counts for incoming neutron spin along and against the target spin-state direction. For a neutron beam of average polarization, P_n, incident on a deuteron target of polarization, P_T, and thickness, x atoms/barn, the longitudinal cross section difference is given by[2]

$$\Delta\sigma_L = \frac{2\epsilon_n}{P_n(P_T x)},\qquad(2)$$

independent of the incident neutron flux and unpolarized total cross section.

We carried out preliminary measurements at the Triangle Universities Nuclear Laboratory, TUNL, accelerator facility. A detailed description of the target and detection apparatus, the TUNL FN Tandem Van de Graaff accelerator facility with an Atomic Beam Polarized Ion Source (ABPIS) used in the $\Delta\sigma_L$ measurements is presented elsewhere[1]. Figure 1 shows the beam line arrangement of the neutron production cell, polarized target, and detectors.

Polarized neutrons of energies 5, 7, and 9 MeV were produced by the well characterized polarization-transfer reaction $^2\mathrm{H}(\vec{d},\vec{n})^3\mathrm{He}$. The ABPIS produced the longitudinally polarized deuteron beam which was accelerated by a 10 MV FN tandem Van de Graaff to the correct energy for neutron production in the center of the gas cell. The production target consisted of a 6.0 cm long and 1.9 cm diameter cell filled with deuterium gas at 2 atm, giving an energy spread of about 200 keV. The neutron yield and beam polarization were dependent upon both the vector (P_z) and tensor (P_{zz}) polarizations of the incident deuteron beam. We measured the deuteron beam polarization both with a spin-filter polarimeter located at the ABPIS,[3] and with a ^{3}He$(\vec{d},p)^4$He reaction polarimeter located in the beam line upstream of the production cell.

For some measurements, the ABPIS transitions were set for maximum vector and tensor polarization, $P_z = P_{zz} = 1$, yielding a neutron beam with $(49\pm4)\%$ polarization. To minimize the sensitivity of the neutron polarization to source fluctuations and reduce the incoming beam asymmetry, we tuned the ABPIS for a deuteron beam with nearly zero tensor polarization. For this configuration, $P_{zz} = 0$ and maximum $P_z = \frac{2}{3}$, the beam intensity was greater, and the neutron beam polarization was reduced to $(33\pm4)\%$. Typical neutron beam current on target was about 1 μA equivalent producing a neutron count rate above threshold of about 70 kHz in the zero-degree detector.

The target was deuterated 1,2-propanediol ($CH_3CHOHCH_2OH$) in which either the six hydrogens bonded to the carbon are replaced with deuterium (D6) or all eight hydrogens are replaced (D8). The D6 target (98% and 99.5% purity) was used for the lower energy experiments, where the proton cross-section difference, $(\Delta\sigma_L)_p$ was small and could reliably be subtracted. At 12 MeV, the deuteron cross-section difference was expected to be small, so we used the D8 target (98%) to reduce the proton contribution to the measured transmission asymmetry. Both targets were chemically doped with an EHBA-Cr^V complex containing small amounts of hydrogen. The D6 target contained 37% hydrogen by number and the D8 target contained 4% hydrogen.

We dynamically polarized the frozen 1 mm diameter deuterated propanediol beads in a Kel-F cube 1.4 cm on a side. The target was cooled to 0.5 K by a ^{3}He refrigerator (of a PSI design[4]) The 69 GHz microwave driven spin-flip transitions transfered polarization from the electrons to the target nuclei.

The target polarization was monitored by NMR techniques using phase sensitive detection with the standard Liverpool module. For the D6 target ($E_n = 5, 7, 9$ MeV), we obtained the deuteron polarization indirectly through the measurement of the polarized proton NMR signal. We applied the equal spin-temperature hypothesis[5] which states that two nuclei in contact with the same spin-reservoir, in our case produced by the chromium-V electrons, will be at the same spin-temperature. Using the measured proton polarization, the magnetic moments of the protons and deuterons, and the applied magnetic field, we determined the polarization of the deuterons. We calibrated the proton NMR signal with thermal equilibrium signals at 1 K and at 0.5 K.

For the D8 target (12 MeV point), we measured the deuteron NMR signal directly and determined the polarization by measuring relative NMR peak heights. In all experiments, we obtained target polarizations of about 13% with an error of $\pm 2\%$ for the direct deuteron NMR measurement and an error of $\pm 1\%$ in the method of measuring the relatively large proton NMR signal.

The target thickness was determined by weighing the target material for a known target volume (2.74 ml) and packing fraction (0.6). We calculated $P_T x$ values for D6 and D8 target materials of $(9\pm 1) \times 10^{-3}$ and $(11\pm 1)\times 10^{-3}$ polarized atoms per barn, respectively.

The target polarization direction was reversed once every two hours. The beam polarization was flipped at 10Hz. Both target and beam polarization directions were reversed in an 8-step sequence $(+ - - + - + + -)$ to remove errors associated with linear and quadratic drifts in time.

For the zero-degree neutron detection, we used a 12.7 cm long $\times$ 12.7 cm diameter liquid organic scintillator located 130 cm from the target. This $0°$ detector was collimated to a 2.5 msr solid angle which illuminated a 1.2 cm $\times$ 1.2 cm cross section of the target. A polyethylene collimator defined the solid angle view of the detector. We used pulse-shape discrimination to separate neutron and gamma-ray events.

To first order, we removed contributions to the measured transmission asymmetry from sources other than the target, including the neutron production asymmetry. A small liquid organic scintillator (11.1 mm $\times$ 22.2 mm $\times$ 25.4 mm thick) located immediately upstream of the target, served as a monitor detector. A glass light-pipe was employed so that the photomultiplier tubes were outside the strong magnetic field region of the target. The neutron asymmetry is the difference in measured asymmetries from the zero-degree neutron detector (ϵ) and the monitor detector (ϵ_m), or $\epsilon_n = \epsilon - \epsilon_m$.

3 Analysis

The measured neutron asymmetry includes contributions from both the deuterons and protons in the target. We first sum the proton (p) and deuteron (d) contributions to the asymmetry (Equation 2). Rearranging then gives

$$(\Delta\sigma_L)_d = \frac{2\epsilon_n}{P_n (P_T x)_d} - \frac{(P_T x)_p}{(P_T x)_d}(\Delta\sigma_L)_p \qquad (3)$$

where the second term is the effective proton cross-section difference correction. The proton longitudinal cross section $(\Delta\sigma_L)_p$, ranges between -50 mb and $+100$ mb in the energy region of interest $(5\text{-}12.5 \text{ MeV})$[1]. The ratio of $P_T x$ values is about 1.5 for the D6 target and 0.18 for the D8 target used when the deuterium asymmetry was small. This leaves a proton correction term that is at most 27% of the deuterium cross-section difference value.

4 Discussion

Our data currently does not agree with the theoretical predictions. Calculations by Witała *et al.* were carried out using several NN potential models with and without the inclusion of 3NF terms[6]. In all cases, the calculations underpredict our results by an unexpected amount. We are reviewing the experimental apparatus and analysis to consider systematics that might have contributed to the present results.

Concurrently, we are improving the target system to increase the deuteron polarization, thereby increasing the asymmetry signal. We have modified the cryostat, converting the ^{3}He evaporation refrigerator to a dilution refrigerator. We expect to repeat the measurements with a target temperature of 200 mK and polarization of about 35%. We are considering placing the passive NMR electronics in the cold region to reduce the d-NMR signal noise, and measure the d-NMR thermal equilibrium signal, thereby decreasing the error in P_T.

In 2002, we plan to repeat the measurements for the above neutron energies and include a low-energy neutron scattering calibration point. At 800 keV, $\Delta\sigma_L = -2.5$ b, a large value that is consistent with scattering length measurements and well-understood calculations.

References

1. J. Walston, *et al.*, Phys. Rev. C **63**, 014004 (2000).
2. W.S. Wilburn, *et al.* Phys. Rev. C **52**, 2351 (1995).
3. T.B. Clegg *et al.*, Nucl. Instr. Meth. **A357**, 200 (1995).
4. B. van den Brandt, J.A. Konter, and S. Mango, Nucl. Instrum. and Methods in Phys. Res. **A289**, 526 (1990).
5. W. de Boer, J. of Low Temp. Phys. **22**, 185 (1976).
6. H. Witała, *et al.* Phys. Lett. B **447** 216 (1999).

Acknowledgments

This work performed under DoE grants DE-FG02-97ER41042 and DE-FG02-97ER41033.

THE GDH EXPERIMENT AT BONN

G. REICHERZ FOR THE GDH COLLABORATION

EP 1 AG , Ruhr - University of Bochum, D-44780 Bochum, Germany

The Gerasimov-Drell-Hearn (GDH) sum rule connects the helicity dependent photoabsorption cross sections with the anomalous magnetic moment of the nucleon. The GDH collaboration is measuring the total cross section of circularly polarized photons on longitudinally polarized protons to check this sum rule experimentally, which will lead to a further progress in the investigation of the spin structure of the nucleon. The experiment has been started using the polarized electron beam of the Mainz accelerator MAMI in the energy range 140 - 800 MeV and is continued at the Bonn accelerator ELSA up to an energy of approximately 3.1 GeV. We have finished the data taking period at MAMI for the proton in august 1998. Our apparatus in Bonn consists of a frozen spin target which had been integrated into the GDH-detector. We obtain polarized photons by helicity transfer in the bremsstrahlung process. The energy has been determined using a tagging system. The degree of polarization of the photon beam has been monitored by a Møller polarimeter. The setup in Bonn and first preliminary results of our proton measurement will be shown.

1 Introduction

The Gerasimov-Drell-Hearn (GDH) sum rule relates the polarized total photoabsorption cross section of the nucleon - an observable of the dynamics - the static characteristics of the nucleon κ and m - the anomalous magnetic moment and the mass of the nucleon - where α represents the fine structure constant [1,2].

$$\int_{m_\pi}^{\infty} \frac{\sigma_{1/2} - \sigma_{3/2}}{\nu} d\nu = -\frac{2\pi^2\alpha}{m^2}\kappa^2 = -204\mu b \qquad (1)$$

$\sigma_{3/2}$ and $\sigma_{1/2}$ are the energy dependent total photoabsorption cross section of circularly polarized photons on longitudinally polarized nucleons with parallel or antiparallel spin orientation, respectively. The experimental verification of the GDH sum rule requires the measurement of the spin dependence of the total photoabsorption cross section, i.e. the left hand side of equation (1). The GDH-Collaboration intends to determine the GDH integrand on the proton and the neutron on the photo energy range from pion threshold 140 MeV up to 3100 MeV at the two accelerators MAMI and ELSA [3,4]. The GDH-detector [5] has been specifically designed and built for the photon energies at ELSA from 700 MeV to 3100 MeV. In this report the experimental set up at Bonn will be explained with the special focus on the polarized solid target.

2   Circularly polarized photons

Circularly polarized photons were produced by bremsstrahlung of longitudinally polarized electrons. A strained GaAs photocathode delivered electrons with a maximum degree of polarization of about 70 % at the GDH - experiment with a current of more than 3 nA. The electron polarization was monitored throughout the experiment with a precision of ±3 % by means of a Møller polarimeter. The polarization direction was flipped randomly every few seconds to minimize systematic effects. The photon polarization was evaluated according to the formula of ref. [8]. The photon energy was determined by the tagging spectrometer which analyses the momenta of the electrons that radiated the bremsstrahlung photon. This detection system is able to tag photons in the range from 700 to 3100 MeV with a resolution of about 5 MeV. The tagging efficiency was continuously monitored throughout the experiment by a e^+e^- pair detector, that was installed downstream of the main hadron detector. The pair detector was regularly calibrated using a 100 % efficient lead glass detector. Using this technique, the photon flux could be determined with a systematic uncertainly of 2 %.

3   Longitudinally polarized protons

The low intensity tagged photon beam demands a solid polarized target to reach reasonable luminosities. A butanol (C_4H_9OH) frozen spin target [9] provided the polarized protons. The target length of 2 cm gave a total thickness of $N_t = 0.9 \cdot 10^{23}$ polarizable protons per cm^2, from which a luminosity results of about 10^{29} sec^{-1}m^{-2}. A polarized solid state target uses the principle of Dynamic Nuclear Polarization (DNP) in which a target material - butanol (C_4H_9OH) - is doped with a dilute assembly of paramagnetic radicals by chemical techniques. The material is cooled to about 300 mK in a magnetic field of 2.5 Tesla and then irradiated by microwaves near to the electron spin resonance frequency to drive the hyperfine transitions, which allow the nucleon spin to be aligned parallel or antiparallel to the applied magnetic field. For the cooling of the target material a horizontal $^3He/^4He$ dilution refrigerator was built, the outer dimensions of which were determined by the inner dimensions of the GDH detector and by the warm bore of the polarizing magnet. Two precooling stages, a separator and an evaporator, and several counter current heat exchangers provide the precooling and the liquefaction of the $^3He/^4He$ gas mixture. The dilution unit consisting of the still, the final heat exchanger and the mixing chamber permits cooling to temperatures below 50 mK. The cryogenic components of the refrigerator are placed symmetrically around the beamline tubing, which is the central part of the refrigerator and serves also as an internal isolation vacuum to reduce the thermal heat load. For the same purpose the refrigerator is also enclosed by an isolation vacuum and surrounded by two heat shields. For the DNP process a magnetic field of 2.5 Tesla with a homogeneity of $\Delta B/B < 10^{-4}$ over the target area is required. This can only be provided by a huge magnet that surrounds the target completely, resulting in a reduction of angular acceptance. To increase the acceptance to a totally open geometry for the scattered particles the 'frozen spin' technique with an internal holding coil is used. In this technique the nucleon polarization is preserved after the polarization process at a temperature of about 50-60 mK and a magnetic field of 0.4 Tesla with a relaxation time of more than one week. This magnetic field is provided by a superconducting, internal holding coil that is integrated in the refrigerator. It consists of four layers of NbTi wire wound on a 300 μm copper layer, which results in a total thickness of 700 mm, such that a significant loss of scattered particles occurs only close to the threshold. As a result of its compact design (length 120 mm, diameter 44 mm) the coil has a weak outer fringe field, so that the influence on the detectors located close to the target area is very small in comparison to that of the outer holding coils formerly used. The coil is cooled to a temperature below 1.2 K by a thermal contact to the liquid helium of the still. At this temperature a maximum field of 0.48 Tesla at 12 A was achieved. The homogeneity of the magnetic field over the target area ($< 5 \cdot 10^{-3}$) allows for the first time an online polarization measurement during 'frozen spin mode' (data taking), so that the systematic error of the polarization measurement is considerably lowered. A maximum polarizatin close to 90 % and relaxation times in the frozen spin mode of about 200 h have been regularly achieved. The degree of the proton polarization has been determined by the NMR technique with a relative precision of less than 2 %.

4   The GDH-Detector

The modular structure of the GDH-Dectector [5] consists of the following major components: The central hadron detector, a threshold Cerenkov detector and a forward hadron detector. The central hadron detector encloses the target and covers the largest fraction of the solid angle. This central hadron detector consists of 6 CAD^a-Modules. The aerogel Cerenkov counter to suppress electromagnetic background is used within a half opening angle of 15° polar angle in forward direction. The forward hadron detector is installed in

---

^{a}Charged particle And Decay detector

the far forwad direction to cover the region of the solid angle left open by the central hadron detector. This forward detector consists of three components: first 1 cm thick plate of plastic scintillator (STP); behind it concentric rings of scintillation detectors which permit resolving polar angles (STAR [10]) and further behind it the FAR -Forward-Wall (FFW [11]) which is constructed like the CAD-Modules of the central detector.

5 Results and Outlook

Data have been taken at five different electron energy settings of 1 GeV, 1.4 GeV, 1.9 GeV, 2.4 GeV and 3.1 MeV, to maximize the photon polarization over the entire energy range. The first setting show a very good link up to the Mainz data. The extraction of the total cross section difference allows the direct determination of the contribution in the energy range investigated to the GDH integral and the forward spin polarizability. The multipole analyses SAID [12] and HDT [13] overestimate the experimental results by about 10 - 20 %.

References

1. S.B. Gerasimov, Sov. J. Nucl. Phys.2 (1966) 430
2. S.D. Drell, A.C. Hearn, Phys. Rev. Lett. 16 (1966) 908
3. G. Anton et al., Prog. Part. nucl, Phys. 34 (1995)
4. J. Ahrens et al. Mainz MAMI Proposal A2/2-95 and
   G. Anton et al., Bonn ELSA Proposal 1992
5. K. Helbing et al., submitted to Nucl. Inst. Meth. A (2001)
6. J. Ahrens et al., Phys. Rev. Letters 84, 5950 (2000)
7. j. Ahrens et al., Phys. Rev. Letters 87, 023022 (2001)
8. H. Olsen and L.C. Maximon, Phys. Rev. 114 (1959) 887
9. C. Bradtke et al., Nucl. Instr. Meth. A 346 (1999) 430
10. A. Fuchs, Diploma thesis, Tübingen (1994)
    M. Sauer et al., Nucl. Instr. Meth. A 378 (1996) 143
11. T. Speckner, Diploma thesis, Erlangen (1998);
    G. Zeitler, Diploma thesis, Erlangen (1998)
12. R.A. Arndt et al., Phys. Rev. C 53 (1996) 430 (solution SM99K)
13. O. Hanstein et al.,Nucl. Phys. A 632 (1999) 561

VIII. Summary Talk

SUMMARY OF THE WORKSHOP ON POLARIZED SOURCES AND TARGETS - PST2001

E. STEFFENS

University of Erlangen-Nürnberg, Germany
Physikalisches Institut, Erwin-Rommel-Str.1, D-91058 Erlangen

The workshops on polarized sources and targets take place every two years and present a broad overview of polarized ion sources, electron sources, solid and gaseous targets and neighbouring fields, like the appli-cation of polarized atoms in applied sciences and medicine. The present meeting organized by the IUCF laboratory was conducted at Nashville, Indiana, near Bloomington. All the necessary tools for spin experiments in nuclear and particle physics were presented and shown to be in very good shape. Impressive progress was reported about polarized proton sources for RHIC and the IUCF cooler. New materials for photo cathodes are under constant development and polarimetry of high-energy electron beams has reached high precision. Neutron polarizers based on ^{3}He cells have opened up new fields in material science and on precision studies of weak interaction in polarized neutron decay. Solid polarized target material for next-generation structure function measurements remains an active field. New materials for special applications are under study. Gaseous hydrogen and deuterium targets have been extensively used at medium and high energy proton and electron storage rings resulting in a wealth of high precision data. RIB facilities in combination with polarization techniques open up new opportunities for the future.

1 Introduction

Workshops of topical and more general nature on the various tools for spin experiments have been organized since the early sixties. These tools were basically developed and used by two communities, the nuclear and particle physics communities. This led to a gradual merging of these fields and to common symposia and workshops. Since 1991, workshops were conducted every two years under the auspices of the International Committee for High Energy Spin Physics, which became the committee for Spin Physics in 1998. The previous workshops were held at Madison (1993), Cologne (1995), Urbana (1997) and Erlangen (1999). At the last meeting PST99 at Erlangen, all relevant tools for Spin Physics were successfully presented at a single meeting:

- Polarized ion sources (protons, deuterons, ...);
- Polarized electron sources;
- Polarized solid targets;
- Polarized gaseous targets of low and high density;
- Polarimetry of hadron and lepton beams;
- Applications in medicine, neutron physics, maerial science and other fields.

This concept was also used for the present meeting, resulting in a comprehensive overview within three and a half days on sources, targets, polarimetry and miscellany. The town of Nashville as meeting has been an excellent choice, as it was a pleasant place with „Indian Summer" wheather conditions and no Email, leading to a creative, but relaxed atmosphere with intense communication among the participants. The social program included a reception on Sunday, Sept.30, a Barbecue dinner on Monday, a tour the the IUCF laboratory and reception on Tuesday, and the Workshop dinner on Wednesday, Oct.3.

In the following, an attempt is made to summarize the status of the field as it was discussed during PST2001, and to indicate future trends. No references are given and the reader is refered to the specific contributions to these proceedings.

2   Polarized Gas Targets for Storage Rings

These targets are pure, have no window and are of low density in order to keep the life time of the stored electron or proton beam at a reasonable level. They can be run in parallel to other storage ring users. A target overview was given in the talk by F. Rathmann (FZ Jülich). A significant number of running or planned installations has been discussed during the meeting:

- the pioneering experiment at VEPP-3 which is now running with an improved deuterium target based on an ABS source;
- the PINTEX target at the IUCF cooler ring which has been converted to a deuterium target for the remaining running time of the ring;
- the HERMES target at the HERA 27.5 GeV electron storage ring which is run for the 2002/03 period in transverse hydrogen mode;
- the targets for the two COSY experiments EDDA (operational since 1998 as jet, storage cell upgrade in progress) and ANKE (first running in 2002);
- the target for the BLAST experiment at MIT-Bates which is a reincarnation of the former AmPS installation;
- an update of the Michigan ultracold Mark-II hydrogen target which was originally planned for the NEPTUN-A experiment at IHEP-Protvino.
- the RHIC polarized hydrogen jet intended as polarimeter standard and possibly the core of a new fixed target experiment at RHIC.

All targets except the last two are based on the successful storage cell concept in order to produce the maximum areal density as seen by the high energy beam from the limited flow rate state-of-the-art sources can provide. The design of these cells differs widely depending on machine and experimental conditions. Cooling to 100K or below is applied for enhancing the density. As reported by HERMES this is also benficial for the build-up of a frozen water layer which has very nice properties concerning recombination - which was found to be negligible for the 2000 deuterium run - and depolarization within the cell.

As target sources for hydrogen or deuterium atomic beam sources (ABS) are employed exclusively in running experiments. The intensities accepted by a standard compression tube which were reported for the various sources fall into a surprisingly narrow interval between 5 and $7 \cdot 10^{16}$ atoms/s. It should be noted that the design of all these ABS sources follows the same principles. The very similar maximum intensities may indicate a fundamental limit of the ABS technology.- Studies on the formation of cold dissociated hydrogen beams can now be compared with simulation results using a DSMC program and show good agreement. Unfortunately, the carrier jet method proposed at Urbana did not work as expected due to a too large admixture of the carrier jet species to the inner atomic beam. - The alternative scheme using spin-exchange by optically pumped alkali vapor (LDS) is still hampered by low atomic fraction and polarization and has not yet been employed in a permanent installation. Development studies are in progress at Erlangen and MIT.

An interesting question has been studied at IUCF, using the existing PINTEX target with some modification: What is the fraction of nuclear polarization surviving wall-recombination of polarized atoms? Precise knowledge of this quantity is requi-red in order to calculate the target polarization e.g. of the HERMES target in the presence of recombination. The result shows that at strong fields about 40% can be conserved. This is in agreement with the picture that one atome comes from a layer of unpolarized hydrogen at the wall and the second out of the - fully polarized - gas phase, plus some losses due to *hfi* within the molecules after they have been formed.

A report on the HERMES target conveyed an overview of the involved diagnostics which is required in order to measure the target polarization precisely using the Breit-Rabi type sampling polarimeter (BRP). A complete understanding of the flow and the depolarization and recombination processes is required in order to calculate the so-called sampling corrections relating the average target polarization as seen by the electron beam and the polarization of the sample beam. Finally it has been reported that for the 2000 run the goal of 3% precision was reached. The 2000 deuterium vector polarization was +0.851 ± 0.031 and -0.840 ± 0.028 with very small variations. - Another interesting diagnostic tool presented at the Workshop is the Lamb-Shift polarimeter developed for the ANKE target. As already demonstrated by the TUNL group it is a versatile instrument to measure the polarization of ions. The production of ions from the target atoms and their transport to the LS polarimeter is well understood that the target polarization can be infered from the measurement.

In conclusion it can be stated that polarized gas targets for storage rings for hydrogen and deuterium are operated exclusively with ABS sources. Electron rings could digest 10x more density as provided with state-of-the-art targets. Spin-exchange targets have not yet reached the required level of quality and reliability.

Optically pumped ^{3}He targets have worked well but there is no actual physics proposal based on such a target. All operational or proposed targets except EDDA and RHIC are employing a storage cell. Probably the largest gains could be realized by squeezing the beam and reducing the cell size as HERMES did in 1999/2000. A 50% gain solely from the reduced cell cross section has been obtained.

3   Polarized Solid Targets

The subject was reviewed by S. Penttilä (Los Alamos). These targets can be run with external beams like e, μ, γ, p or n. The limited intensity of muon beams can be com-pensated for by rather thick targets more than 1m long as demonstrated for the SMC experiment. The standard „dynamical nuclear polarization" (DNP) is produced by microwaves at low temperature and high magnetic field (e.g. 1K, 2.5T). Here new studies are underway at Bochum where both irradiated d-buthanol and ^{6}LiD are in-vestigated in order to improve polarization and hardness. 70% deuteron polarization have been achieved with buthanol at 5T. Such a target is envisaged for the GDH experiment at Bonn and Mainz on the neutron. ^{6}LiD is the ultimate choice for the COMPASS experiment (CERN) on the gluon contribution to the nucleon spin. The new large target magnet has not been commissioned yet due to fabrication problems. Test results with the SMC magnet, which is used as fall-back solution showed a deuterium polarization up to about 50%.

At TUNL, a deuterium target for reactions induced by polarized neutrons is employed. A frozen-spin target for a GDH experiment at the Duke FEL facility is being considered.

For an inverse-kinematics experiment with radioactive ions at RIKEN, a polari-zed proton target with low magnetic field is required in order not to deflect low-energy recoils. As a new DNP material, doped naphthalene has been studied. Polarization of more than 15% at 100K and 0.3T have been obtained.

Frozen „brute-force polarized" HD targets for photo-nuclear reactions are stu-died by a Syracuse-BNL collaboration for the SPHice experiment. Polarizing takes place at 15mK and 17T within a time period of one week. High purity is required and a small H_2 admixture plays a crucial role for the relaxation process. A 0.7T holding field is sufficient to keep the target polarized. The test of the production and transfer process is in progress.

An interesting study combining „small angle neutron scattering" (SANS) and DNP techniques is performed at PSI, yielding information on the location of protons with repect to radicals in large biomolecules.

4   High-Pressure Polarized Noble Gas Targets

The two established methods were discussed:

- Optical pumping of a small Alkali (e.g. Rb) fraction and subsequent spin-exchange to ^{3}He or ^{129}Xe gas <u>at full pressure</u> up to several bar;
- Direct optical pumping of metastable ^{3}He atoms followed by metastability exchange collosions <u>at low pressure</u>. In a second step the gas is compressed using a „nonmagnetic" mechanical compressor up to several bar.

There are several application (i) as target in - mostly electron - scattering experiments, (ii) for MRI - magnetic resonance imaging - in medicine, and (iii) as neutron spin filter for polarizing and analyzing thermal and cold neutron beams.

- a) A ^{3}He target based on spin-exchange for the study of nucleon structure is in operation at Jefferson Lab.
- b) MRI in medicine is a very active field. ^{3}He is not solvable in blood; it can be nicely used to study ventilation of the lung. ^{129}Xe is solvable and can monitore the blood supply e.g. of the brain.
- c) There were a number of high-pressure ^{3}He cells from various labs presented. No clear preference for one of the two methods is visible up to now. Their appli-cation on neutron beams as polarizers or analyzers is discussed in Section 8.

5   Polarized Electron Sources

The subject was reviewed by K. Aulenbacher (Mainz). Compared with the specs of an „ideal source" presented by Aulenbacher there were still some open issues. The key technology is the art of making photocathodes. Strained GaAlAs - InGaAlAs superlattices as developed at Petersburg may lead to cathodes with high QE **and** polarization at the same time.

At Bates, strained GaAs is illuminated from high-power diode laser arrays using an extended spot size. A 'monster' Pockels cell is employed to switch the thick beam, and NEG pumping close to the cathode is applied. For such a system the life time problem seems to be much less severe.

At JLab, a lot of effort is put into a careful calibration of the electron polarimeters of different type, located at various places but all of them looking at beams from the same source and accelerator. This so-called 'Spin Dance 2000' revealed clear problems of old design Möller polarimeters due to the uncertainty of electron polarization in magnetized foils on the 'couple of percent' level. A new Möller design with very strong magnetizing field seems to give the correct electron polarization. - At the same Lab, a photon polarimeter based on pair production is under construction. The calibration was done at the Spring-8 laser-backscattering beam line with known photon polarization.

6   Polarized Ion Sources

An overview was given by T. Clegg (TUNL). Two completely different schemes are employed: (i) the OPPIS principle which is based on pick-up of polarized electrons from optically pumped alkali vapor - and (ii) the ABS principle where atoms are polarized using Stern-Gerlach separation combined with various types of ionizers. Deuterium ABS sources can provide a more flexible scheme of vector and tensor polarized beams.

OPPIS sources run at the TRIUMF H^- cyclotron (about 0.5 mA in dc mode) and at the RHIC injector complex. For the BNL project the former KEK source has been upgraded to deliver 1.1 mA with 0.04% duty cycle. Further improvement is expec-ted by exchanging the ECR proton source with a plasma source of 10 A peak current. The present source has been successfully used during the RHIC polarized proton commissioning runs.

About 15 ABS sources are operational but many of them not continuously. The TUNL and IUCF sources are in operation for more than 180 d /y. A close collabo-ration between IUCF and INR has boosted the performance figures of pulsed H^- sources to more than 2 mA. Ionization is performed using a negative -ion plasma source. The pulse length is limited to about 100 µs.

Other ionizing techniques were reported, too. In the new Munich source as well as at TUNL, ECR ionization to positive ions is applied, followed by double charge exchange on Cesium to negative ions. Thermal polarized hydrogen atoms are converted directly to negative ions using a fast neutral Cs beam. At COSY this scheme led to H^- currents of 30 µA at 1% duty cycle resulting in up to 2.5×10^{10} stored protons. At full momentum of 3.2 GeV/c a polarization of 0.7 has been achieved.

7   Proton Polarimetry

The precise measurement of the polarization of multi-MeV and -GeV proton beams is mandatory for precision experiments. Two facilities were discussed: (i) a high precision scanning polarimeter for the TRIUMF parity violation experiment and (ii) polarimetry for the RHIC polarized proton beams beyond 1 GeV.

The report on the scanning polarimeter for TRIUMF revealed the strong requirements imposed by the measurement of a 10^{-7} longitudinal parity violating A_z component in the presence of a transverse 'allowed' A_y component which is stronger by more than 6 orders of magnitude. This requires the measurement and minimization of first moments of the transverse polarization in the 221 MeV proton beam to high precision. Using photodiode readout in the current mode, realization of this challenging goal may be within reach.

At RHIC, for beam tuning and extraction of physics quantities the proton polarization has to be known to better than 5% at various energies. Different methods are under study. The traditional methods like pC inclusive or pp elastic scattering have small analyzing powers at AGS energies and beyond. Inclusive π production exhibits large effects at all relevant energies, but the error in the analyzing power is too big. A possible solution is a polarized hydrogen jet target with known polarization. Such a target is presently under study.

8 Polarized Neutrons

The technique of polarizing and analyzing of slow neutrons was reviewed by T. Gentile (NIST). Different neutron polarizers are employed: (i) Supermirrors and Heusler alloy mirrors for reflecting one spin state; (ii) polarized proton targets and (iii) high pressure polarized ^{3}He targets as spin filters transmitting one spin state. In particular, ^{3}He cells are very well suited as broad band 'transmission' polarizers and analyzers.

Two methods based on optical pumping are applied (see section 4) in order to provide the required ^{3}He cells with large aperture and limited thickness. At NIST, both methods are applied with no clear preference for one of them. At IUCF and HMI-Berlin, ^{3}He filling stations with compression were built. An interesting application is polarized neutron small angle scattering (SANS). Beautiful scattering data taken at the HMI were shown illustrating the potential of ^{3}He spin filters in material research.

A TU-Munich - ILL group is applying 'opaque' spin filters to polarized neutron decay aiming at precision tests of the Standard Model in the weak sector. Using opaque filters, i.e. at the limit of high density $P \rightarrow P_\infty$, reduces the error of the polari-zation of the transmitted neutrons to the 0.1% level. Combined with other observab-les, this boost in precision will result in more sensitivity and possibly the detection of a significant deviation from the SM.

9 Polarized Radioactive Beams

At MPI-Heidelberg, a Marberg university group employs thermal polarized ^{8}Li atoms to study spin relaxation on surfaces since more than 10 years. ^{8}Li is produced by a beam of ^{7}Li ions from the tandem accelerator via the ^{2}H(^{7}Li,^{8}Li) reaction and thermalized in a graphite stopper, followed by laser optical pumping. The polariza-tion of ^{8}Li adsorbed at the surface is determined by the β decay

362

asymmetry. A study of the modification of a Si surface structure as function of Li coverage has been presented at the meeting.

The advent of on-line isotope separators combined with techniques from polari-zed ion sources enables the production of a broad variety of polarized radioactive beams. As an example, a 30 keV polarized ^{8}Li+ beam from the ISAC facility at TRIUMF was described. Polarization is obtained by optical pumping of the fast neutrals after a neutralizer cell which are then re-ionized on He gas. Beams of ^{8}Li$^+$, ^{11}Li$^+$ and ^{20}Na$^+$ ions of variable energy will be employed for the study of thin films.

10 Conclusions

Despite the work of several decades, the development of 'spin tools' is still a very active field. Instead of inventing new methods, most of the activity goes into adaptation and refinement of known principles to various experiments and applications. A considerable spin-off takes place into applied sciences. Examples discussed at the meeting are MRI with ^{3}He or ^{129}Xe in medicine, ^{3}He neutron spin filters for the study of magnetic materials and ^{8}Li for the study of surfaces or thin films.

Our host laboratory, the Indiana University Cyclotron Facility, has always been in the forefront of developing and utilizing such tools. I am convinced that even in view of the forthcoming changes to the laboratory they will continue along these lines and come up with new plans, facilities and results. Our thanks are due to the organizers of this wonderful and productive meeting, in particular to Laddie, Janet and the local organizers. We are all looking forward to PST2003 in two years from now!

The Ninth International Workshop on Polarized Sources and Targets
PST 2001
SEPTEMBER 30TH TO OCTOBER 4TH, 2001
Nashville, IN USA

MONDAY, OCTOBER 1st, 2001

8-8:30 AM	Registration, Coffee and Continental Breakfast

Session Chair: Hans Meyer
8:30 AM Welcome - John Cameron
8:40 AM **Overview of Gas Target Development** - *Frank Rathmann*
9:20 AM *P.Lenisa*, The Gaseous Polarized H-D Target of the HERMES Experiment
9:40 AM *V.Luppov*, Status of the Michigan Ultra-Cold Polarized Hydrogen Jet Target
10:00 AM *H. Kolster*, The BLAST Polarized Hydrogen/Deuterium Target

10:20 AM-10:40 AM
Coffee Break

Session Chair: Mike Lowry
10:40 AM **Overview of Solid Target Development** - *Seppo Penttilä*
11:20 AM *D. Crabb (for G. Court)*, High Precision Measurements of Target Polarization using NMR Techniques
11:40 AM *St. Goertz*, On the Hunt for the Most Efficient Paramagnetic Centers for the DNP Process

12:00 PM - 1:30 PM
Lunch

Session Chair: Dieter Eversheim
1:30 PM **Polarization of Molecules** - *Tom Wise*
2:10 PM *R. Engels*, A Lamb-Shift Polarimeter for the Polarized Gas Target at ANKE
2:30 PM *A. Nass*, Studies on Beam Formation in the HERMES - ABS

2:50 PM - 3:10 PM
Coffee Break

Session Chair: Frank Rathmann
3:10 PM **Polarized Noble Gases in Medical Imaging** - *T. Chupp*

3:50 PM	*M. Mikirtytchiants*, The Polarized Gas Target for ANKE at COSY/Julich
4:10 PM	*B. v. Przewoski*, A Vector and Tensor Polarized Internal Deuterium Target for the IUCF Cooler
4:30 PM	*D. Toporkov*, Status of the Novosibirsk Polarized Deuteron Target
4:50 PM	*K. Kramer*, The Jefferson Lab Polarized ^{3}He Target

6:00 PM

Barbecue Dinner @ Brown County Inn

TUESDAY, OCTOBER 2nd, 2001

8-8:30 AM	Coffee and Continental Breakfast
<u>Session Chair:</u>	<u>Kurt Aulenbacher</u>
8:30 AM	**An Overview of Electron Polarimeters and Results of an Intercomparison** - *J. Grames*
9:10 AM	*M. Wood*, A High-Energy Photon Polarimeter Based on Pair Production
9:30 AM	*D. Orlov*, Cold Electrons from GaAs(Cs,O)

9:50 AM

Coffee Break

10:10 AM - 12:00 AM   Poster Session

C. Crawford	Laser Driven Target at MIT-Bates
V.P.Ershov	Cryogenic Source of Polarized Deuterons POLARIS for the JINR Accelerator Nuclotron
D. Eversheim	Design Criteria for Polarized Sources
M. Glende	The Storage Cell for the TRI-Experiment at COSY-Jülich
D. Haase	Design of a Frozen Spin Deuteron Target for the HIGS/DFELL Facility
A. Hatakeyama	Detailed Calculations for Optical Pumping at ISAC Polarizer
H. Kolster	Novel Approach to the Design of a Six-pole Magnet System for an Atomic Beam Source
R. Raymond	The University of Michigan Polarized Proton Target
C. Schröter	Pulsed Polarized Electron Source for Atomic Physics Experiments
E. Stephenson	Development of a Polarized Deuteron Beam for the IUCF Cooler
A. Subashiev	Emission from Semiconductor Photocathodes at High Intensity Inhomogeneous Excitation
A. Subashiev	Charge Limitation Effects in Photoemission from GaAsP Strained Layer Cathode

J. Wilbert	The Erlangen Laser Driven Source for Polarized Hydrogen and Deuterium
T. Wise	Design of a Jet Polarimeter for RHIC

12:00 PM - 1:45 PM
Lunch

Session Chairs:	Seppo Penttilä
1:45 PM	**Interaction of Polarized Atoms with Solid Surfaces** - *D. Fick*
2:25 PM	*P. Levy*, Production of a Highly Polarized $^8Li^+$ Ion Beam at ISAC
2:45 PM	*D. Markoff*, $\vec{n} \to \vec{d}$ Scattering Measurements With a Dynamically Polarized Deuteron Target at TUNL
3:05 PM	*G. Reicherz*, The GDH Experiment at Bonn

3:25 PM - 3:50 PM
Coffee

Session Chair:	Claus Dieter Schröter
3:50 PM	**Overview of Polarized Electron Source Progress** - *Kurt Aulenbacher*
4:30 PM	*A. Subashiev*, Polarized Emission from Strained Superlattices
4:50 PM	*E. Tsentalovich*, Photoemission from Strained GaAsP with High Power Diode Laser at MIT-Bates Accelerator Center

6:30 PM
Tour of IUCF and reception. Sponsored by Vacuumschmelze Corporation.

WEDNESDAY, OCTOBER 3rd, 2001

8-8:30 AM	Coffee and Continental Breakfast
Session Chair:	Don Crabb
8:30 AM	**HD Target at BNL** - *Mike Lowry*
9:10 AM	*P. Hautle*, Polarized Protons Domains in Matter
9:30 AM	*T. Wakui*, Development of Polarized Solid Proton Target for RI Beam Experiment

9:50 AM - 10:20 AM
Coffee Break

Session Chair:	Oliver Zimmer
10:20 AM	**Overview of Polarized Neutron Sources** - *Tom Gentile*
11:00 AM	*A. Rupp*, Development and Tests of ^3He Neutron Spin Filters and Construction of a Filling Station

11:20 AM	*H. Sakai*, Polarized Neutron Facility at RCNP

11:40 AM - 1:10 PM
Lunch

<u>Session Chair:</u>	<u>Tom Gentile</u>
1:10 PM	**Neutron Polarimeters** - *Oliver Zimmer*
1:50 AM	*M. Snow*, Development of Polarized ^{3}He-based Neutron Spin Filters at IUCF
2:10 PM	**Polarized Ion Source Progress** - *Tom Clegg*

2:50 PM - 3:20 PM
Coffee Break

<u>Session Chair:</u>	<u>Tom Clegg</u>
3:20 PM	*A. Zelenski*, A New Polarized Ion Source for RHIC Spin Physics
3:40 PM	*R. Gebel*, New Developments at the Polarized Ion Source of COSY-Julich
4:00 PM	*A. Belov*, Development of Pulsed Polarized Ion Source with Resonant Charge-Exchange Plasma Ionizer at INR Moscow
4:20 PM	*V. Derenchuk*, Recent Improvements in CIPIOS Intensity and Operation
4:40 PM	*R. Hertenberger*, The Munich Stern-Gerlach Source for Intense and Bright Beams of Negative Polarized Hydrogen Ions

6:00 PM Banquet

THURSDAY, OCTOBER 4th, 2001

8-8:30 AM	Coffee and Continental Breakfast
<u>Session Chair:</u>	<u>Tom Wise</u>
8:30 AM	**Proton Polarimetry at 1 GeV and Beyond** - *Yousef Makdisi*
9:10 AM	*D. Ramsay*, A High Precision Scanning Polarimeter for the TRIUMF Proton-Proton Parity Violation Experiment
9:30 AM	*E. Stephenson*, A Polarimeter for Nanosecond Bunch Length Protons
9:50 AM	*I. Alekseev*, RHIC pC CNI Polarimeter, Current Status and Future Plans

10:10 AM - 10:30 AM
Coffee Break

10:30 AM	**Workshop Summary** - *E. Steffens*

Adjournment

LIST OF PARTICIPANTS

ALEKSEEV, Igor
ITEP
Polar. Research
B. Cheremushkinskaya 25
Moscow Region, 117259 RUSSIA
Igor.Alekseev@itep.ru
Telephone: 7 (095) 1299629
FAX: 7 (095) 8839601

AULENBACHER, Kurt
Institut fur Kernpysik der Uni Mainz
B2
Insitut fur Kernphysik, Universitat
Mainz
J.J Becherweg 45
Mainz, RLP 55099 BRD
aulenbac@kph.uni-mainz.de
Telephone: 0049-6131-3925804
FAX: 0049-6131-3922964

BACHER, Andrew
Indiana University Cyclotron Facility
2401 Milo B. Sampson Lane
Bloomington, IN 47408 USA
bacher@iucf.indiana.edu
Telephone: 812-855-9365
FAX: 812-855-6645

BAYLAC, Maud
Jefferson Lab
Accelerator Division / Injector Group
12000 Jefferson Avenue
MS 5A
Newport News, VA 23606 USA
baylac@jlab.org
Telephone: 757-269-7703
FAX: 757-269-5520

BEAUMEL, Didier
I.P.N. Orsay
15 Rue Georges Clemenceau
Orsay, 91406 FRANCE
beaumel@ipno.in2p3.fr
Telephone: 1 69 15 73 27
FAX: 1 69 15 64 70

BELOV, Alexander
Indiana University Cyclotron Facility
2401 Milo B. Sampson Lane
Bloomington, IN 47408 USA
belov@iucf.indiana.edu
Telephone: 812-855-2887
FAX: 812-855-6645

CHUPP, Tim
University of Michigan
1049 Randall Lab.
500 E. University Ave.
Ann Arbor, MI 48109
chupp@umich.edu
Telephone: 734-647-2514
FAX: 734-763-9694

CLEGG, Thomas B.
UNC-Chapel Hill and TUNL
Physics and Astronomy
278 Phillips Hall
University of North Carolina
Chapel Hill, NC 27599-3255 USA
clegg@tunl.duke.edu
Telephone: 919-843-8168
FAX: 919-962-0480

CRABB, Donald
University of Virginia
Physics
382 McCormick Road
Charlottesville, VA 22903 USA
dgc3q@virginia.edu
    Telephone: 434-924-6790
        FAX: 434-924-4576

CRAWFORD, Christopher
MIT
Physics
25 Evergreen Lane
Arlington, MA 02474 USA
chris2@lns.mit.edu
    Telephone: 617-253-6734
        FAX: 617-258-5440

DERENCHUK, Vladimir
Indiana University Cyclotron Facility
2401 Milo B. Sampson Lane
Bloomington, IN 47408 USA
laddie@iucf.indiana.edu
    Telephone: 812-855-9130
        FAX: 812-855-6645

DOSKOW, Jack
Indiana University Cyclotron Facility
2401 Milo B. Sampson Lane
Bloomington, IN 47408 USA
doskow@iucf.indiana.edu
    Telephone: 812-855-9365
        FAX: 812-855-6645

ENGELS, Ralf
Universitat zu Koln
Institut fuer Kernphysik
Zulpicher Strasse 77
Koeln, D-50937 GERMANY
engels@ikp.uni-koeln.de
    Telephone: 49-221-4703457
        FAX: 49-221-4705168

ERSHOV, Vladimir Petrovich
Res. Eng. Of the Joint Inst. Nucl.
Research
Cryogenic
Str. 14, app. 10
Dubna, Moscow 141980 RUSSIA
pilipen@sunhe.jinr.ru
    Telephone: 007 096 21 64931
        FAX: 007 096 21 65180

EVERSHEIM, Dieter
University Bonn
Institut fuer Kernphysik
Nussallee 14-16
Bonn, NRW D-53115 GERMANY
evershei@iskp.uni-bonn.de
    Telephone: 49 228 73 5299
        FAX: 49 228 73 2505

FICK, Dieter
Philipps-Universitaet
Fachbereich Physik
Marburg,   D-35032 GERMANY
fick@physik.uni-marburg.de
    Telephone: 49 6421 282 2017
        FAX: 49 6421 282 8993

FOSTER, Ryan
North Carolina State University
Physics
3523 N. Roxboro Street
Apartment 2A
Durham, NC 27704 USA
rdfoster@tunl.duke.edu
    Telephone: 919-660-2613

GEBEL, Ralf
Forschungszentrum Juelich
IKP COSY
Leo Brandt Strasse
IKP 7.2
Juelich, NRW 52428 GERMANY
r.gebel@fz-juelich.de
    Telephone: 49 2461 613097
        FAX: 49 2461 612854

GENTILE, Thomas
NIST
Physics Laboratory
Stop 8461, NIST
Gaithersburg, MD 20899 USA
thomas.gentile@nist.gov
    Telephone: 301-975-5431
        FAX: 301-926-1604

GLENDE, Martin
Forschungszentrum Juelich
IKP
Leo Brandt Strasse
Juelich, D-52425 GERMANY
m.glende@fz-juelich.de
    Telephone: 49 2461 61 8039
        FAX: 49 2461 61 8039

GOERTZ, Stefan
Ruhr-Universitaetsstr. 150
Institute for Experimental Physics
Universitaetsstr. 150
NB 2/30
Bochum, NRW 44780 GERMANY
goertz@epl.ruhr-uni-bochum.de
    Telephone: 49 234 322 3570
        FAX: 49 234 321 4170

GRAMES, Joseph
Thomas Jefferson National
Accelerator Facility
Accelerator Division MS-5A
12000 Jefferson Avenue
Newport News, VA 23606 USA
grames@jlab.org
    Telephone: 757 269 7097
        FAX: 757-269-7363

HAASE, David
North Carolina State University
Physics
Physics Dept., Box 8202
Raleigh, NC 27695-8211 USA
david_haase@ncsu.edu
    Telephone: 919-515-6118
        FAX: 919-515-7545

HATAKEYAMA, Atsushi
TRIUMF
4004 Wesbrook Mall
Vancouver, BC V6T 2A3
CANADA
hatakeya@triumf.ca
    Telephone: 604-222-7509
        FAX: 604-222-1074

HAUTLE, Patrick
Paul Scherrer Institute
WLGA B25
Villigen PSI, AG 5232
SWITZERLAND
patrick.hautle@psi.ch
    Telephone: 41-56-310-32-10
        FAX: 41-56-310-32-94

HERTENBERGER, Ralf
University of Munich LMU
Sektion Physik
Am Coulombwall 1
Garching, 85748 GERMANY
ralf.hertenberger@physic.uni-
muenchen.de
  Telephone: 89 289 14130
      FAX: 89 289 14103

IBALD, Rolf
Indiana University Cyclotron Facility
2401 Milo B. Sampson Lane
Bloomington, IN 47408 USA
ibald@iucf.indiana.edu
  Telephone: 812-855-5195
      FAX: 812-855-6645

KOLSTER, Hauke
MIT
Laboratory for Nuclear Science
77 Massachusetts Ave.
Cambridge, MA 02139 USA
kolster@mit.edu
  Telephone: 617-253-2693
      FAX: 617-258-5440

KRAMER, Kevin
College of William and Mary
Physics
973 Banks Lane
Newport News, VA 23608 USA
kramer@physics.wm.edu
  Telephone: 757-890-8098
      FAX: 757-269-3500

LENISA, Paolo
INFN - Sez. Di Ferrara and
University
Physics
Via Paradiso, 12
Ferrara, 44100 ITALY
lenisa@hermes.desy.de
  Telephone: 0049-40-8998-1963

LEVY, Phillip
TRIUMF/Manitoba
4004 Wesbrook Mall
Vancouver, BC V6T 2A3
CANADA
levy@triumf.ca
  Telephone: 604-222-7424
      FAX: 604-222-1074

LOWRY, Michael
Brookhaven National Laboratory
Physics
Bldg. 510
Upton, NY 11934 USA
lowry@bnl.gov
  Telephone: 631-344-3771
      FAX: 631-344-3181

LOZOWSKI, Bill
Indiana University Cyclotron Facility
2401 Milo B. Sampson Lane
Bloomington, IN 47408 USA
lozowski@iucf.indiana.edu
  Telephone: 812-855-2928
      FAX: 812-855-6645

LUPPOV, Vladimir
University of Michigan
Physics
500 East University
Ann Arbor, MI 48109-1120 USA
vluppov@umich.edu
Telephone: 743-764-5111
FAX: 734-763-9027

MAKDISI, Yousef
Brookhaven National Laboratory
Collider-Accelerator
Building 911 B
Upton, NY 11973 USA
makdisi@bnl.gov
Telephone: 631-344-4932
FAX: 631-344-5954

MARECHAL, Francois
Institut de Recherches Subatomiques
23, Rue du Loess
Strasbourg, 67037 FRANCE
francois.marechal@ires.in2p3.fr
Telephone: 33-3-88-10-62-82
FAX: 33-3-88-10-65-34

MARKOFF, Diane
North Carolina State University /
TUNL
Physics
TUNL Box 90308
Durham, NC 27708 USA
markoff@tunl.duke.edu
Telephone: 919-660-2624
FAX: 919-660-2634

MEYER, Hans
Indiana University Cyclotron Facility
2401 Milo B. Sampson Lane
Bloomington, IN 47408 USA
meyer@iucf.indiana.edu
Telephone: 812-855-2883
FAX: 812-855-6645

MIKIRTYTCHIANTS, Maxim
Forschungszentrum Juelich
IKP
Leo Brandt Strasse
Juelich, D-52428 GERMANY
m.mikirtytchiants@fz-juelich.de
Telephone: (49) 2461 612147
FAX: (49) 2461 613930

NANN, Hermann
Indiana University Cyclotron Facility
2401 Milo B. Sampson Lane
Bloomington, IN 47408 USA
nann@iucf.indiana.edu
Telephone: 812-855-2884
FAX: 812-855-6645

NASS, Alexander
University of Erlangen - Nuremburg
Physikalisches Institut
E. -Rommel-Str. 1
Erlangen, Bavaria 91058
GERMANY
nass@pi2p7.physik.uni-erlangen.de
Telephone: 49 40 8998 4682
FAX: 49 40 8998 4034

ORLOV, Dmitry
MPI fuer Kernphysik
Saupfercheckweg 1
Heidelberg, 69117 GERMANY
Dmitry.Orlov@mpi-hd.mpg.de
    Telephone: 49 6221 516504
        FAX: 49 6221 516602

PENTTILÄ, Seppo
LANL
P-23
MS-H803
Los Alamos, NM 87545 USA
penttila@lanl.gov
    Telephone: 505-665-0641
        FAX: 505-665-4121

PILIPENKO, Yuri
Joint Institute for Nuclear Research
Cryogenic
Moscow reg.
Dubna, 141980 RUSSIA
pilipen@sunhe.jinr.ru
    Telephone: 7 096 21 65044
        FAX: 7 096 21 65180

von PRZEWOSKI, Barbara
Indiana University Cyclotron Facility
2401 Milo B. Sampson Lane
Bloomington, IN 47408 USA
przewoski@iucf.indiana.edu
    Telephone: 812-855-2913
        FAX: 812-855-6645

RAMSAY, Des
TRIUMF/Manitoba
4004 Wesbrook Mall
Vancouver, BC V6T 2A3
CANADA
rams@triumf.ca
    Telephone: 604-222-1047
        FAX: 604-222-1074

RATHMANN, Frank
Forschungszentrum Juelich
Institut fuer Kernphysik
Leo Brandt Strasse
Julich, 52425 GERMANY
f.rathmann@fz-juelich.de
    Telephone: 0049-2461-614-558
        FAX: 0049-2461-613-930

RAYMOND, Richard
University of Michigan
Physics
Randall Lab of Physics
Ann Arbor, MI 48109 USA
rraymond@umich.edu
    Telephone: 734-764-5113

REICHERZ, Gerhard
Ruhr-Universitaet Bochum
Experimental Physik
Universitaetsstr. 150
NB 2/30
Bochum, D-44780 GERMANY
reicherz@ep1.ruhr-uni-bochum.de
    Telephone: 492343223542
        FAX: 493214170

RINCKEL, Tom
Indiana University Cyclotron Facility
2401 Milo B. Sampson Lane
Bloomington, IN 47408 USA
rinckel@iucf.indiana.edu
    Telephone: 812-855-5197
        FAX: 812-855-6645

RUPP, Axel
Hahn-Meitner-Institut Berlin
SF2
Glienicker Strasse 100
Berlin, D-14109 GERMANY
rupp-a@hmi.de
Telephone: 49/30-8062-3177
FAX: 49/30-8062-2999

SAKAI, Hideyuki
University of Tokyo
Physics
7-3-1 Hongo, Bunkyo
Toyko, 113-0033 JAPAN
sakai@phys.s.u-tokyo.ac.jp
Telephone: 81 3 5841 4237
FAX: 81 3 5841 7642

SCHROETER, Claus Dieter
Max-Planck-Institut fuer Kernphysik
Saupferchechweg 1
Heidelberg, D-69117 GERMANY
Claus.Schroeter@mpi-hd.mpg.de
Telephone: 49-6221-516-513
FAX: 49-6221-516-604

SNOW, Mike
Indiana University Cyclotron Facility
2401 Milo B. Sampson Lane
Bloomington, IN 47408 USA
snow@iucf.indiana.edu
Telephone: 812-855-7914
FAX: 812-855-6645

STEFFENS, Erhard
University of Erlangen
Physik. Institut
ERWIN-ROMMEL-STR.1
Erlangen, D-91058 GERMANY
steffens@physik.uni-erlangen.de
Telephone: 49-9131-852-7093
FAX: 49-9131-15249

STEPHENSON, Ed
Indiana University Cyclotron Facility
2401 Milo B. Sampson Lane
Bloomington, IN 47408 USA
stephenson@iucf.indiana.edu
Telephone: 812-855-5469
FAX: 812-855-6645

SUBASHIEV, Arsen
St. Petersburg State Technical
University
Experimental Phsyics
Politekhnicheskaya 26
St. Petersburg, 195251 RUSSIA
arsen@subashiev.hop.stu.neva.ru
Telephone: 78125527790

TENCKHOFF, Georg
Universitaet zu Koln
Institut fuer Kernphysik
Zulpicher Strasse 77
Koeln, D-50937 GERMANY
gt@ikp.uni-koeln.de
Telephone: 49 (0) 221 470 3623
FAX: 49 (0) 221 470 5168

TOPORKOV, Dmitri
Budker Institute of Nuclear Physics
BINP, Pr. Lavrentieva 11
Novosibirsk, 630090 RUSSIA
toporkov@inp.nsk.su
Telephone: 7-3832-394910
FAX: 7-3832-342163

TSENTALOVICH, Evgeni
MIT/Bates
21 Manning Road
Middleton, MA 01949 USA
evgeni@mit.edu
Telephone: 617-253-9507
FAX: 617-253-9599

WAKUI, Takashi
RIKEN
Hirosawa 2-1
Wako, Saitama 351-0198 JAPAN
wakui@rarfaxp.riken.go.jp
    Telephone: 81-48-467-7932
        FAX: 81-48-462-4689

WEIGEL, Udo M.
Max-Planck-Institut fuer Kernphysik
Saupfercheckweg 1
Heidelberg, 69117 GERMANY
udo.weigel@mpi-hd.mpg.de
    Telephone: 49 6221 516504
        FAX: 49 6221 516602

WHITAKER, T.J.
Indiana University Cyclotron Facility
2401 Milo B. Sampson Lane
Bloomington, IN 47408 USA
whitaket@iucf.indiana.edu
    Telephone: 812-855-5195
        FAX: 812-855-6645

WISE, Tom
University of Wisconsin
Department of Physics
1150 University Ave.
Madison, WI 53706 USA
wise@uwnuc0.physics.wisc.edu
    Telephone: 608-262-6555
        FAX: 608-262-3598

WOOD, Michael
University of South Carolina
Physics
12000 Jefferson Avenue
Newport News, VA 23606 USA
mikewood@jlab.org
    Telephone: 757-269-5551
        FAX:

WOOTEN, Gary
VACUUMSCHMELZE
4027 Will Rogers Parkway
Oklahoma City, OK 73108 USA
garyw@vaccorp.com
    Telephone: 405-246-7015
        FAX: 405-246-7115

ZELENSKI, Anatoli
Brookhaven National Laboratory
CAD
B830
Upton, NY 11973 USA
zelenski@bnl.gov
    Telephone: 631-344-8387
        FAX: 631-344-5011

ZIMMER, Oliver
Technische Universitat Muenchen
Physik-Dep. E 18
James-Franck-Strasse
Garching, 85748 GERMANY
oliver.zimmer@ph.tum.de
    Telephone: 49 -89-28912573
        FAX: 49-89-28912570

ZISKIN, Vitaliy
LNS
MIT-Bates
21 Manning Rd.
Middleton, MA 01949 USA
vziskin@mit.edu
    Telephone: 617-253-9530
        FAX:

AUTHOR INDEX